STAT CITY

Understanding Statistics through Realistic Applications

The Irwin Series in Quantitative Analysis for Business

Consulting Editor   Robert B. Fetter   *Yale University*

STAT CITY

Understanding Statistics through Realistic Applications

HOWARD S. GITLOW
University of Miami

ROSA OPPENHEIM
Rutgers University

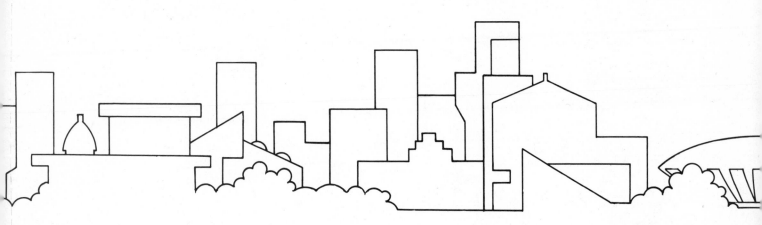

1986   Second Edition

IRWIN
Homewood, Illinois 60430

© RICHARD D. IRWIN, INC., 1982 and 1986

ISBN 0-256-03382-X

Library of Congress Catalog Card No. 85–81463

Printed in the United States of America

1 2 3 4 5 6 7 8 9 0 ML 3 2 1 0 9 8 7 6

To Shelly and Ali
To Alan, Adam, and David

our rainbows in the sky

PREFACE

Stat City is a supplementary textbook which emphasizes the interpretation and communication of statistics. The objectives of *Stat City* are to:

1. Provide readers with *complete* statistical problems so that they can grasp the totality of statistical studies, from inception through memorandum.
2. Provide readers with *unified* statistical problems; all problems in *Stat City* are related to a Stat City data base and are performed for Stat City interest groups (for example, the mayor, the telephone company, the electric company, and so on).
3. Provide readers with a role model for the performance of statistical studies.

Stat City is comprised of six parts. Part One introduces Stat City, a fictitious town in the United States. Stat City is comprised of families, businesses, schools, parks, nonprofit organizations, and places of worship. Each dwelling unit in Stat City has 5 identifiers and 22 pieces of descriptive information. It is critical that Part One be mastered by the reader before he/she proceeds onto Parts Two through Six; understanding Part One (Chapter 1) is the key to successfully using *Stat City*. Part Two introduces descriptive statistics, the basics of inferential statistics, and the fundamentals of statistical memorandum writing. The third part of the text introduces probability theory, probability distributions, and expands the fundamentals of statistical memorandum writing. Parts Four and Five formally introduce inferential statistics and complete the basics of statistical memorandum writing. Part Six presents how computers [in particular, SPSS (Statistical Package for the Social Sciences) and Minitab] can be used to facilitate statistical studies. All of the above parts are linked together through the Stat City data base.

Most chapters end with a brief summary and a series of problems which are designed to provide a vehicle for the reader to determine if he/she has grasped the essential concepts in the chapter. Assignment problems are designed to be completed and turned in to the instructor.

Due to the nature of the Stat City data base, some topics often covered in basic statistics textbooks have been omitted—index numbers, nonparametric statistics, time series, and decision theory. We believe these omissions are not a limitation because of the supplementary nature of this text and because these topics are adequately covered in most basic texts.

Stat City has been classroom tested for three years in basic and advanced courses at both the undergraduate and graduate levels. The students' responses to the text have been extremely enthusiastic. A frequent comment is that the complete and unified examples greatly facilitate the learning of statistics.

We would like to express thanks to our colleagues and students at the University of Miami and at Rutgers University, especially Paul Sugrue, who was instrumental in the conception of this book; David Orden and Marian McDuffie, for their invaluable guidance in the use of statistical computer packages; Margaret Updike and Mary McKenry, graduate assistants who spent many hours searching for errors in the manuscript; and copy editor Virginia Lee Fogle.

We are indebted to Professor Alan Oppenheim (Montclair State College) for his significant contributions to the text and to Professors Paul D. Berger (Boston University), James F. Horrell (University of Oklahoma), Ernest Kurnow (New York University), Donald S. Miller (Emporia State University), Barbara Price (Wayne State University), John C. Shannon (Suffolk University), and Howard J. Williams (Wilkes College) for their reviews and comments on the manuscript.

We are grateful to the Literary Executor of the late Sir Ronald A. Fisher, F.R.S., to Dr. Frank Yates, F.R.S., and to Longman Group Ltd., London, for permission to reprint Table III from their book *Statistical Tables for Biological, Agricultural and Medical Research* (6th edition, 1974).

Most important, we want to thank Shelly Gitlow and Alan Oppenheim for their encouragement and patience, and our parents, Abraham and Beatrice Gitlow and Aaron and Esther Blitzer, for their unwavering support and confidence.

Howard S. Gitlow
Rosa Oppenheim

CONTENTS

Can be A $CI_{\bar{x}}$

Can be $CI_{\bar{x}}$ →

PART THREE

Chapter 4   Probability concepts   49

Chapter 5   Probability Distributions   67

B, Normal

Binomial

Normal

Normal
Binomial
Poisson

Normal

Normal) (handwritten)

PART FOUR

CI x̄ (handwritten)
CI x̄ (handwritten)
Proportion. (handwritten)
CI x̄ (handwritten)
CI x̄ (handwritten)

Z x̄ (handwritten)
x̄ (handwritten)
Percentage and proportion. (handwritten)
t (handwritten)

Proportion (handwritten)

PART FIVE

(handwritten margin notes: r, y', y', y', Multiple R, y', y', x', Multiple R, y')

PART ONE

STAT CITY

CHAPTER 1

Statistics involves methods and theory as they are
applied to numerical data or observations with
the objective of making rational decisions in
the face of uncertainty.[1]

Kurnow, Ottman, and Glasser

Stat City is a fictitious town in the United States. Stat City was created to
help you learn how to *use* statistics, not to teach you about mathematical
statistics. Once you have mastered using statistics to explain the characteristics of Stat City for decision-making purposes, you will be prepared to
use statistics on any real-world problem.

Remember, statistics is a way of thinking and attacking tough problems; it does not have to be torture.

Stat City is comprised of families (dwelling units), businesses, schools,
parks, nonprofit organizations, and places of worship. The map inside the
back cover of this book gives you a pictorial view of Stat City. As you can
see, the business district is located in approximately the center of town.
The residential sections of Stat City surround the business district and are
divided into four zones. Table 1–1 depicts the breakdown of Stat City
dwelling units (families) into each residential zone as of January 1985.

Each zone in Stat City is subdivided into blocks. A block is a geograph-
ical area separated by streets and avenues. The subdivision of zones into
blocks gives you an extremely detailed way of breaking down Stat City
into component parts for analysis. For example, by grouping blocks you
could compare the incomes of families in Zone 3 east of 10th Avenue with
the incomes of families in Zone 3 west of 10th Avenue.

Each dwelling unit in Stat City has three identifiers: the last name of the
head of the household, a street address, and an identification number.
Each dwelling unit has a unique street address, except in the case of
apartment buildings where families have the same street address. If more

INTRODUCTION

**Table 1–1 Dwelling units
(families) in Stat City by
residential zone (January 1985)**

Zone	Number of units
1	130
2	157
3	338
4	748
Total	1,373

---

[1] *Statistics for Business Decisions* (Homewood, Ill.: Richard D. Irwin, 1959). © 1959 by
Richard D. Irwin, Inc.

detail is needed to isolate a particular dwelling unit (family), their identification number can be used.

Aside from name, address, block, identification number, and zone, 22 other pieces of information were collected for each dwelling unit (family) in Stat City. Exhibit A–1 in Appendix A at the back of this book lists all the information collected on Stat City families.

The information listed in Exhibit A–1 was collected through a mail survey (questionnaire) sent out on January 30, 1985—one questionnaire to each dwelling unit. The head of each household was instructed that he or she alone should complete the questionnaire and answer all questions relative to January 1985. A copy of the questionnaire can be seen in Exhibit A–2, Appendix A.

Once all 1,373 questionnaires were completed and checked, the data was compiled into the Stat City Data Base (see Appendix C). The information is ordered by identification number (variable 4). Zone 1 contains the dwelling units with identification numbers 1 through 130. Zone 2 contains the dwelling units with identification numbers 131 through 287. Zone 3 contains the dwelling units with identification numbers 288 through 625. Finally, Zone 4 contains the dwelling units with identification numbers 626 through 1,373. Table 1–2 depicts how the identification numbers are assigned.

To interpret the collected information for a particular dwelling unit (a line of data), simply read across the Stat City Data Base and use Exhibit A–1 as a reference guide for the descriptions of each piece of information. For example, an interpretation of the information collected for the Kilfoyle family as of January 1985 (the first line of data in the Stat City Data Base) appears in Exhibit 1–1.

The Kilfoyles reside at 406 6th Street, on block 43, in Zone 1, in a house with a monthly mortgage payment of $502 and an assessed value of $88,264. The house has 10 rooms, an average yearly heating bill of $1,199, an average monthly electric bill of $88, and an average monthly telephone bill of $90. The Kilfoyle family income was $56,419 in 1984. There are six people in the Kilfoyle family. The Kilfoyles own three cars, spend an average of $165 bimonthly on gas ($55 per car), average six trips to the gas station per month, and have most of their car repairs done at Howie's Gulf Station (gas station 2). The Kilfoyles average about six trips to the hospital per year. The Kilfoyles spend an average of $128 per week on food ($21 per person) in the A&P (supermarket 3). Their many food purchases include a weekly supply of two six-packs of diet soda, no regular soda, and five six-packs of beer.

Table 1–2 Identification number assignments

| | Identification number | |
Zone	From	To
1	1	130
2	131	287
3	288	625
4	626	1,373

WHAT IS STATISTICS?

Statistics can be broadly defined as the study of numerical data to better understand the characteristics of a population. Statistics may help in making some rational decision about a population. Statistics can also be thought of as the *art* of extracting a clear picture out of numerical information. We refer to statistics as an art because frequently statisticians must be creative if they are going to be helpful in solving problems. The creative portion of a statistician's job makes his/her work exciting.

The use of statistics in business and industry is pervasive. Some examples of applications of statistics are:

Opinion polling to predict the outcome of an election.

Determining the proper dosage level for a drug.

Surveying consumers to establish appeal and nutritional acceptability of a new food product.

Forecasting the population of southern Florida so that an adequate mass transportation system can be built.

Exhibit 1–1

NUMBER	NEUMONIC	DESCRIPTION	
1	NAME	Last name of the head of the household	Kilfoyle
2	ADDR	Street address	406 6th Street
3	BLOCK	Block location of dwelling unit	43
4	ID	Dwelling unit's identification code	0001
5	ZONE	Residential housing zone	1
6	DWELL	Type of dwelling unit 0 = Apartment 1 = House	1
7	HCOST	Monthly housing cost as of January 1985 Rent if apartment Mortgage if house	$502
8	ASST	Assessed value of home ($0.00 if apartment) as of January 1985	$88,264
9	ROOMS	Number of rooms in dwelling unit as of January 1985	10
10	HEAT	Average total yearly heating bill as of January 1985 (includes all types of heat—electric, gas, etc.)	$1,199
11	ELEC	Average monthly electric bill as of January 1985	$88
12	PHONE	Average monthly telephone bill as of January 1985	$90
13	INCOM	Total family income for 1984	$56,419
14	PEPLE	Number of people in household as of January 1985	6
15	CARS	Number of cars in household as of January 1985	3
16	GAS	Average bimonthly automobile gas bill as of January 1985	$165
17	GASCA	Average bimonthly automobile gas bill per car as of January 1985	$55
18	GASTR	Average number of trips to the gas station per month as of January 1985	6
19	REPAR	Favorite place to have automotive repairs performed as of January 1985 0 = Performs own repairs 1 = Repairs done by service station 2 = Repairs done by dealer	1

Exhibit 1–1 (*concluded*)

NUMBER	NEUMONIC	DESCRIPTION	
20	FAVGA	Favorite gas station as of January 1985 1 = Paul's Texaco (11th St. & 7th Ave.) 2 = Howie's Gulf (11th St. & Division St.)	2
21	HOSP	Average yearly trips to the hospital by all members of dwelling unit as of January 1985	6
22	EAT	Average weekly supermarket bill as of January 1985	$128
23	EATPL	Average weekly supermarket bill per person as of January 1985	$21
24	FEAT	Favorite supermarket as of January 1985 1 = Food Fair 2 = Grand Union 3 = A&P	3
25	LSODA	Average weekly purchase of six-packs of diet soda as of January 1985	2
26	HSODA	Average weekly purchase of six-packs of regular soda as of January 1985	0
27	BEER	Average weekly purchase of six-packs of beer as of January 1985	5

Controlling the quality of goods from a production line.

Estimating the population in each state in the United States to disburse federal monies.

An aspect of statistics that makes it exciting is that it can be applied in almost any field of endeavor—from controlling rat populations in Bombay to allocating salespeople's commissions in Detroit. Statistical skills are transferable. A world of opportunity is open to the statistician.

POPULATIONS AND SAMPLES

One of the fundamental concepts you will encounter when using statistics is the difference between a POPULATION and a SAMPLE.

A POPULATION (or Universe) is the total number of items under study. All the dwelling units in Stat City form a population.

A SAMPLE is a part of the population under investigation selected so that information can be drawn from it about the population. For example, 100 randomly selected Stat City dwelling units form a sample.

It is important to keep in mind that the dwelling units in one residential zone of Stat City could be considered a population or a sample. Despite the general differentiation between population and sample, it is important to remember that a population is what a client defines it to be, no more, no less.[2] A population may include all dwelling units in a city, or it may only include dwelling units in a given area. Your client is responsible for defining clearly the population of interest. For example, if you are working for the Stat City Chamber of Commerce and are asked to do a citywide analysis of dwelling units, the population must be defined as all dwelling units in Stat City. However, if you are working for the Zone 3 Community Planning Board and are asked to do an analysis of Zone 3 dwelling units, all the dwelling units in Zone 3, and only the dwelling units in Zone 3, must be defined as the relevant population.

Let us look at another facet of the same issue. If you assume that the relevant population is all the dwelling units in Stat City and you select all the dwelling units in Zone 4 as your sample, you may very well commit a severe error. The error occurs when the dwelling units in Zone 4 are in some way unique (as the French Quarter is to New Orleans or as Harlem is to New York City). You may then be reporting facts about the dwelling units in Zone 4 which do not pertain to the other dwelling units in Stat City. Be careful!

Population parameters versus sample statistics

Decision makers frequently use two types of information to understand a population, parameters and statistics.

A PARAMETER is a measure of a population characteristic. For example, the average[3] monthly telephone bill computed from all dwelling units in Stat City or the average monthly electric bill computed from all dwelling units in Zone 2 (if your relevant population is Zone 2) are parameters.

A STATISTIC is an estimate of a population characteristic which is computed from a sample. For example, if 10 dwelling units were randomly selected from Stat City and the average number of rooms computed, this average would be called a statistic. This sample average would be an estimate of the average number of rooms per dwelling unit in Stat City.

How to select a random sample of Stat City dwelling units

There are 1,373 dwelling units in Stat City. To draw a random sample of, say, 10 Stat City dwelling units, the following steps should be followed.

1. Specify the objective in drawing the sample. For example, you may want to estimate the average number of rooms (variable 9) per dwelling unit in Stat City.
2. Number the dwelling units in Stat City from 0001 through 1373. (This has already been done if you will look at variable 4 in the Stat City Data Base.)
3. Select a page in a table of random numbers. Exhibit A–3, Appendix A, is a random number table.
4. On the selected page in Exhibit A–3 (for example, the first page), select a column of numbers with as many digits in it as are in your largest ID number, 1373 (4 digits). For example, begin in the first column of the first page in Exhibit A–3 with 5347.
5. Select the first 10 four-digit numbers in the chosen column on that page which are between 0001 and 1373, inclusive.
   If a number is encountered which is smaller than 0001 (0000) or larger than 1373 (5374, for example), discard the number and continue down the column. If an acceptable number appears more than once, ignore every repetition and continue moving down the column until 10

---

[2] In many instances the statistician and the client are the same person.

[3] Arithmetic mean.

unique numbers have been selected. Finally, if you get to the bottom of the page before you have obtained all 10 random numbers, go to the top of the page and move down the next four-digit column. Table 1–3 depicts the 10 random numbers drawn, the family names of those dwelling units selected, and the number of rooms in each selected dwelling unit.

Table 1–3

Variable 4 Identification number (selected random numbers)	Variable 1 Name	Variable 9 Number of rooms
933	Althouse	6
481	Comas	7
576	Pomer	10
1318	Shapiro	5
1348	MacMaster	8
966	Slater	5
1215	Lehman	6
573	Reeves	7
997	Singer	5
1144	Halbrooks	6

6. Finally, analyze the information on the variable(s) of interest. The estimated average number of rooms per household is 6.5 rooms, based on the sample of 10 dwelling units.

An important point to remember is that different samples of size 10 may yield different room estimates. We will deal with this frustrating but critical point later on in this book.

The remaining sections of this book will be filled with cases, problems, and assignments for you to tear out and turn in to your instructor, all dealing with populations and samples to help you learn how to use statistics in the business world.

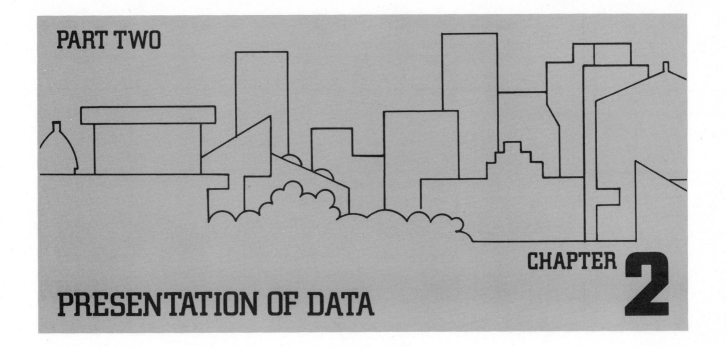

PART TWO

PRESENTATION OF DATA

CHAPTER 2

INTRODUCTION

Business people are frequently confronted with huge amounts of data which must be summarized into a workable format so that decisions can be made. In this chapter, you will learn how to reduce large amounts of data to manageable proportions for decision-making purposes.

A common method of presenting statistical data is by means of a chart. Charts can be used to summarize large amounts of statistical data in a form which is readily interpreted and easily understood. Charts are widely used in presenting data in newspapers, magazines, and business reports.

The most common types of statistical charts are: bar charts, line charts, pie charts, pictographs, and statistical maps. An example of each type of chart will be illustrated as a needed tool to solve a problem for a Stat City interest group. Each problem in this chapter will present parameters for you to use as input in constructing charts and writing business memos. Don't worry about actually computing the parameters now. We will deal with calculating parameters in later chapters. Before we get involved in calculations, let us get acquainted with writing business memorandums that use statistical data.

Table 2–1 Automobiles in Stat City by residential zone (January 1985)

Zone	Number of automobiles per zone
1	323
2	365
3	874
4	1,405
Total	2,967

Example 2–1 *Stat City Beacon* article on overnight parking

Ms. Donna Nelson, a reporter for the *Stat City Beacon,* is preparing an article on recent problems with overnight parking on public streets within Stat City. She has decided that it would be appropriate to describe clearly in the article the number of automobiles in each zone of the city. She has chosen to present a bar chart which would enable the readers of the article to comprehend readily the preponderance of automobiles in Zones 3 and 4 as compared with Zones 1 and 2. She obtained the information shown in Table 2–1 from the Stat City Department of Traffic as of January 1985.

You have been asked by Ms. Nelson to construct the appropriate bar chart. Type a business memorandum presenting the bar chart to Ms. Nelson (see Exhibit 2–1).

Problem

Solution

Exhibit 2–1

HOWARD S. GITLOW, PH.D.
STATISTICAL CONSULTANT

MEMORANDUM

TO: Donna Nelson, Reporter
         Stat City Beacon

FROM: Howard Gitlow

DATE: February 21, 1985

RE: Bar chart for article on overnight parking on public
         streets in Stat City as of January 1985

I have taken the Department of Traffic data you supplied
to me concerning the total number of automobiles per zone in
Stat City and have constructed the desired bar chart per your
request.

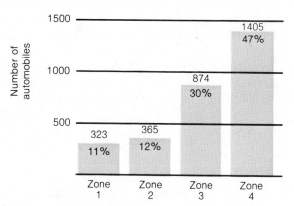

Note: Percentages have been rounded so they
will total 100 percent.
    Source: Stat City Department of Traffic,
January 1985.

The chart indicates that more than 75 percent of the
automobiles in Stat City are owned by families in Zones 3
and 4.

If you have any further questions, please do not hesitate
to call me at 305-999-9999.

The reporter for the *Stat City Beacon* has decided that a pictograph would be understood more widely by her readership. She has asked you once again to type a memorandum presenting her with the pictograph (see Exhibit 2-2).

Exhibit 2-2

HOWARD S. GITLOW, PH.D.
STATISTICAL CONSULTANT

MEMORANDUM

TO: Donna Nelson, Reporter
        Stat City Beacon

FROM: Howard Gitlow

DATE: March 4, 1985

RE: Pictograph for article on overnight parking on
        public streets in Stat City as of January 1985

    As per your request, I have constructed a pictograph based
upon data supplied by the Department of Traffic concerning
the total number of automobiles per zone in Stat City as of
January 1985. The pictograph appears below.

Exhibit 2–2 (*concluded*)

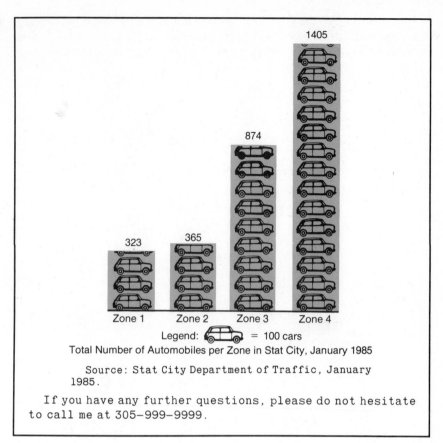

Legend: = 100 cars

Total Number of Automobiles per Zone in Stat City, January 1985

Source: Stat City Department of Traffic, January 1985.

If you have any further questions, please do not hesitate to call me at 305-999-9999.

Example 2–2   Purchasing books for the Stat City Library

Problem

Ms. Elissa Gitlow, chief librarian for the Stat City Library, will increase the number of books purchased by the library if the percentage change in the Stat City population was more than 10 percent between January 1981 and January 1985. The Stat City Chamber of Commerce has supplied Ms. Gitlow with the Stat City population for January 1981 through January 1985. The population figures are shown in Table 2–2.

Ms. Gitlow has decided that a line chart supplemented by the percentage increase in population for January 1985 over January 1981 would be required to make the correct purchase decision. You have been asked by Ms. Gitlow to construct the line chart and to compute the percentage increase in population. Type a memorandum to Ms. Gitlow reporting the requisite information (see Exhibit 2–3).

Table 2–2

Month	Year	Population
January	1981	5,127
January	1982	5,279
January	1983	5,499
January	1984	5,640
January	1985	5,814

ROSA OPPENHEIM, PH.D.
STATISTICAL CONSULTANT

MEMORANDUM

TO: Ms. Elissa Gitlow, Chief Librarian
        Stat City Library

FROM: Rosa Oppenheim

DATE: March 11, 1985

RE: Purchasing books for the Stat City Library

Based upon the data you supplied from the Stat City
Chamber of Commerce, I have constructed a line chart to
illustrate the population increase in Stat City from
January 1981 to January 1985.

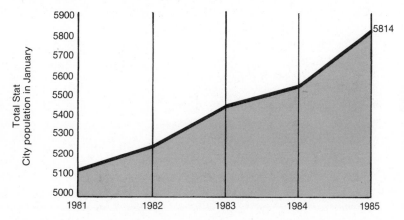

Source: Stat City Chamber of Commerce, January 1981–January
1985.

The percentage increase in the population between
January 1981 and January 1985 is approximately 13.4 percent
[(5,814 − 5,127 = 687) ÷ 5,127 = .13399 = +13.4%]. This
increase exceeds your 10.0 percent decision criteria
established for the expansion.
    If you have any further questions, please do not hesitate
to call me at 305–999–9999.

Example 2–3   Department of Public Works population survey

Problem

Table 2–3   Population of Stat City by zone, January 1985

Zone	Population
1	504
2	546
3	1,411
4	3,353
Total	5,814

Mr. Joseph Moder, division head of the Stat City Department of Public Works, is preparing a report for the mayor on the need to expand the sewage system in Zone 4. This expansion is required because of the rapid growth in the population of Zone 4. Mr. Moder has decided to enclose a pie chart in his report to illustrate the population inequities existing in the four zones. He obtained the information shown in Table 2–3 from the Stat City Chamber of Commerce.

You have been asked to type a memorandum presenting the pie chart to Mr. Moder.

Solution

The degrees of the angles needed to construct the pie chart are computed below.

Zone 1: $504/5,814 = .0867 =$ percentage of population in Zone 1.
$.0867 (360°) = 31.212° =$ number of degrees in the angle describing Zone 1's slice of the pie.

Zone 2: $546/5,814 = .0939 =$ percentage of population in Zone 2.
$.0939 (360°) = 33.804° =$ number of degrees in the angle describing Zone 2's slice of the pie.

Zone 3: $1,411/5,814 = .2427 =$ percentage of population in Zone 3.
$.2427 (360°) = 87.372° =$ number of degrees in the angle describing Zone 3's slice of the pie.

Zone 4: $3,353/5,814 = .5767 =$ percentage of population in Zone 4.
$.5767 (360°) = 207.612° =$ number of degrees in the angle describing Zone 4's slice of the pie.

The pie chart appears in Exhibit 2–4.

Exhibit 2–4

HOWARD S. GITLOW, PH.D.
STATISTICAL CONSULTANT

MEMORANDUM

TO: Joe Moder, Division Head
        Stat City Department of Public Works

FROM: Howard Gitlow

DATE: March 5, 1985

RE: Pie chart presenting population by zone in Stat City
        as of January 1985

I have taken the Chamber of Commerce data concerning the
total population in each zone of Stat City and have
constructed the desired pie chart per your request. The pie
chart appears below.

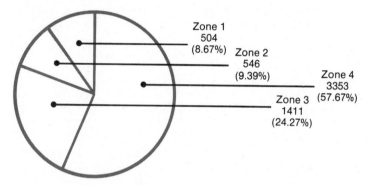

Source: Stat City Chamber of Commerce, January 1985.

Zone 4, accounting for almost 60 percent of Stat City's
population, has more than twice the population of the next
largest area, Zone 3.
        If you have any further questions, please do not
hesitate to call me at 305–999–9999.

Example 2–4   Block analysis of the number of dwelling units in Stat City

Ms. Sharon Vigil, chairperson of the Stat City Real Estate Board, **Problem**
would like a map depicting the density of dwelling units by block. She is
interested in obtaining the locations of all blocks with 6 or less dwelling
units, 7 to 10 dwelling units, 11 to 29 dwelling units, and 30 or more
dwelling units.
        You have been hired by Ms. Vigil to construct a statistical map of Stat
City reflecting the density of dwelling units per block. (Hint: Use a copy
of the Stat City map to construct the statistical map.)

The density categories and the gradations of gray used to represent **Solution**
those categories are:

Number of dwelling units

     1 to 6   7 to 10   11 to 29   30 or more

The map appears as part of the memorandum shown in Exhibit 2–5. It was constructed by counting the number of ID numbers (variable 4) on each block from the Stat City map.

Exhibit 2–5

ROSA OPPENHEIM, PH.D.
STATISTICAL CONSULTANT

MEMORANDUM

TO: Ms. Sharon Vigil, Chairperson
        Stat City Real Estate Board

FROM: Rosa Oppenheim

DATE: March 12, 1985

RE: Statistical map depicting dwelling unit density by
        block in Stat City as of January 1985

  The statistical map requested by your office is herewith
enclosed. If you have any further questions, please call me
at 305-999-9999.

Exhibit 2–5 (concluded) Stat City map depicting dwelling unit density

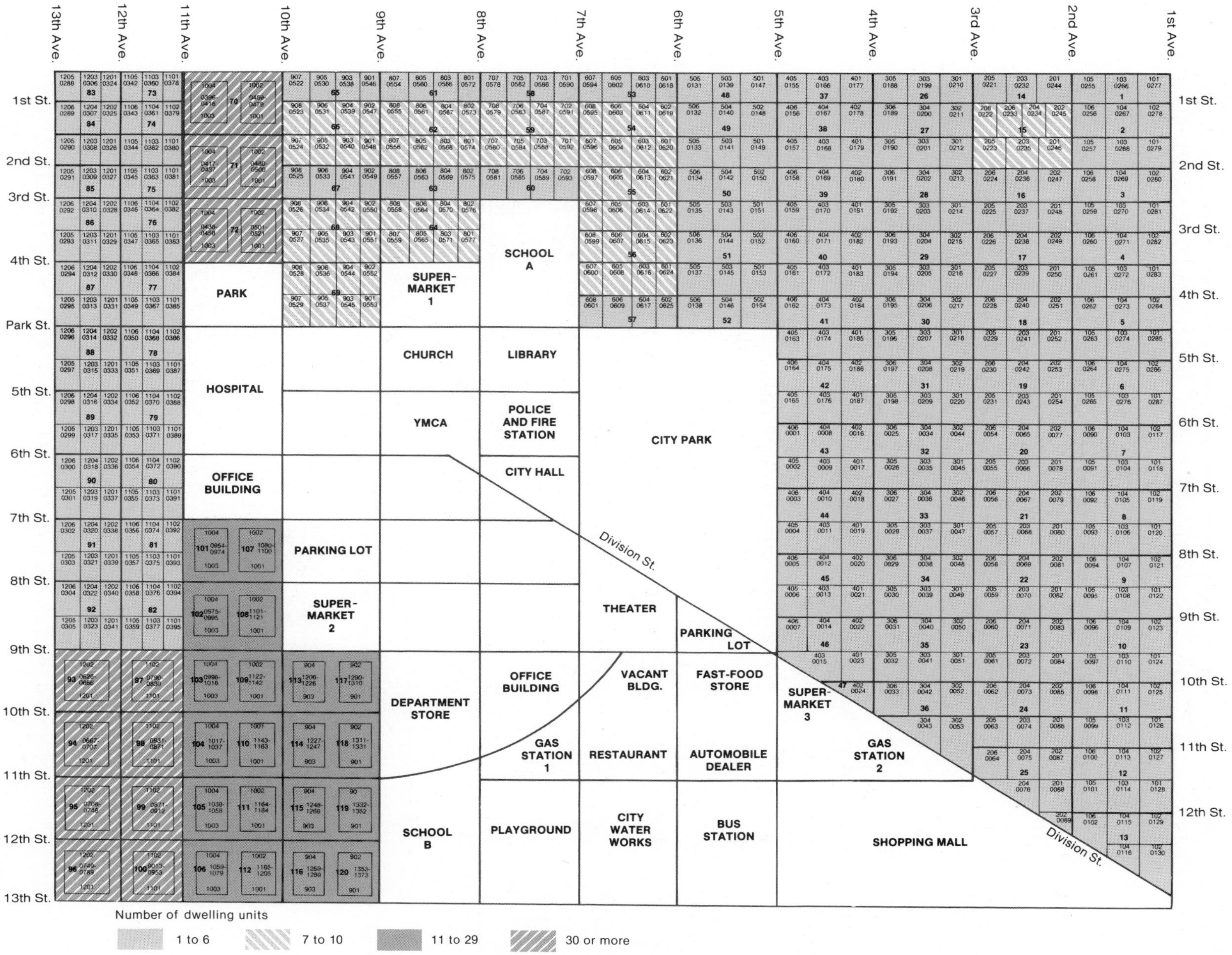

2–5 Analysis of telephone bills by zone in Stat City   The chairman of the Stat City Telephone Company, Jack Davis (ID = 232), wants you to construct a bar chart for his "Zonal Use" study indicating the total of all average monthly telephone bills for each residential zone as of January 1985. Mr. Davis had his secretary extract the following information from the telephone company's accounting records:

Zone	Total of average monthly telephone bills by zone as of January 1985
1	$ 4,613
2	5,704
3	12,346
4	27,732

Type a memorandum to Mr. Davis presenting the bar chart.

2–6 Promotional campaign for beer in Stat City   The Stat City Supermarket Association is considering a special promotional campaign for beer if the rate of growth in the number of six-packs purchased per year was less than 15 percent between 1981 and 1984. The Supermarket Association collected the following data:

Year	Number of six-packs of beer purchased in Stat City
1981	2,720
1982	2,760
1983	2,780
1984	2,800

The association has asked you to prepare a line chart of beer purchases for 1981 through 1984 and to compute the percentage change in beer purchases between 1981 and 1984. Type a memorandum to the Supermarket Association reporting the required information.

2–7 Analysis of electric bills by zone in Stat City The chief financial officer of the Stat City Electric Company, Saul Reisman (ID = 223), is preparing the annual report for stockholders. He has decided to include either a bar or pie chart or pictograph describing the total of all the average monthly electric bills in each residential zone of Stat City as of January 1985. He has extracted the following information from the electric company's accounting records:

Zone	Total of average monthly electric bills by zone as of January 1985
1	$ 9,150
2	11,189
3	22,016
4	34,197
Total	$76,552

Type a memorandum to Mr. Reisman presenting the charts. Use the diagrams below to present the charts.

Total of All Average
Monthly Electric Bills
as of January 1985

Total of All Average
Monthly Electric Bills
as of January 1985

Total of All Average
Monthly Electric Bills
as of January 1985

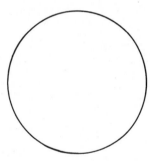

Zone   1   2   3   4

Zone   1   2   3   4

Legend: = $10,000

Ms. Arlene Davis, director **ASSIGNMENT PROBLEM**
of the Park View Hospital, located on Park Street and 10th Avenue, has
decided that the hospital will set up an outpatient clinic in any zone that
generates more than 2,000 patient visits per year on average. You have
been asked by Park View management to construct a pie chart and a
pictograph illustrating the average yearly number of hospital visits per
zone as of January 1985.

A scrupulous study of the hospital's patient records reveals the following information:

Zone	*Average yearly number of hospital visits by zone as of January 1985*
1	414
2	379
3	1,036
4	2,523
Total	4,352

Using the diagrams below, type of memorandum to Ms. Davis presenting the pie chart and pictograph.

Average Yearly Number of
Hospital Visits by Zone
as of January 1985

Average Yearly Number of
Hospital Visits by Zone
as of January 1985

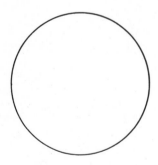

Legend: ⚲ = 100 Hospital Visits

2–9 Block analysis of the population in Stat City Mr. Lee Kaplowitz, the mayor of Stat City, is preparing a report for the town council and needs a chart depicting the number of people residing in each block of Stat City. The town council is interested in knowing the location of blocks that have fewer than 10 people, between 10 and 19 people, between 20 and 29 people, between 30 and 39 people, and 40 people or more.

Mr. Kaplowitz obtained a table from the Stat City Chamber of Commerce indicating the population on each block in Stat City. The table is as follows:

ASSIGNMENT PROBLEM

Block	Population	Block	Population	Block	Population
1	8	41	24	81	30
2	17	42	28	82	27
3	16	43	28	83	9
4	26	44	24	84	23
5	16	45	18	85	16
6	25	46	23	86	27
7	18	47	4	87	25
8	26	48	10	88	28
9	22	49	24	89	37
10	15	50	22	90	31
11	20	51	21	91	18
12	22	52	12	92	29
13	22	53	13	93	181
14	11	54	25	94	196
15	17	55	33	95	192
16	20	56	32	96	193
17	20	57	20	97	175
18	19	58	15	98	192
19	23	59	35	99	200
20	23	60	17	100	180
21	22	61	15	101	86
22	28	62	30	102	89
23	25	63	23	103	101
24	26	64	32	104	85
25	24	65	18	105	97
26	10	66	36	106	96
27	17	67	15	107	102
28	17	68	34	108	102
29	23	69	34	109	93
30	18	70	175	110	91
31	23	71	162	111	89
32	31	72	172	112	98
33	20	73	12	113	80
34	23	74	25	114	82
35	22	75	11	115	88
36	18	76	22	116	98
37	11	77	30	117	90
38	21	78	26	118	92
39	24	79	27	119	94
40	21	80	24	120	91

You have been asked by the mayor to use this table in preparing a statistical map of Stat City reflecting the block populations according to the previously stated population categories. Use the map on the back of this page and prepare an accompanying memorandum for Mr. Kaplowitz.

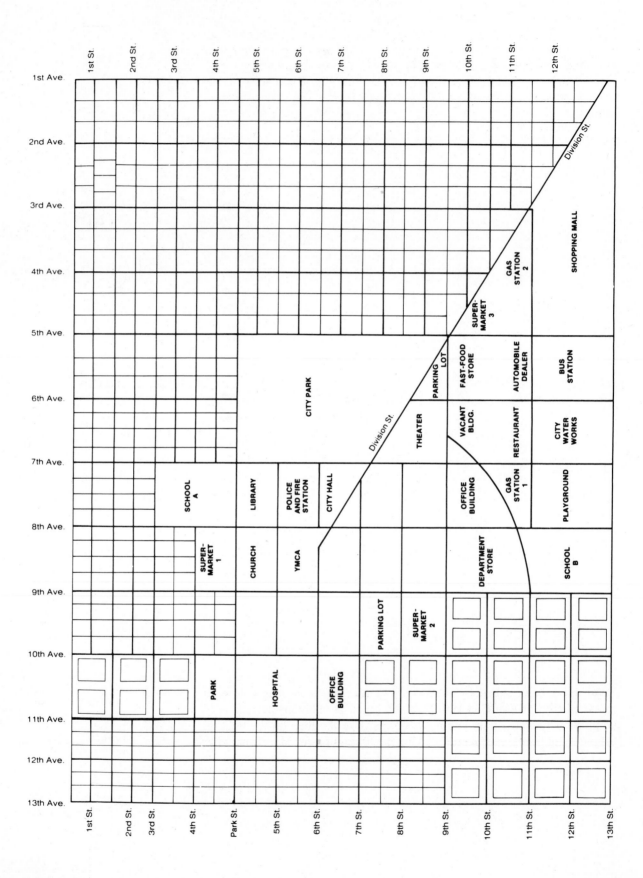

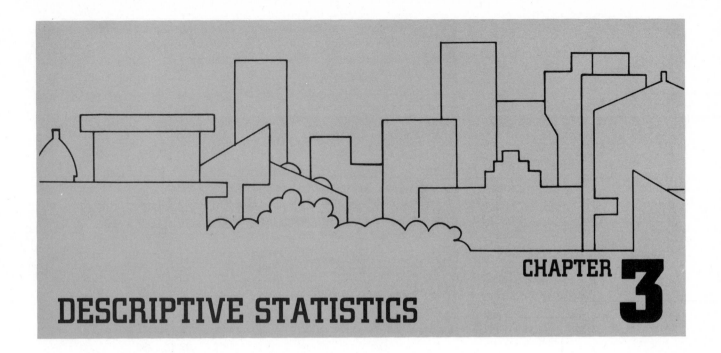

DESCRIPTIVE STATISTICS

CHAPTER 3

INTRODUCTION

In this chapter, new techniques will be presented for reducing large amounts of data to manageable dimensions for decision-making purposes. You will learn how to display data using frequency distributions, histograms, and frequency polygons. You will also learn how to use numerical measures to summarize data. These fall into three distinct categories: measures of location, measures of dispersion, and measures of shape. Once you have mastered these tools on Stat City, you will begin to feel the tremendous insight that can be gained about a population by using statistics.

The solutions to some of the examples in this chapter will have two parts: a population section and a sample section. The population section will show you how to compute parameters. The sample section will show you how to estimate parameters by drawing a random sample and computing statistics. Finally, you will be asked to determine if your statistics are good estimators of the parameters.

Once you have confidence in your ability to sample, you will no longer be required to compute parameters. In fact, it is important that you do not develop a dependence on parameters because they are rarely obtainable in real business situations.

Example 3–1 Image study of Howie's Gulf Station

Problem

Mr. Howie, the owner of Howie's Gulf Station, wishes to change the image of his station in the minds of Zone 1 residents. He believes a more favorable, service-conscious image would improve business. Before attempting the image alteration, Mr. Howie wants to know how many Zone 1 families think of Howie's Gulf as their favorite station and how many Zone 1 families think of Paul's Texaco as their favorite station. Mr. Howie thinks this information is important in determining whether he should attempt the image change.

You have been retained by Mr. Howie to perform the above study. Conduct the study and prepare a typed memorandum for him (see Exhibit 3–1).

Note to the student: Make two surveys. First, conduct a survey investigating all 130 dwelling units in Zone 1. Second, conduct a survey investigating a random sample of 30 dwelling units in Zone 1. Finally, compare the results of your two surveys. Do you see anything surprising in the comparison?

Solution

Part 1 The population frequency distribution can be seen in Table 3–1.

Part 2 A sample frequency distribution based on 30 randomly selected dwelling units is constructed as shown.

1. Draw 30 random numbers from a random number table using the rules set out in Chapter 1 (see Table 3–2).

Table 3–1

Classes	Frequency	Percent
Howie's Gulf	80	61.5
Paul's Texaco	50	38.5
Total	130	100.0

Table 3–2

good example of sample & population

Begin here →

6067	7516	2451	1510	²⁴0201	1437
4541	9863	8312	9855	0995	6025
6987	4802	8975	2847	4413	5997
¹0376	8636	9953	4418	2388	8997
8468	5763	3232	1986	7134	4200
9151	4967	3255	8518	2802	8815
²1073	4930	1830	2224	2246	²⁷1000
5487	1967	5836	2090	3832	0002
4896	4957	6536	7430	6208	3929
9143	7911	¹³0368	¹⁸0541	2302	5473
9256	2956	4747	6280	7342	²⁸0453
4173	⁸1219	7744	9241	6354	4211
2525	7811	5417	7824	0922	8752
9165	⁹1156	6603	2852	8370	0995
³0014	8474	6322	5053	5015	6043
5325	7320	8406	5962	6100	3854
2558	1748	5671	4974	7073	3273
⁴0117	1218	¹⁴0688	2756	7545	5426
8353	1554	4083	2029	8857	4781
1990	9886	3280	6109	9158	3034
9651	7870	2555	3518	2906	4900
9941	5617	1984	2435	5184	0379
7769	5785	9321	2734	2890	3105
3224	8379	9952	¹⁹0515	2724	4826
⁵1287	7275	6646	1378	6433	0005
6389	4191	4548	5546	6651	8248
1625	4327	2654	4129	3509	3217
7555	3020	4181	7498	4022	9122
4177	1844	3468	1389	3884	6900
⁶0927	¹⁰0124	8176	0680	²⁵1056	1008
8505	1781	7155	3635	9751	5414
8022	8757	6275	1485	3635	2330
8390	8802	5674	2559	7934	4788
3630	5783	7762	²⁰0223	5328	7731
9154	6388	6053	9633	2080	7269
1441	3381	7823	8767	9647	4445
8246	¹¹0778	¹⁵0993	6687	7212	9968
2730	3984	¹⁶0563	9636	7202	0127
9196	8276	¹⁷0233	²¹0879	3385	2184
5928	9610	9161	²²0748	3794	9683
⁷1042	9600	7122	2135	7868	5596
6552	4103	7957	0510	5958	²⁹0211
5968	4307	9327	3197	0876	8480
4445	¹²1018	4356	4653	9302	³⁰0761
8727	8201	5980	7859	6055	1403
9415	9311	4996	2775	8509	7767
2648	7639	9128	²³0341	6875	8957
3707	3454	8829	6863	²⁶1297	5089

2. List the 30 random numbers and the favorite gas station for each dwelling unit to which that random number (identification number—variable 4) belongs (see Table 3–3).

3. Construct a frequency distribution from the sample data and estimate the population frequency distribution (see Table 3–4).

Table 3–4

Class	Sample frequency	Sample percentage	Estimated population frequency
Howie's Gulf	20	66.7	87*
Paul's Texaco	10	33.3	43†
Total	30	100.0	130

* 87 = 130 × .667.
† 43 = 130 × .333.

4. The memorandum appears in Exhibit 3–1.

Exhibit 3–1

HOWARD S. GITLOW, PH.D.
STATISTICAL CONSULTANT

MEMORANDUM

TO: Mr. Howie, Owner
        Howie's Gulf Station

FROM: Howard Gitlow

DATE: March 29, 1985

RE: Image study of Howie's Gulf Station

As per your instructions and using commonly accepted statistical techniques, I have found that approximately 87 Zone 1 dwelling units (66.7 percent of the dwelling units) consider Howie's Gulf Station their favorite. The remaining 43 Zone 1 dwelling units (33.3 percent of the dwelling units) prefer Paul's Texaco Station.*

I direct your attention to the following table for a summary of my findings.

	Percent	Estimated frequency
Howie's Gulf	66.7	87
Paul's Texaco	33.3	43
Total	100.0	130

If you have any questions, please do not hesitate to call me at 305-999-9999.

* The above information was obtained through selecting a random sample of 30 Stat City dwelling units.

Table 3–3

Random identification number (variable 4)	Favorite gas station (variable 20)*
37	1
107	2
1	2
11	1
128	1
92	2
104	2
121	1
115	1
12	2
77	2
101	2
36	2
68	2
99	2
56	1
23	1
54	2
51	2
22	1
87	1
74	2
34	2
20	2
105	2
129	2
100	2
45	1
21	2
76	2

* 1 = Paul's Texaco.
  2 = Howie's Gulf.

Comments: A comparison of the actual and estimated population frequency distributions are shown in Table 3–5.

Table 3–5

Class	Actual	Estimated	Difference
Howie's Gulf	80	87	−7
Paul's Texaco	50	43	+7
Total	130	130	

It is easy to see that the actual and estimated population frequency distributions are different. This difference is due to sampling error—error introduced by using only 30 Zone 1 dwelling units rather than all 130 Zone 1 dwelling units.

It is important to realize that random samples and statistical procedures yield correct answers only on average and that a particular sample's statistics may be very different from the parameters the sample was drawn to estimate. On average means that if many random samples were drawn and a statistic was computed from every sample, the average of all the statistics would yield the parameter the statistics were estimating.

Just for practice, construct another sample frequency distribution based on a different sample of 30 dwelling units. The second sample frequency distribution is constructed as shown.

1. Draw 30 random numbers from a random number table. Use Table 3–6.

Table 3–6

Begin here →

5347	8111	9803	*15*1221	5952	4023	4057	3935	4321	6925
9734	7032	5811	9196	2624	4464	8328	9739	9282	7757
6602	3827	7452	7111	8489	1395	9889	9231	6578	5964
9977	7572	*11*0317	4311	8308	8198	1453	2616	2489	2055
3017	4897	9215	3841	4243	2663	8390	4472	6921	6911
8187	8333	1498	9993	1321	3017	4796	9379	8669	9885
1983	9063	7186	9505	5553	6090	8410	5534	4847	6379
1 0933	3343	5386	5276	1880	2582	9619	6651	7831	9701
3115	5829	4082	4133	2109	9388	4919	4487	4718	8142
6761	5251	*12*0303	8169	1710	6498	6083	8531	4781	0807
6194	4879	*13*1160	8304	2225	*26*1183	0434	9554	2036	5593
2 0481	6489	9634	7906	2699	4396	6348	9357	8075	9658
3 0576	3960	5614	2551	8615	7865	0218	2971	0433	1567
7326	5687	4079	1394	9628	9018	4711	6680	6184	4468
5490	*6* 0997	7658	*16* 0264	3579	4453	6442	3544	2831	9900
4258	3633	6006	*17*0404	2967	1634	4859	2554	6317	7522
2726	2740	9752	2333	3645	3369	2367	4588	4151	0475
4984	71144	6668	3605	3200	7860	3692	5996	6819	6258
2931	4046	2707	6923	5142	5851	4992	0390	2659	3306
3046	2785	6779	1683	7427	*05*79	0290	6349	0078	3509
2870	8408	6553	4425	3386	8253	9839	2638	0283	3683
1318	5065	9487	2825	7854	5528	3359	6196	5172	1421
6079	7663	3015	4029	9947	2833	1536	4248	6031	4277
1348	4691	6468	*18*0741	7784	*27*0190	4779	6579	4423	7723
3491	9450	3937	3418	5750	2251	0406	9451	4461	1048
2810	*04*81	8517	8649	3569	*28* 0348	5731	6317	7190	7118
5923	4502	*14* 0117	*19*0884	8192	7149	9540	3404	0485	6591
8743	8275	7109	3683	5358	2598	4600	4284	8168	2145
2904	*8* 0130	5534	6573	7871	4364	4624	5320	9486	4871
6203	7188	9450	1526	6143	*29*1036	4205	6825	1438	7943
3885	8004	5997	7336	5287	4767	4102	8229	2643	8737
4066	4332	8737	8641	9584	2559	5413	9418	4230	0736
4058	9008	3772	*20*0866	3725	2031	5331	5098	3290	3209
7823	8655	5027	2043	*23*0024	*30* 0230	7102	4993	2324	0086
9824	6747	7145	6954	*05*16	0332	6701	9254	9797	5272
6997	7855	6543	3262	2831	6181	1459	7972	5569	9134
3984	2307	4081	*21* 0371	2189	9635	9680	2459	2620	2600
6288	8727	9989	9996	3437	4255	1167	9960	9801	4886
5613	6492	2945	5296	8662	6242	3016	7618	9531	3926
9080	5602	4899	6456	6746	6018	1297	0384	6258	9385
4 0966	4467	7476	3335	6730	8054	9765	1134	7877	4501
3475	5040	7663	*22*1276	3222	3454	1810	5351	1452	7212
5 1215	7332	7419	2666	7808	5363	5230	0000	0570	6353
6938	*9* 0773	9445	7642	1612	0930	6741	6858	8793	3884
9335	6456	4376	4504	4493	6997	1696	0827	6775	6029
3887	3554	9956	8540	*24*0491	6254	7840	0101	8618	2207
5831	6029	7239	6966	*25*1247	9305	0205	2980	6364	1279
8356	*10* 1022	9947	7472	2207	1023	2157	2032	2131	5712
2806	9115	4056	3370	6451	0706	6437	2633	7965	3114
*05*73	7555	9316	8092	5587	5410	3480	8315	0453	8136

2. List the 30 random numbers and the favorite gas station for each dwelling unit to which that random number (identification number—variable 4) belongs (Table 3–7).

Table 3–7

Random identification number (variable 4)	Favorite gas station (variable 20)*
93	2
48	2
57	2
96	1
121	1
99	2
114	2
13	2
77	2
102	2
31	1
30	1
116	2
11	1
122	2
26	1
40	1
74	2
88	2
86	1
37	1
127	2
2	1
49	1
124	1
118	2
19	2
34	2
103	1
23	1

* 1 = Paul's Texaco.
2 = Howie's Gulf.

3. Construct a frequency distribution from the sample data and estimate the population frequency distribution (Table 3–8).

Table 3–8

Class	Sample frequency	Sample percentage	Estimated population frequency
Howie's Gulf	16	53.3	69*
Paul's Texaco	14	46.7	61†
Total	30	100.0	130

* 69 = 130 × .533.
† 61 = 130 × .467.

Comments: A comparison of the two sample percentages (percent of dwelling units favoring Howie's Gulf) with the population percentage demonstrates that sample statistics differ from sample to sample (66.7 percent versus 53.3 percent) and from the population percentage (61.5 percent). Not all statistics accurately reflect parameters. For example, the first sample percentage (66.7 percent) was reasonably close to the population percentage (61.5 percent). But the second sample percentage (53.3 percent) reflected a wide variation from the population percentage (61.5 percent). Of course in actual business situations you don't know the value of the population parameter (if you did, you wouldn't have to sample). and so, unlike this case, you don't know how close a particular sample statistic is to the parameter.

Later in this course you will learn about techniques for controlling the accuracy of statistics when estimating parameters.

Very often you may want to describe data by means of a numerical measure in addition to or instead of a frequency chart. Next you will learn how to calculate and use such measures of location, dispersion, and shape.

Measures of location

There are many measures of location that can be used to summarize data: for example, the arithmetic mean, the geometric mean, the harmonic mean, the weighted mean, the mode, the median, percentages, and quantiles. In this chapter, you will only work with the arithmetic mean, the mode, the median (for ungrouped data), and proportions (or percentages).

The arithmetic mean of ungrouped data is the average value.[1] The mode is the most frequently occurring number in a population or a sample. The median is the middle data point in a population or sample in which the numbers have been ranked from lowest to highest.[2]

A proportion can be computed as a ratio:

$$\pi = \frac{\text{Number of times a number (or an event) occurs in a population}}{\text{Size of the population (number of elementary units in the population)}}$$

$$\bar{p} = \frac{\text{Number of times a number (or an event) occurs in a sample}}{\text{Size of the sample (number of elementary units in the sample)}}$$

If you multiply the fraction obtained by 100, you have a percentage.

Measures of dispersion and shape

The most commonly used measures of dispersion are the range, the interquartile range, the variance, the standard deviation, the coefficient of variation, sums of squares, and the average deviation. In this chapter, you will work with the range and the standard deviation of ungrouped data. The range of ungrouped data is the difference between the maximum and minimum values. The standard deviation of ungrouped data is shown in Table 3–9.

This book will cover measures of shape, such as skewness and kurtosis, in the "Computational Exercises" section.

We continue with several Stat City related examples worked out in detail and several additional problems to solve for Stat City interest groups.

Table 3–9

Population formulas
$\sigma = \sqrt{\dfrac{\sum_{i=1}^{N}(X_i - \mu)^2}{N}}$
or
$\sigma = \sqrt{\dfrac{\sum_{i=1}^{N}(X_i^2) - \dfrac{(\sum_{i=1}^{N} X_i)^2}{N}}{N}}$
Sample formulas
$s = \sqrt{\dfrac{\sum_{i=1}^{n}(x_i - \bar{x})^2}{n-1}}$
or
$s = \sqrt{\dfrac{\sum_{i=1}^{n}(x_i^2) - \dfrac{[\sum_{i=1}^{n}(x_i)]^2}{n}}{n-1}}$

---

[1] The formula for the arithmetic mean is:

$$\mu = \sum_{i=1}^{N} X_i/N \quad \text{for a population, and}$$

$$\bar{x} = \sum_{i=1}^{n} x_i/n \quad \text{for a sample.}$$

[2] The median's position in a ranked sequence of numbers is:

$$\frac{N+1}{2} \quad \text{for a population, and}$$

$$\frac{n+1}{2} \quad \text{for a sample.}$$

when N is odd

Example 3–2 Distribution of rooms per dwelling unit in Zone 2

Problem   The Stat City Tax Assessor's Office is planning to reassess all dwelling units in Zone 2. The tax office needs information about the number of rooms per dwelling unit to set the new assessment rates. With the above information, the tax office can set assessment guidelines to yield a target total tax.

You have been hired by the tax office to determine the distribution of rooms per dwelling unit in Zone 2 as well as the mean, median, minimum, maximum, range, and standard deviation of the number of rooms per dwelling in Zone 2. Conduct the necessary survey twice (once for the population and once for a sample of 30 randomly selected dwelling units) and write a summary of your findings to the tax office (see Exhibit 3–2). Include a relative frequency polygon in your memorandum.

Solution   **Part 1**   The population frequency distribution can be seen in Table 3–10.

Table 3–10

Number of rooms	Frequency	Percent
5	7	4.5
6	15	9.6
7	37	23.6
8	49	31.2
9	36	22.9
10	10	6.4
11	3	1.9
Total......	157	100.0*

* Percentages do not total 100 percent due to rounding.

The population parameters can be seen below:

Mean*...........................	7.854
Mode............................	8.000
Median†........................	8.000
Minimum	5.000
Maximum.......................	11.000
Range‡	6.000
Standard deviation§..............	1.286

* Mean $= \mu = \dfrac{\Sigma X}{N} = \dfrac{1,233}{157} = 7.854$.

† Median position $= \dfrac{157 + 1}{2} = 79$.

  79th number is 8.

‡ Range $= 11 - 5 = 6$.

§ $\sigma = \sqrt{\dfrac{\Sigma X^2 - (\Sigma X)^2/N}{N}}$

$\quad = \sqrt{\dfrac{9,943 - (1,233)^2/157}{157}} = 1.286$.

Part 2 A sample frequency distribution based on 30 randomly selected dwelling units is constructed as shown.

1. Draw 30 random numbers from a random number table using the rules set down in Chapter 1 (use Table 3–11).

Table 3–11

Begin here →

```
 5347  8111  9803  1221     5952  4023  4057  3935  4321  6925
 9734  7032  5811  9196  21 2624  4464  8328  9739  9282  7757
 6602  3827  7452  7111     8489  1395  9889  9231  6578  5964
 9977  7572  0317  4311     8308  8198  1453  2616  2489  2055
 3017  4897  9215  3841     4243  2663  8390  4472  6921  6911

 8187  8333 // 1498  9993  22 1321  3017  4796  9379  8669  9885
/1983  9063  7186  9505     5553  6090  8410  5534  4847  6379
 0933  3343  5386  5276  23 1880  2582  9619  6651  7831  9701
 3115  5829  4082  4133  24 2109  9388  4919  4487  4718  8142
 6761  5251  0303  8169  25 1710  6498  6083  8531  4781  0807

 6194  4879  1160  8304  26 2225  1183  0434  9554  2036  5593
 0481  6489  9634  7906  27 2699  4396  6348  9357  8075  9658
 0576  3960  5614 /3 2551  8615  7865  0218  2971  0433  1567
 7326  5687  4079 /4 1394  9628  9018  4711  6680  6184  4468
 5490  0997  7658     0264  3579  4453  6442  3544  2831  9900

 4258  3633  6006     0404  2967  1634  4859  2554  6317  7522
2 2726 8 2740  9752 /5 2333  3645  3369  2367  4588  4151  0475
 4984  1144  6668     3605  3200  7860  3692  5996  6819  6258
 2931  4046 /2 2707  6923  5142  5851  4992  0390  2659  3306
 3046 9 2785  6779 /6 1683  7427  0579  0290  6349  0078  3509

3 2870  8408  6553  4425  3386  8253  9839  2638  0283  3683
4 1318  5065  9487 /7 2825  7854  5528  3359  6196  5172  1421
 6079  7663  3015     4029  9947  2833  1536  4248  6031  4277
5 1348  4691  6468     0741  7784  0190  4779  6579  4423  7723
 3491  9450  3937     3418  5750  2251  0406  9451  4461  1048

6 2810  0481  8517  8649  3569  0348  5731  6317  7190  7118
 5923  4502  0117  0884  8192  7149  9540  3404  0485  6591
 8743  8275  7109  3683  5358  2598  4600  4284  8168  2145
 2904  0130  5534  6573  7871  4364  4624  5320  9486  4871
 6203  7188  9450 /8 1526  6143  1036  4205  6825  1438  7943

 3885  8004  5997  7336  5287  4767  4102  8229  2643  8737
 4066  4332  8737  8641  9584  2559  5413  9418  4230  0736
 4058  9008  3772  0866  3725  2031  5331  5098  3290  3209
 7823  8655  5027 /9 2043  0024  0230  7102  4993  2324  0086
 9824  6747  7145  6954  0116  0332  6701  9254  9797  5272

 6997  7855  6543  3262  28 2831  6181  1459  7972  5569  9134
 3984 /0 2307  4081  0371  29 2189  9635  9680  2459  2620  2600
 6288  8727  9989  9996     3437  4255  1167  9960  9801  4886
 5613  6492  2945  5296     8662  6242  3016  7618  9531  3926
 9080  5602  4899  6456     6746  6018  1297  0384  6258  9385

 0966  4467  7476  3335  6730  8054  9765  1134  7877  4501
 3475  5040  7663  1276  3222  3454  1810  5351  1452  7212
 1215  7332  7419 20 2666  7808  5363  5230  0000  0570  6353
 6938  0773  9445  7642  30 1612  0930  6741  6858  8793  3884
 9335  6456  4376  4504  4493  6997  1696  0827  6775  6029

 3887  3554  9956  8540  0491  6254  7840  0101  8618  2207
 5831  6029  7239  6966  1247  9305  0205  2980  6364  1279
 8356  1022  9947  7472  2207  1023  2157  2032  2131  5712
7 2806  9115  4056  3370  6451  0706  6437  2633  7965  3114
 0573  7555  9316  8092  5587  5410  3480  8315  0453  8136
```

Table 3–12

Random identification number (variable 4)	Number of rooms (variable 9)
198	6
272	9
287	9
131	6
134	7
281	9
280	8
274	7
278	10
230	8
149	8
270	5
255	7
139	9
233	8
168	7
282	9
152	8
204	8
266	7
262	5
132	7
188	8
210	8
171	10
222	8
269	7
283	8
218	7
161	8

2. List the 30 random numbers and the number of rooms for each dwelling unit to which that random (identification number—variable 4) number belongs (Table 3–12).

3. Construct a frequency distribution from the sample data and estimate the population frequency distribution (Table 3–13).

Table 3–13

Number of rooms	Sample frequency	Sample percentage	Estimated population frequency
5	2	6.7	10
6	2	6.7	10
7	8	26.7	42
8	11	36.7	58
9	5	16.7	26
10	2	6.7	10
Total	30	100.0*	157†

* Percentages do not add up to 100 due to rounding.
† Estimated frequencies do not add up to 157 due to rounding.

We calculate the required sample statistics as shown below.

Mean*	7.700
Mode	8.000
Median†	8.000
Minimum	5.000
Maximum	10.000
Range‡	5.000
Standard deviation§	1.236

* Mean = $\bar{x} = \dfrac{\Sigma x}{n} = \dfrac{231}{30} = 7.7.$

† Median position = $\dfrac{30 + 1}{2} = 15.5.$

5, 5, 6, 6, 7, 7, 7, 7, 7, 7, 7, 8, 8, (8, 8,) 8, 8, 8, 8, 8, 8, 9, 9, 9, 9, 9, 10, 10

Median = $\dfrac{8 + 8}{2} = \dfrac{16}{2} = 8.$

‡ Range = 10 − 5 = 5.

§ $s = \sqrt{\dfrac{\Sigma(x - \bar{x})^2}{n - 1}} = \sqrt{\dfrac{44.3}{30 - 1}}$

$= \sqrt{1.5276} = 1.236.$

(See Table 3–14.)

Table 3–14

Random identification number	Number of rooms (x)	$(x - \bar{x}) = (x - 7.7)$	$(x - 7.7)^2$
198	6	−1.7	2.89
272	9	1.3	1.69
287	9	1.3	1.69
131	6	−1.7	2.89
134	7	−.7	.49
281	9	1.3	1.69
280	8	.3	.09
274	7	−.7	.49
278	10	2.3	5.29
230	8	.3	.09
149	8	.3	.09
270	5	−2.7	7.29
255	7	−.7	.49
139	9	1.3	1.69
233	8	.3	.09
168	7	−.7	.49
282	9	1.3	1.69
152	8	.3	.09
204	8	.3	.09
266	7	−.7	.49
262	5	−2.7	7.29
132	7	−.7	.49
188	8	.3	.09
210	8	.3	.09
171	10	2.3	5.29
222	8	.3	.09
269	7	−.7	.49
283	8	.3	.09
218	7	−.7	.49
161	8	.3	.09

$$\Sigma(x - \bar{x})^2 = 44.3$$

Comments: A comparison of the actual and estimated population characteristics is shown in Table 3–15.

Table 3–15

Number of rooms	Actual frequency	Estimated population frequency	Difference = Actual − Estimated
5	7	10	−3
6	15	10	+5
7	37	42	−5
8	49	58	−9
9	36	26	+10
10	10	10	0
11	3	0	+3
Total	157	157*	

	Population	Sample
Mean	7.854	7.700
Mode	8.000	8.000
Median	8.000	8.000
Minimum	5.000	5.000
Maximum	11.000	10.000
Range	6.000	5.000
Standard deviation	1.286	1.236

* Estimated frequencies do not add up to 157 due to rounding.

It is apparent that the actual and estimated population frequency distributions are different. This difference is once again due to sampling error. However, you should notice that eight rooms is the most commonly occurring class in both distributions and the number of rooms trails off on either side of eight rooms in both distributions. Consequently, the final picture we get from the sample is not as inaccurate as we had originally thought. Both the actual and the estimated frequency distributions tell us that the typical number of rooms per dwelling unit in Zone 2 is approximately eight.

Similarly, it is apparent that the population parameters and the sample statistics are different due to sampling error. Again however, the final picture we get from the sample is not an inaccurate one; the sample measures of location indicate a typical number of rooms of about eight, and the sample range and standard deviation are close to the population measures of dispersion.

Remember, random samples and statistical procedures yield correct answers only on average, and a particular sample's statistics may be very different from the parameters the sample was drawn to estimate.

The memorandum appears in Exhibit 3–2.

Exhibit 3–2

ROSA OPPENHEIM, PH.D.
STATISTICAL CONSULTANT

MEMORANDUM

TO: Stat City Tax Assessor's Office

FROM: Rosa Oppenheim

DATE: October 18, 1985

RE: Distribution of rooms per dwelling unit in Zone 2
        of Stat City as of January 1985

As per your instructions and using commonly accepted statistical techniques, I have estimated the distribution of rooms per dwelling unit in Zone 2 of Stat City. I direct your attention to the attached chart for a summary of my findings.

I have also computed several summary measures concerning the number of rooms per dwelling unit in Zone 2.[1] My research indicates that the typical Zone 2 dwelling unit has approximately eight rooms.[2] Further, the number of rooms per dwelling unit in Zone 2 ranges between 5 and 10,[3] indicating some degree of variability in dwelling unit size.[4]

Exhibit 3–2 (*concluded*)

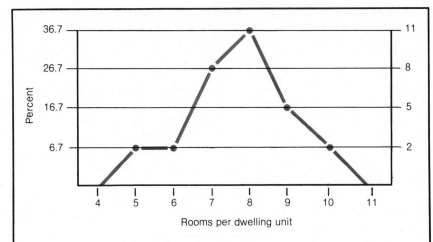

The above information was obtained through selecting a sample of 30 Zone 2 dwelling units.

If you have any further questions, please do not hesitate to call me at 305–999–9999.

Footnotes

[1] The summary statistics reported are based on a random sample of 30 Zone 2 dwelling units.

[2] Typical refers to the fact that the average number of rooms per dwelling unit is 7.7, and the mode and median for the number of rooms per dwelling unit in Zone 2 are both 8.

[3] Dwelling units range in size from two bedrooms, one bathroom, one kitchen, one living room up to four bedrooms, three bathrooms, one kitchen, one den, one living room.

[4] The degree of variability in the number of rooms per dwelling unit in Zone 2 of Stat City is also evidenced by the fact that the standard deviation of the number of rooms per dwelling unit in Zone 2 is 1.236 rooms. This number indicates that approximately 95 percent of Zone 2 dwelling units have between 5.28 and 10.12 rooms. The above statement is true if the sample mean and standard deviation accurately reflect the true, but unknown, population mean and standard deviation and the distribution of rooms in Zone 2 is bell–shaped. I believe that this assumption is tenable; consequently, the standard deviation can be interpreted as stated above.

Please do not be confused by the discrepancy between 95 percent of the dwelling units having between 5.28 and 10.12 rooms and that the number of rooms per dwelling unit in Zone 2 ranges between 5 and 10. The discrepancy can be easily explained by realizing that all the statistics presented are only sample estimates. If the true, but unknown, minimum and maximum number of rooms per dwelling unit in Zone 2 and the true, but unknown, standard deviation of the number of rooms per dwelling unit in Zone 2 were available, this discrepancy would no longer exist. Briefly expressed, the interval created by subtracting the minimum from the maximum number of rooms per dwelling unit would be larger than the interval created in which 95 percent of Zone 2 dwelling units would fall.

Many of the memorandums in this book will have footnotes like the ones in Exhibit 3–2. You will see that these are important in clarifying your results and procedures as well as in serving as a partial disclaimer for the errors which can occur when computing statistics from a particular sample. In Exhibit A–4, Appendix A, you will find a listing of the standard footnotes presented and referred to in some of these memorandums.

Example 3–3   Analysis of electric bills in Stat City

Problem   Mr. Saul Reisman, the chief financial officer of the Stat City Electric Company, needs information concerning average monthly electric bills for Stat City families. In particular, Mr. Reisman needs to know the mean, the range, and the number of Stat City dwelling units with average monthly electric bills in the following categories: $10 to less than $40, $40 to less than $70, $70 to less than $100, and $100 to less than $130.

Conduct a survey and write a summary of your findings to Mr. Reisman (see Exhibit 3–4). Use a histogram and frequency polygon in your report and include the mean and range for the sample.

Note to the student: Draw a random sample of 30 Stat City dwelling units to perform your appointed task. In this problem you are *not* being asked to compute parameters; consequently, you must rely on statistics for decision-making purposes. From now on, we will work with statistics and will draw inferences about the unknown parameters. Reliance on statistics, and nonreliance on parameters, is a big step into the world of statistics.

It will be assumed from now on that you know how to use a random number table. Consequently, in most problems you will not be required to actually select the random identification numbers used in obtaining your sample. Instead, the random numbers to be used in drawing your random samples will be listed in the problems. The only reason for doing this is to save time and effort that could more wisely be invested elsewhere.

Table 3–16 lists the random sample of 30 dwelling units' identification numbers (variable 4) to be used in solving Mr. Reisman's problem. Remember, the numbers are just random identification numbers. You must still draw the sample of 30 average monthly electric bills (variable 11) from Stat City.

Table 3–16

4,	37,	163,	164,	191,
208,	391,	395,	408,	485,
511,	540,	541,	632,	780,
805,	815,	923,	954,	968,
973,	1038,	1060,	1063,	1153,
1188,	1206,	1209,	1281,	1291.

Solution   A sample frequency distribution based on the 30 randomly selected dwelling units is constructed as shown.

1.   List the 30 random identification numbers and the average monthly electric bills for each dwelling unit to which those random identification numbers belong (Table 3–17).

Table 3–17

Random identification number (variable 4)	Average monthly electric bill (variable 11)
4	$78
37	83
163	66
164	70
191	41
208	75
391	73
395	97
408	48
485	64
511	47
540	72
541	80
632	24
780	40
805	37
815	31
923	46
954	54
968	48
973	43
1038	38
1060	54
1063	34
1158	29
1188	38
1206	56
1209	34
1281	37
1291	35

2. Construct a sample frequency distribution and estimate the population frequency distribution, population histogram, and population frequency polygon (Table 3–18 and Exhibit 3–3).

Table 3–18

Class	Sample frequency	Sample percentage	Estimated population frequency
$ 10 to less than $ 40	10	33.3	.333 × 1,373 = 458
40 to less than 70	12	40.0	.400 × 1,373 = 549
70 to less than 100	8	26.7	.267 × 1,373 = 366
100 to less than 130	0	0.0	0
Total	30	100.0	1,372

Exhibit 3–3

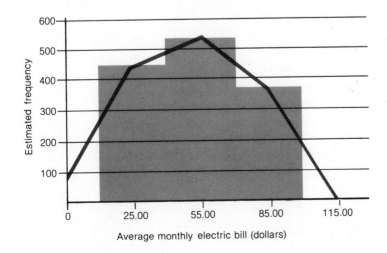

Average monthly electric bill (dollars)

3. Calculate the arithmetic mean and range.

$$\text{Mean} = \bar{X} = \frac{\Sigma x}{n} = \frac{1,572}{30} = 52.4$$

$$\text{Range} = 97 - 24 = 73$$

4. The memorandum appears in Exhibit 3–4.

Exhibit 3–4

HOWARD S. GITLOW, PH.D.
STATISTICAL CONSULTANT

MEMORANDUM

TO: Mr. Saul Reisman, Chief Financial Officer
        Stat City Electric Company

FROM: Howard Gitlow

DATE: April 15, 1985

RE: Average monthly electric bills in Stat City as of
        January 1985

As per your instructions and using commonly accepted statistical techniques, I have estimated the distribution of average monthly electric bills in Stat City. I direct your attention to the statistics and charts below for a summary of my findings. I am enclosing two pictorial representations of the same information so that you may select the one which best meets your needs.

My research indicates that the average monthly electric bill in Stat City can be estimated as $52.40; the range of bills in $73, from a low of $24 a month to a high of $97 a month.

Exhibit 3–4 (concluded)

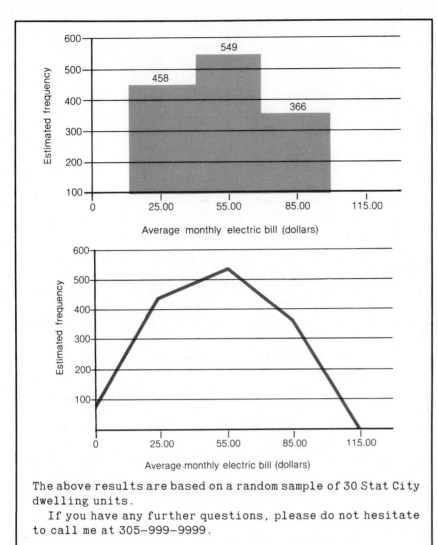

The above results are based on a random sample of 30 Stat City dwelling units.

If you have any further questions, please do not hesitate to call me at 305-999-9999.

Comments: This is the first time you do not have parameters to compare with your statistics. Consequently, you do not know if your random sample of 30 dwelling units is giving you a true picture of Stat City average monthly electric bills. WELCOME TO THE WORLD OF DECISION MAKING!

Example 3–4   Construction of an emergency clinic in Zone 1

Problem

Ms. Arlene Davis, director of the Stat City Hospital, is considering the advisability of locating a new emergency clinic in Zone 1 of Stat City. Ms. Davis decided that if the average number of trips to the hospital per year in Zone 1 is more than three, the emergency clinic should be constructed.

She has asked you to provide her with appropriate statistics. Conduct the required survey and type a memorandum to Ms. Davis (see Exhibit 3–5). Make sure you include information concerning the mean, mode, minimum, maximum, range, and standard deviation for the relevant variable.

Please use the following list of 19 randomly selected identification numbers (ID) in solving the problem:

8, 17, 29, 36, 37, 44, 55, 57, 65, 67, 69, 74, 80, 85, 90, 98, 110, 122, and 130.

Solution   The hospital statistics are shown below.

Mean*	3.316
Mode	1 and 2
Median†	3.000
Minimum	0.000
Maximum	8.000
Range‡	8.000
Standard deviation§	2.237

Yearly trips to hospital	Frequency	Percent
0	1	5.3
1	4‖	21.1
2	4‖	21.1
3	1	5.3
4	3	15.8
5	2	10.5
6	3	15.8
8	1	5.3
Total	19	100.0‖

* Mean = $\bar{x} = \dfrac{\Sigma x}{n} = \dfrac{63}{19} = 3.316$.

† Median position = $\dfrac{19 + 1}{2} = \dfrac{20}{2} = 10$.

0, 1, 1, 1, 1, 2, 2, 2, ③, 4, 4, 4, 5, 5, 6, 6, 6, 8
Median = 3.

‡ Range = 8 − 0 = 8.

§ $s = \sqrt{\dfrac{\Sigma(x - \bar{x})^2}{n - 1}}$

$s = \sqrt{\dfrac{90.106}{19 - 1}} = \sqrt{5.0059} = 2.237$.

(See Table 3–19.)

‖ Percentages do not total 100 percent due to rounding.

Table 3–19

Random identification number (variable 4)	Average yearly trips to hospital per dwelling unit (variable 21) x	$(x - \bar{x})$ $(x - 3.316)$	$(x - \bar{x})^2$ $(x - 3.316)^2$
8	5	1.684	2.8359
17	4	.684	.4679
29	2	1.316	1.7319
36	4	.684	.4679
37	2	1.316	1.7319
44	2	1.316	1.7319
55	0	−3.316	10.9959
57	5	1.684	2.8359
65	2	1.316	1.7319
67	3	−.316	.0999
69	8	4.684	21.9399
74	1	−2.316	5.3639
80	6	2.684	7.2039
85	1	−2.316	5.3639
90	1	−2.316	5.3639
98	1	−2.316	5.3639
110	6	2.684	7.2039
122	6	2.684	7.2039
130	4	.684	.4679

$\Sigma(x - \bar{x})^2 = 90.106$

The memorandum appears in Exhibit 3–5.

Exhibit 3–5

ROSA OPPENHEIM, PH.D.
STATISTICAL CONSULTANT

MEMORANDUM

TO: Ms. Arlene Davis, Director
        Stat City Hospital

FROM: Rosa Oppenheim

DATE: April 30, 1985

RE: Construction of an emergency clinic in Zone 1

As per your request and using commonly accepted statistical techniques, I have computed several summary measures concerning the average number of trips to the hospital per year for Zone 1 families (dwelling units)[1] My research indicates that the most common number of trips to the hospital per year for a Zone 1 family is 1 or 2.[2] However, the average number of yearly trips to the hospital per family is 3.316.[3] Further, the number of trips to the hospital per year ranges between 0 and 8, indicating some degree of variability in yearly trips to the hospital.[4]

If you have any further questions, please do not hesitate to call me at 305–999–9999.

Footnotes

[1] The summary statistics reported are based on a random sample of 19 Zone 1 dwelling units.

[2] The distribution of yearly trips to the hospital is bimodal, with one and two trips being the modes.

[3] The mean yearly number of trips to the hospital per year is 3.316. The mean is greater than the modes indicating that the distribution of yearly trips to the hospital is skewed to the right; that is, most families make a few trips while a few families make many trips.

[4] The variability is evidenced by the fact that the standard deviation of average yearly trips to the hospital is 2.237 trips. This number indicates that at least 75 percent of Zone 1 families make between 0 and 7.79 trips to the hospital per year. The above statement is true if the sample mean and standard deviation accurately reflect the true, but unknown, population mean and standard deviation. The 75 percent was derived by invoking Chebyshev's inequality.

Example 3–5 Construction of an automotive parts outlet in Ṣtat City

Problem   Sears, Roebuck & Co. is considering building an automobile parts outlet in Stat City. Sears management decided that if more than 30 percent of the families living in Stat City perform their own automotive repairs, they will construct the automobile parts outlet. Sears's management has asked you to conduct a survey and type a memorandum presenting them with your findings (see Exhibit 3–6).

Please use the following list of 30 randomly selected identification numbers in solving the problem:

794, 616, 1326, 131, 55, 1306, 516, 493, 734, 94,
859, 223, 1283, 378, 57, 1339, 517, 65, 541, 451,
1337, 1104, 93, 1311, 797, 238, 1261, 533, 383, 770.

Remember, the above numbers are just random identification numbers. You must still draw the sample of 30 favorite places for car repairs (variable 19) from the Stat City data base.

Solution   The "favorite place for car repairs" variable has to be redefined as follows: 1 = does own repair (original 0) and 0 = other (original 1 and 2). The "favorite place for car repairs" statistics are shown below:

$$\bar{p} = \frac{6}{30} = .2 = \text{Proportion which does own repair.}$$

Mode = 0.

Standard deviation* = .407.

Favorite place for car repair	Frequency
Do own repairs (1)	6
Other (0)	24
Total	30

$$* \; s = \sqrt{\frac{\Sigma(x - \bar{x})^2}{n - 1}} = \sqrt{\frac{4.8}{29}} = .407.$$

or

$$s = \sqrt{\frac{n\bar{p}(1 - \bar{p})^2}{n - 1}} = \sqrt{\frac{30(.2)(.8)}{29}} = .407.$$

(See Table 3–20.)

Table 3–20

Random identification number (variable 4)	Favorite place for car repairs (variable 19)	$(x - \bar{x})$ $(x - .2)$	$(x - \bar{x})^2$ $(x - .2)^2$
794	0 (1)	−.2	.04
616	0 (2)	−.2	.04
1326	0 (2)	−.2	.04
131	0 (2)	−.2	.04
55	0 (2)	−.2	.04
1306	1 (0)	.8	.64
516	0 (2)	−.2	.04
493	0 (1)	−.2	.04
743	0 (1)	−.2	.04
94	0 (1)	−.2	.04
859	1 (0)	.8	.64
223	0 (2)	−.2	.04
1283	1 (0)	.8	.64
378	0 (1)	−.2	.04
57	0 (2)	−.2	.04
1339	1 (0)	.8	.64
517	0 (1)	−.2	.04
65	0 (1)	−.2	.04
541	0 (1)	−.2	.04
451	0 (1)	−.2	.04
1337	0 (1)	−.2	.04
1104	0 (2)	−.2	.04
93	0 (1)	−.2	.04
1311	1 (0)	.8	.64
797	0 (1)	−.2	.04
238	0 (1)	−.2	.04
1261	0 (1)	−.2	.04
533	1 (0)	.8	.64
383	0 (1)	−.2	.04
770	0 (1)	−.2	.04

$$\Sigma(x - \bar{x})^2 = 4.8$$

The memorandum appears in Exhibit 3–6.

Exhibit 3–6

HOWARD S. GITLOW, PH.D.
STATISTICAL CONSULTANT

MEMORANDUM

TO: Sears, Roebuck & Co.

FROM: Howard Gitlow

DATE: May 1, 1985

RE: Construction of an automotive parts outlet in Stat City

I have estimated that 20 percent of Stat City families perform their own automotive repairs using commonly accepted statistical techniques. Based on your decision criteria (that more than 30 percent of the families living in Stat City must perform their own repairs to open the outlet), the automotive parts outlet should not be opened.

If I can be of any further assistance, please do not hesitate to call me at 305-999-9999.

3–6 Image study of the Grand Union Mr. Hector Chavez (ID = 251), manager of the Stat City Grand Union, is interested in conducting an image study of consumers' perceptions of Stat City supermarkets so he can prepare his annual marketing plan. Basically, Mr. Chavez wants to know the number of Stat City dwelling units that favor the Grand Union. Also, Mr. Chavez wants to know the number of Stat City dwelling units that favor the Food Fair and the A&P.

You have been hired as a marketing research consultant by Mr. Chavez to perform the above study. Type a memorandum summarizing your results for Mr. Chavez.

In order that everyone in your class will arrive at the same results, use the following list of randomly selected dwelling unit identification numbers (variable 4):

> 634, 933, 1008, 47, 178, 579, 1039, 278, 420, 1007, 222,
> 484, 1363, 620, 709, 1299, 1279, 573, 628, 552, 819, 783,
> 433, 809, 939, 990, 972, 535, 1353, 575.

Remember, the above numbers are just random identification numbers. You must still draw the sample of 30 favorite supermarket designations (variable 24) from the Stat City Data Base.

3–7 Heating bill survey Mr. Saul Reisman, chief financial officer of the Stat City Electric Company, needs to know the distribution of average yearly heating bills for Stat City residents. In particular, Mr. Reisman needs to know the percentage of dwelling units with average yearly heating bills in the following categories: under $500, $500–$899, and $900 and over. If any category contains more than 45 percent of the dwelling units in Stat City, the electric company must change their master plan for providing service to Stat City residents.

Conduct a survey and type a summary of your findings to Mr. Reisman.

In order that everyone in your class will arrive at the same results, use the following randomly selected list of dwelling unit identification numbers (variable 4):

> 279, 1131, 985, 222, 127, 212, 643, 1224, 881, 154,
> 254, 1330, 287, 441, 1317, 543, 70, 920, 1119, 957,
> 859, 1180, 1138, 536, 808, 1367, 280, 489, 1178, 996.

Remember, the above numbers are just random identification numbers. You must still draw the sample of 30 average yearly heating bills (variable 10) from the Stat City data base.

3–8 Widening 12th Street between 9th and 13th Avenues The Stat City Chamber of Commerce is contemplating widening 12th Street between 9th and 13th Avenues in answer to residents' complaints concerning congestion caused by on-street parking.

You have been retained by the Stat City Chamber of Commerce to determine how many cars are owned per family in the apartment buildings bordering the proposed construction. If more than 20 percent of the dwelling units along the proposed construction area own two or more cars, then the construction should be performed. Conduct the necessary survey and write a summary of your findings to the Stat City Chamber of Commerce.

Note to the student: This problem uses addresses to define the relevant population.

In order that everyone in your class will come up with the same sample frequency distribution, use the random number table (shown in margin) to draw 20 random identification numbers (variable 4).

3-9 Distribution of telephone bills in Stat City Mr. Jack Davis, chairman of the Stat City Telephone Company, needs to know the distribution of average monthly telephone bills for Stat City families. In particular, Mr. Davis needs to know the number of Stat City dwelling units with average monthly telephone bills in the following categories: under $30, $30 to less than $60, $60 to less than $90, $90 to less than $120, and $120 to less than $150. If any category contains more than 500 of the dwelling units in Stat City, the telephone company must change the capacity of their switching stations.

Conduct a survey and type a summary of your findings to Mr. Davis. Use a histogram and a frequency polygon in your report.

Please use the following random sample of 30 dwelling units' identification numbers (variable 4):

72,	86,	180,	195,	271,
285,	322,	340,	426,	432,
468,	520,	611,	616,	620,
759,	853,	855,	941,	1061,
1078,	1104,	1180,	1184,	1217,
1222,	1240,	1246,	1264,	1331.

Remember, the above numbers are just random identification numbers. You must still draw the sample of 30 average monthly telephone bills (variable 12) from the Stat City data base.

3-10 Federal funding for Stat City Mr. Lee Kaplowitz, the mayor of Stat City, is attempting to obtain federal funding for several projects. To apply for the funds, the mayor must prepare a statistical summary concerning the number of persons per dwelling unit in Stat City as of January 1980.

Mr. Kaplowitz has asked you to conduct a survey and type a memorandum indicating the mean, mode, minimum, maximum, range, and standard deviation of family size for Stat City dwelling units.

Please use the following random sample of 10 dwelling units' identification numbers (variable 4):

694,	286,	917,	457,	319,
817,	1105,	363,	233,	268.

Remember, the above numbers are just random identification numbers. You must still draw the sample of 10 family sizes (variable 14) from the Stat City data base.

3-11 Housing cost survey for the Stat City Real Estate Board Ms. Sharon Vigil (ID = 231), chairperson of the Stat City Real Estate Board, wants to obtain some information concerning the housing costs (mortgages and rents) of Stat City families. She needs this information so the board can revise the prospectus they give to potential Stat City residents.

You have been hired by Ms. Vigil to conduct a survey of Stat City housing costs. Use the following randomly selected identification numbers (variable 4) to draw your sample:

25, 61, 306, 310, 337, 370, 407, 488, 597, 609, 633, 669, 673, 720, 734, 768, 795, 849, 902, 916, 923, 1014, 1015, 1039, 1061, 1106, 1150, 1196, 1216, 1321.

0918	7703	6832	1273
2827	0907	8327	4883
2184	3554	1263	3477
2202	7955	1969	2525
0904	2077	0908	0916
0731	7112	0742	2051
0922	0901	2025	0931
2260	5671	2178	6878
3669	0751	3427	6549
5980	6998	7180	6498
5647	0905	5851	9540
0756	3487	8320	0313
7448	3399	0917	9258
1979	2248	9423	8105
1194	2785	6052	2414
0921	6063	6570	7788
6927	2320	6103	8426
8365	0740	0737	4108
7861	5110	6682	4043
1498	7415	7821	3898

Type a memorandum to Ms. Vigil indicating the mean, median, minimum, maximum, range, and standard deviation of housing costs.

3–12 Diet soda survey Mr. Marc Cooper, manager of the A&P, would like to determine the average weekly purchase of six-packs of diet soda per dwelling unit in Stat City. This information will be helpful to Mr. Cooper in establishing an A&P inventory policy.

You have been hired by Mr. Cooper to conduct a survey to supply him with the required information. Use the following random sample.

Random ID number (variable 4)	Average weekly purchase of six-packs of diet soda (variable 25)	Random ID number (variable 4)	Average weekly purchase of six-packs of diet soda (variable 25)
5	2	519	2
58	1	560	2
68	1	563	0
150	2	586	2
198	1	604	0
201	1	624	1
210	0	662	0
213	0	684	0
216	0	727	0
217	0	741	0
226	2	776	2
278	1	850	0
279	2	947	2
320	0	962	2
340	2	988	0
379	0	996	0
388	2	1089	0
397	2	1144	1
425	2	1145	1
465	2	1152	2
467	2	1223	2
492	0	1236	0
499	0	1255	2
504	0	1333	0
511	0	1338	0

Type a memorandum to Mr. Cooper indicating the mean, mode, minimum, maximum, range, percentage breakdown, and standard deviation of average weekly purchases of six-packs of diet soda in Stat City as of January 1985.

3–13 Family income and costs in Stat City Mr. Kaplowitz, the mayor of Stat City, is preparing a report describing several characteristics of Stat City families. The characteristics to be included in the report are: housing cost (rent or mortgage) as of January 1985 (HCOST), average monthly telephone bill as of January 1985 (PHONE), total 1984 family income (INCOM), average bimonthly automobile gasoline bill as of January 1985 (GAS), and average weekly supermarket bill as of January 1985 (EAT). The mayor has hired you to conduct a survey of 50 Stat City dwelling units and to construct frequency distributions, histograms, and frequency polygons for each of the aforementioned variables.

The mayor has specified the following class limits for each variable:

HCOST   ($0 to less than $250)
($250 to less than $500)
($500 to less than $750)
($750 to less than $1,000)
($1,000 to less than $1,250)
($1,250 to less than $1,500)

PHONE   ($0 to less than $25)
($25 to less than $50)
($50 to less than $75)
($75 to less than $100)
($100 to less than $125)
($125 to less than $150)

INCOM   ($0 to less than $25,000)
($25,000 to less than $50,000)
($50,000 to less than $75,000)
($75,000 to less than $100,000)
($100,000 to less than $125,000)
($125,000 to less than $150,000)
($150,000 to less than $175,000)

GAS   ($0 to less than $50)
($50 to less than $100)
($100 to less than $150)
($150 to less than $200)
($200 to less than $250)

EAT   ($0 to less than $50)
($50 to less than $100)
($100 to less than $150)
($150 to less than $200)
($200 to less than $250)

Use the accompanying simple random sample of 50 Stat City dwelling units.

CASE-N	ID	HCOST	PHONE	INCOM	PEPLE	GAS	GASTR	REPAR	EAT	LSODA	HSODA
1	6.	340.	24.	45454.	2.	133.	3.	1.	50.	0.	0.
2	47.	610.	20.	51671.	4.	162.	5.	2.	80.	0.	0.
3	53.	435.	25.	43451.	1.	65.	6.	1.	27.	2.	0.
4	68.	305.	21.	41836.	1.	65.	4.	2.	27.	1.	0.
5	88.	359.	19.	40283.	2.	99.	2.	2.	41.	1.	0.
6	92.	629.	19.	52548.	5.	190.	8.	1.	97.	0.	1.
7	164.	827.	25.	68773.	3.	175.	3.	1.	71.	0.	1.
8	167.	975.	47.	91910.	5.	211.	5.	2.	112.	1.	5.
9	291.	505.	68.	21086.	4.	176.	8.	2.	80.	2.	0.
10	459.	424.	90.	42806.	6.	120.	6.	0.	124.	0.	0.
11	469.	370.	16.	36907.	5.	197.	6.	1.	99.	0.	0.
12	477.	379.	20.	29803.	6.	222.	11.	1.	111.	0.	0.
13	490.	433.	16.	27809.	4.	163.	6.	1.	84.	0.	0.
14	498.	424.	19.	41587.	4.	231.	4.	0.	117.	0.	1.
15	528.	315.	34.	41557.	5.	199.	9.	1.	102.	2.	0.
16	538.	589.	15.	42033.	7.	165.	5.	2.	170.	2.	0.
17	539.	279.	17.	49245.	3.	142.	6.	0.	77.	2.	0.
18	588.	396.	21.	44824.	4.	160.	7.	1.	88.	0.	1.
19	597.	346.	67.	48075.	4.	183.	6.	1.	84.	0.	4.
20	599.	366.	19.	34509.	5.	173.	6.	2.	90.	2.	0.
21	616.	484.	48.	27071.	6.	217.	5.	2.	138.	0.	2.
22	657.	450.	20.	28604.	4.	214.	2.	0.	94.	0.	2.
23	684.	667.	20.	12539.	5.	65.	4.	1.	117.	0.	4.
24	705.	425.	80.	16726.	4.	65.	6.	1.	74.	0.	0.
25	724.	518.	20.	17943.	2.	120.	6.	1.	53.	0.	0.
26	764.	143.	81.	8102.	2.	120.	6.	0.	55.	0.	0.
27	788.	286.	63.	7548.	7.	120.	3.	0.	99.	2.	0.
28	795.	333.	73.	9558.	5.	65.	4.	1.	107.	2.	0.
29	797.	278.	19.	13705.	4.	65.	3.	1.	94.	0.	5.
30	816.	603.	16.	25718.	4.	179.	8.	1.	97.	0.	5.
31	835.	483.	12.	13661.	3.	120.	7.	0.	69.	0.	1.
32	854.	395.	23.	12097.	7.	65.	5.	0.	150.	0.	0.
33	873.	485.	25.	11509.	6.	120.	7.	1.	97.	2.	0.
34	897.	340.	26.	11969.	7.	65.	4.	1.	138.	0.	0.
35	962.	365.	10.	7852.	6.	120.	3.	0.	139.	2.	0.
36	970.	337.	14.	21258.	6.	223.	12.	1.	126.	0.	0.
37	996.	491.	10.	18157.	4.	65.	4.	1.	102.	2.	0.
38	1073.	359.	71.	7306.	6.	120.	7.	0.	138.	2.	0.
39	1086.	222.	23.	20539.	5.	191.	5.	1.	88.	0.	0.
40	1092.	525.	13.	26576.	4.	149.	6.	2.	78.	0.	0.
41	1134.	468.	17.	17395.	8.	65.	5.	1.	163.	2.	0.
42	1136.	424.	105.	23923.	8.	120.	5.	2.	162.	0.	5.
43	1192.	405.	16.	5786.	2.	65.	2.	1.	124.	2.	0.
44	1215.	538.	20.	21094.	5.	207.	7.	0.	123.	2.	0.
45	1244.	352.	23.	16095.	3.	65.	6.	1.	78.	2.	0.
46	1279.	313.	36.	5555.	2.	120.	6.	1.	45.	0.	1.
47	1297.	583.	15.	18922.	5.	65.	6.	0.	109.	0.	0.
48	1299.	305.	76.	29004.	4.	160.	7.	2.	85.	0.	6.
49	1320.	450.	10.	19193.	7.	120.	4.	1.	142.	0.	0.
50	1348.	611.	19.	19695.	4.	65.	3.	0.	72.	2.	0.

3–14 Demographic profile of Stat City Mr. Paul Lund, chairman of the Stat City Chamber of Commerce, is preparing a demographic profile of Stat City families. The demographic variables that Mr. Lund has chosen to include in the profile are: total 1984 family incomes (INCOM), assessed value of a home as of January 1985 (ASST), average total yearly heating bill as of January 1985 (HEAT), average monthly electric bill as of January 1985 (ELEC), average monthly telephone bill as of January 1985 (PHONE), number of cars in the household as of January 1985 (CARS), average bimonthly automobile gasoline bill as of January 1985 (GAS), and average number of trips to the gas station per month as of January 1985 (GASTR). Mr. Lund has provided you with enough funds to survey 10 randomly selected Stat City families. Further, he wants you to compute the mean, standard deviation, variance, minimum, maximum, range, skewness, and kurtosis for each of the above variables.

Use the following simple random sample of 10 Stat City families.

CASE-N	ID	ASST	HEAT	ELEC	PHONE	INCOM	CARS	GAS	GASTR
1	258.	92419.	1136.	73.	24.	116203.	2.	127.	3.
2	295.	23773.	538.	42.	65.	25328.	1.	99.	6.
3	386.	39558.	792.	72.	23.	32667.	2.	160.	6.
4	415.	0.	413.	37.	22.	26972.	2.	65.	6.
5	416.	0.	457.	37.	17.	41549.	3.	174.	7.
6	736.	0.	392.	38.	14.	18404.	2.	120.	12.
7	858.	0.	256.	21.	21.	17052.	2.	120.	5.
8	1016.	0.	512.	46.	19.	24800.	3.	185.	5.
9	1149.	0.	651.	69.	27.	19670.	2.	120.	5.
10	1242.	0.	471.	47.	22.	24743.	3.	168.	8.

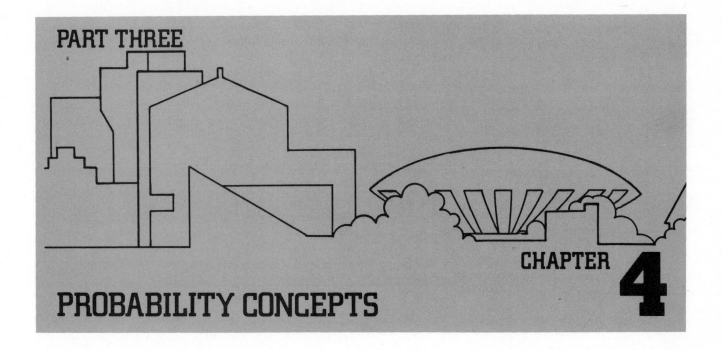

PART THREE

PROBABILITY CONCEPTS

CHAPTER 4

Part Two of this book was devoted to investigating various characteristics of Stat City; for example, the distribution of average monthly telephone bills. This chapter will introduce probability concepts to determine the chances of various phenomena occurring. For example, you will learn how to determine the likelihood of a randomly selected family in Zone 4 earning over $30,000 a year or the chances of a family with children going to the hospital at least once during a year.

The basic probability concepts you learn in this chapter will increase your capabilities of making sound inferences about a population based only on a sample. The topics discussed in this chapter are:

Cross-tabulation, or contingency tables.

Marginal, joint, and conditional probabilities.

Addition and multiplication rules.

Independent and dependent events.

Probabilities indicate the relative frequency of the occurrence of events. For example, the probability of randomly selecting a Stat City family that lives in Zone 1 is 0.0947 (9.47% = 130/1,373); there are 130 dwelling units in Zone 1 and 1,373 dwelling units in Stat City. Probabilities give decision makers a handle on the likelihood that a particular event will occur. If the probability of an event's occurring is small, then the event is a rare event and the chances of it occurring are not likely. On the other hand, if the probability of an event's occurring is large, then the event is a common event and the chances of it occurring are likely. It is important to realize that the definitions of rare and common are subjective judgments on the part of the decision maker, not the statistician.

Example 4–1 Revision of Howie's Gulf Station marketing plan

The owner of Howie's Gulf Station, Mr. Howie, wants to know if there is a relationship between "a family's favorite gas station as of January 1985 (FAVGA)" and "a family's favorite place for car repairs as of Janu-

Problem

ary 1985 (REPAR)." If such a relationship does exist, Mr. Howie will revise his marketing plan to cater to the types of customers he is currently not receiving.

You have been retained by Mr. Howie to determine if a meaningful relationship exists between FAVGA and REPAR. Mr. Howie has provided you with funds to draw a random sample of 30 Stat City dwelling units. Conduct the survey and type a memorandum to Mr. Howie reporting your findings (see Exhibit 4–1). Use a cross-tabulation (contingency) table in your memorandum.

Please use the following randomly selected identification numbers (variable 4) to draw your sample:

8, 12, 22, 103, 142, 229, 258, 260, 321, 352, 363, 446, 452,
517, 687, 693, 729, 745, 828, 932, 967, 1013, 1075, 1100, 1112,
1138, 1250, 1297, 1305, 1341.

Solution   The random sample is shown in Table 4–1.

Table 4–1

ID	REPAR	FAVGA	ID	REPAR	FAVGA
8.	2.	2.	693.	1.	2.
12.	1.	2.	729.	0.	1.
22.	2.	1.	745.	2.	2.
103.	2.	1.	828.	2.	1.
142.	2.	2.	932.	1.	2.
229.	1.	2.	967.	1.	2.
258.	2.	2.	1013.	1.	2.
260.	2.	1.	1075.	1.	2.
321.	2.	2.	1100.	1.	2.
352.	1.	1.	1112.	1.	2.
363.	1.	1.	1138.	0.	2.
446.	2.	1.	1250.	0.	2.
452.	0.	2.	1297.	0.	1.
517.	1.	1.	1305.	2.	2.
687.	0.	1.	1341.	2.	1.

Set up a contingency table using the above data (Table 4–2).

Table 4–2

	Performs own repair (0)	Repairs done by service station (1)	Repairs done by dealer (2)	Total
Paul's Texaco Station (1)	3 (10%)*	3 (10%)	6 (20%)	12 (40%)
Howie's Gulf Station (2)	3 (10%)	9 (30%)	6 (20%)	18 (60%)
Total	6 (20%)	12 (40%)	12 (40%)	30 (100%)

* All percents are computed by dividing the cell count by 30.

Use the sample cell percentages to estimate the population cell frequencies (Table 4–3).

Table 4–3

	Performs own repairs (0)	Repairs done by service station (1)	Repairs done by dealer (2)	Total
Paul's Texaco Station	137*	137	275	549
Howie's Gulf Station	137	412	275	824
Total	274	549	550	1,373

* All estimated cell counts are computed as follows: Estimated cell count = 1,373 × Sample cell percent.

The memorandum appears in Exhibit 4–1.

Exhibit 4–1

HOWARD S. GITLOW, PH.D.
STATISTICAL CONSULTANT

MEMORANDUM

TO: Mr. Howie, Owner
        Howie's Gulf Station

FROM: Howard Gitlow

DATE: May 15, 1985

RE: Revision of marketing plan based on customer's
        favorite place of car repair.

As per your request and using commonly accepted statistical techniques, I have investigated the relationship between "a family's favorite gas station as of January 1985 (FAVGA)" and "a family's favorite place for car repairs as of January 1985 (REPAR)." The following chart shows the estimated breakdown of Stat City residents by FAVGA and REPAR.[1]

	Performs own repairs (0)	Repairs performed by service station (1)	Repairs done by dealer (2)	Total
Paul's Texaco Station	137	137	275	549
Howie's Gulf Station	137	412	275	824
Total	274	549	550	1,373

Exhibit 4–1 (*concluded*)

As you can see, there is little relationship between FAVGA and REPAR. Howie's and Paul's split the "performs own repairs" and "repairs done by dealer" customers. However, Howie's Gulf Station does have a significant inroad to the "repairs done by service station" customers. Consequently, according to your marketing strategy, Howie's Gulf Station should maintain the current marketing plan for the "repairs performed by service station" customers and create a new marketing plan to attract the "performs own repairs" and "repairs done by dealer" customers.

If you have any further questions, please do not hesitate to call me at 305-999-9999.

Footnotes

[1] Insert standard footnote 6, using a sample size of 30.

Example 4–2 Investigation of territorial shopping behavior

Problem The Stat City Supermarket Association would like to investigate the "territorial shopping behavior" of Stat City residents. "Territorial shoppers" are consumers who shop as close to their neighborhoods as possible. This would mean that (1) most of Zone 1 residents would shop at the A&P (supermarket 3), (2) most of Zone 2 residents would shop at either the Food Fair (supermarket 1) or the A&P (supermarket 3), (3) most of Zone 3 residents would shop either at Foot Fair (supermarket 1) or the Grand Union (supermarket 2), and (4) most of Zone 4 residents would shop at the Grand Union (supermarket 2). The association would like you to indicate the percent of Stat City families favoring each supermarket and the percent of Stat City families favoring a particular supermarket given their residential housing zone.

The Stat City Supermarket Association insists (and provides you with the funds) that you draw a random sample of 100 Stat City families to investigate the relationship between "residential housing zone (ZONE)" and "a family's favorite supermarket as of January 1985 (FEAT)." Conduct the survey and type a memorandum to the Stat City Supermarket Association reporting your findings (see Exhibit 4–2).

Please use the randomly selected sample of Stat City dwelling units given in Table 4–4.

Table 4–4

ID	ZONE	FEAT	ID	ZONE	FEAT
6.	1.	3.	734.	4.	2.
8.	1.	3.	740.	4.	2.
21.	1.	3.	747.	4.	2.
76.	1.	3.	765.	4.	2.
79.	1.	1.	785.	4.	2.
85.	1.	3.	787.	4.	2.
92.	1.	3.	802.	4.	2.
112.	1.	3.	808.	4.	2.
123.	1.	3.	825.	4.	2.
140.	2.	2.	858.	4.	2.
150.	2.	3.	861.	4.	2.
153.	2.	1.	867.	4.	2.
161.	2.	1.	881.	4.	2.
228.	2.	3.	885.	4.	2.
233.	2.	3.	896.	4.	2.
243.	2.	1.	902.	4.	2.
246.	2.	1.	913.	4.	2.
253.	2.	1.	914.	4.	2.
289.	3.	1.	916.	4.	2.
291.	3.	2.	921.	4.	2.
325.	3.	3.	929.	4.	2.
322.	3.	2.	931.	4.	3.
327.	3.	2.	951.	4.	2.
337.	3.	1.	953.	4.	2.
398.	3.	1.	956.	4.	2.
390.	3.	1.	994.	4.	2.
414.	3.	2.	996.	4.	2.
429.	3.	2.	1005.	4.	2.
440.	3.	1.	1007.	4.	2.
460.	3.	2.	1020.	4.	2.
468.	3.	2.	1030.	4.	2.
482.	3.	2.	1037.	4.	2.
492.	3.	1.	1046.	4.	2.
517.	3.	2.	1094.	4.	2.
545.	3.	2.	1122.	4.	2.
561.	3.	2.	1141.	4.	1.
585.	3.	1.	1156.	4.	2.
586.	3.	1.	1159.	4.	2.
622.	3.	1.	1171.	4.	2.
629.	4.	2.	1186.	4.	2.
630.	4.	2.	1227.	4.	2.
636.	4.	2.	1239.	4.	2.
638.	4.	2.	1243.	4.	2.
668.	4.	2.	1245.	4.	2.
675.	4.	1.	1263.	4.	2.
678.	4.	2.	1277.	4.	2.
692.	4.	2.	1282.	4.	2.
699.	4.	2.	1295.	4.	2.
721.	4.	2.	1297.	4.	2.
730.	4.	1.	1365.	4.	2.

Solution   The memorandum appears in Exhibit 4–2.

Exhibit 4–2

HOWARD S. GITLOW, PH.D.
STATISTICAL CONSULTANT

MEMORANDUM

TO: Stat City Supermarket Association

FROM: Howard Gitlow

DATE: May 27, 1985

RE: Investigation of territorial shopping behavior

In accordance with the association's request, I have investigated the "territorial shopping behavior" of Stat City residents using commonly accepted statistical techniques. The following table depicts the estimated breakdown of Stat City families into housing zone and favorite supermarket as of January 1985.[1]

Favorite supermarket	Zone				Total*
	1	2	3	4	
Food Fair	14 (1%)	69 (5%)	124 (9%)	41 (3%)	248 (18%)
Grand Union	0 (0%)	14 (1%)	151 (11%)	796 (58%)	961 (70%)
A&P	110 (8%)	40 (3%)	14 (1%)	0 (0%)	164 (12%)
Total	124 (9%)	123 (9%)	289 (21%)	837 (61%)	1,373 (100%)

* Numbers in this table are rounded to sum to 1,373.

The above table indicates that 18 percent of Stat City families favor the Food Fair, 70 percent favor the Grand

Exhibit 4-2 (*concluded*)

Union, and 12 percent favor the A&P. Approximately 11 percent of Zone 1 families frequent the Food Fair (14/124 = 11.1 percent), while the majority of Zone 1 residents (88.9 percent) frequent the A&P (110/124). Approximately 56 percent of Zone 2 families prefer to shop at the Food Fair (69/123 = 55.6 percent), 11 percent shop at the Grand Union (14/123 = 11.1 percent), while 33 percent choose the A&P (40/123 = 32.5 percent). Approximately 43 percent of Zone 3 families market at Food Fair (124/289 = 42.9 percent), 52 percent at Grand Union (151/289 = 52.4 percent), and only 5 percent favor the A&P (14/289 = 4.8 percent). Approximately 5 percent of Zone 4 families prefer the Food Fair (41/837 = 4.9 percent), but the Grand Union is the overwhelming favorite of the rest (796/837 = 95.1 percent).

All of the above statistics indicate that Zone 1 families primarily shop in the A&P, Zone 2 families mainly shop in the Food Fair or the A&P, Zone 3 families generally shop in the Grand Union or the Food Fair, and Zone 4 families almost exclusively shop at the Grand Union. Consequently, our analysis verifies the existence of Stat City "territorial shopping" behavior.

I hope that this memorandum satisfies your informational needs. If you have any further questions, please do not hesitate to call me at 305-999-9999.

_____ _____

Footnotes

[1] Insert standard footnote 6, using a sample size of 100.

Example 4-3   Relationship between alcohol consumption and income

Problem

Dr. Marsha Cox, chief psychiatrist for the Stat City Hospital, believes that poorer families (families earning under $20,000 per year) consume more alcohol than wealthier families (families earning $20,000 or more per year). She believes that understanding the relationship between alcohol consumption and income will enhance her staff's ability to detect families with drinking problems.

If Dr. Cox's belief is correct, then the Stat City Hospital psychological counselors should be alerted for possible alcohol-related problems when working with poorer families. Before issuing a directive to her staff, she would like to sample Stat City families with reference to this problem. Basically, she needs comparative statistics depicting the relationship between the consumption of alcohol of poor and wealthy families.

She has decided that the most "practical" measure of alcohol consumption is if a family purchases at least one six-pack of beer per week.

You have been asked by Dr. Cox to perform a study which will shed light upon her "income-alcohol" premise. She has provided you with funds to draw a random sample of 50 Stat City families. Conduct the survey and type a memorandum reporting your findings to Dr. Cox (see Exhibit 4-3).

Table 4–5

ID	INCOM	BEER
5.	40603.	1.
82.	17046.	4.
94.	36165.	3.
122.	62205.	3.
123.	56479.	0.
134.	63330.	0.
140.	23576.	4.
211.	96618.	5.
292.	48054.	0.
316.	28784.	0.
320.	31665.	4.
326.	22232.	4.
408.	33300.	0.
439.	25971.	5.
487.	23527.	9.
501.	26669.	0.
568.	51522.	3.
581.	41974.	0.
637.	16917.	0.
655.	23675.	5.
689.	14369.	7.
712.	13527.	0.
718.	19359.	3.
723.	13377.	3.
743.	16473.	0.
767.	7809.	0.
778.	6476.	0.
793.	18811.	10.
851.	12642.	0.
862.	17522.	0.
884.	11445.	0.
897.	11969.	0.
912.	16601.	3.
931.	8987.	5.
942.	3938.	1.
967.	26820.	0.
972.	21765.	0.
992.	33790.	0.
1018.	14985.	5.
1024.	12822.	5.
1032.	18651.	0.
1078.	6894.	0.
1090.	24361.	2.
1127.	18341.	1.
1155.	14018.	0.
1242.	24743.	2.
1247.	21735.	1.
1299.	29004.	0.
1358.	9207.	0.
1368.	8603.	4.

Solution Use the randomly selected sample of Stat City dwelling units shown in Table 4–5.

The memorandum appears in Exhibit 4–3.

Exhibit 4–3

ROSA OPPENHEIM, PH.D.
STATISTICAL CONSULTANT

MEMORANDUM

To: Dr. Marsha Cox, Chief Psychiatrist
        Stat City Hospital

FROM: Rosa Oppenheim

DATE: July 16, 1985

RE: Income–alcohol premise

In order to conduct your requested investigation I used the following definitions for income status and alcohol consumption.

1. Income status =
   - 0 if income was less than \$20,000 in 1984 (poor family)
   - 1 if income was \$20,000 or more in 1984 (wealthy family)

2. Alcohol consumption =
   - 0 if a family's average weekly purchase of beer as of January 1985 was zero six–packs (nondrinking family)
   - 1 if a family's average weekly purchase of beer as of January 1985 was one or more six–packs (drinking family)

The following table, which was constructed using commonly accepted statistical techniques, depicts the estimated breakdown of Stat City families by alcohol consumption and income status.[1]

Alcohol consumption (average weekly purchase of beer as of January 1985)	Income status		Total
	Under \$20,000 in 1984 — 0	\$20,000 or more in 1984 — 1	
0 six-packs	357 (26%)	330 (24%)	687 (50%)
1 or more six-packs	329* (24%)	357 (26%)	686 (50%)
Total	686 (50%)	687 (50%)	1,373 (100%)

* This number is rounded so cells will add to 1,373.

Exhibit 4-3 (*concluded*)

From the above table we note that:

1. The chances of randomly selecting a beer purchasing family out of all poor Stat City families are about 48 in 100 (.48 = 329/686) times.
2. The chances of randomly selecting a beer purchasing family out of all wealthy Stat City families are about 52 in 100 (.52 = 357/687) times.

Consequently the "income-alcohol premise" does not hold up upon statistical investigation. However, I would take note that 50 percent of Stat City families do drink so that you may want to issue a directive to your staff telling them to be on guard for alcohol-related problems with all families they counsel.

If you have any further questions, please do not hesitate to call me at 305-999-9999.

Footnotes

[1] Insert standard footnote 6, using a sample size of 50.

Example 4-4   Soda purchase survey

Mr. Marc Cooper, manager of the A&P, would like to determine: **Problem**

a. The percentage of Stat City families that did not purchase any soda as of January 1985.
b. The percentage of Stat City families that purchased either diet or regular soda (or both) as of January 1985.
c. The percentage of Stat City families that purchased only diet soda (no regular soda) as of January 1985.
d. The percentage of Stat City families that purchased only regular soda (no diet soda) as of January 1985.
e. If diet soda purchases were related to regular soda purchases as of January 1985.

You have been asked by Mr. Cooper to conduct a study to answer the above questions. Mr. Cooper has provided you with sufficient funds to draw a simple random sample of 120 Stat City families. Conduct the survey and type a memorandum reporting your findings to Mr. Cooper (see Exhibit 4-4).

Please use the randomly selected sample of 120 Stat City dwelling units given in Table 4-6.

Table 4-6

CASE-N	ID	LSODA	HSODA	CASE-N	ID	LSODA	HSODA	CASE-N	ID	LSODA	HSODA
1	3.	1.	0.	41	469.	0.	0.	81	917.	0.	0.
2	12.	2.	0.	42	477.	0.	0.	82	922.	0.	0.
3	23.	0.	0.	43	494.	0.	1.	83	924.	1.	0.
4	45.	0.	1.	44	496.	2.	0.	84	943.	0.	0.
5	65.	1.	0.	45	504.	0.	1.	85	945.	0.	0.
6	75.	0.	0.	46	529.	0.	1.	86	946.	0.	1.
7	93.	2.	0.	47	531.	0.	0.	87	949.	2.	0.
8	103.	0.	0.	48	533.	0.	0.	88	967.	0.	2.
9	123.	0.	0.	49	546.	0.	0.	89	975.	1.	0.
10	128.	0.	1.	50	571.	0.	5.	90	978.	0.	5.
11	132.	0.	1.	51	588.	0.	1.	91	982.	0.	0.
12	138.	0.	0.	52	608.	1.	0.	92	985.	0.	0.
13	145.	0.	1.	53	611.	1.	0.	93	997.	0.	1.
14	154.	0.	6.	54	613.	0.	0.	94	998.	0.	0.
15	161.	1.	0.	55	627.	0.	6.	95	1007.	2.	0.
16	162.	1.	0.	56	650.	0.	0.	96	1033.	0.	0.
17	183.	0.	0.	57	654.	0.	0.	97	1043.	0.	3.
18	190.	0.	5.	58	666.	2.	0.	98	1077.	0.	1.
19	194.	0.	0.	59	676.	0.	0.	99	1108.	2.	0.
20	209.	2.	0.	60	690.	2.	0.	100	1121.	0.	0.
21	222.	2.	0.	61	695.	0.	3.	101	1126.	0.	1.
22	247.	2.	0.	62	700.	0.	4.	102	1130.	2.	0.
23	249.	2.	0.	63	713.	2.	0.	103	1131.	2.	0.
24	260.	0.	1.	64	720.	0.	0.	104	1133.	0.	0.
25	261.	0.	5.	65	725.	0.	2.	105	1171.	1.	0.
26	266.	0.	0.	66	735.	0.	0.	106	1204.	0.	1.
27	279.	2.	0.	67	737.	2.	0.	107	1206.	0.	3.
28	284.	1.	0.	68	744.	0.	0.	108	1207.	0.	0.
29	327.	0.	1.	69	747.	0.	0.	109	1211.	0.	1.
30	329.	0.	0.	70	776.	2.	0.	110	1231.	0.	4.
31	339.	0.	0.	71	784.	0.	4.	111	1238.	2.	0.
32	346.	0.	1.	72	801.	0.	0.	112	1246.	0.	0.
33	356.	2.	0.	73	818.	0.	1.	113	1251.	2.	0.
34	367.	0.	3.	74	823.	0.	0.	114	1274.	0.	0.
35	382.	0.	0.	75	826.	0.	0.	115	1276.	0.	0.
36	383.	0.	4.	76	832.	0.	2.	116	1291.	2.	1.
37	400.	0.	0.	77	836.	0.	3.	117	1317.	1.	0.
38	423.	0.	1.	78	858.	2.	0.	118	1356.	2.	0.
39	424.	0.	0.	79	865.	0.	1.	119	1359.	0.	4.
40	445.	0.	0.	80	868.	2.	0.	120	1367.	0.	1.

First, Mr. Cooper's questions must be restated statistically.

Mr. Cooper's question	Statistical restatement of Mr. Cooper's question
a. Determine the percentage of Stat City families that did not purchase any soda as of January 1985.	Determine the probability of randomly selecting a Stat City family that did not purchase diet soda and regular soda as of January 1985. $P(\text{LSODA} = 0 \cap \text{HSODA} = 0)$
b. Determine the percentage of Stat City families that purchased either diet or regular soda (or both) as of January 1985.	Determine the probability of randomly selecting a Stat City family that purchased diet soda or regular soda or both as of January 1985. $P(\text{LSODA} > 0 \cup \text{HSODA} > 0)$
c. Determine the percentage of Stat City families that purchased only diet soda (no regular soda) as of January 1985.	Determine the probability of randomly selecting a Stat City family that purchased only diet soda (no regular soda) as of January 1985. $P(\text{LSODA} > 0 \cap \text{HSODA} = 0)$
d. Determine the percentage of Stat City families that purchased only regular soda (no diet soda) as of January 1985.	Determine the probability of randomly selecting a Stat City family that did purchase only regular soda (no diet soda) as of January 1985. $P(\text{LSODA} = 0 \cap \text{HSODA} > 0)$
e. Determine if diet soda purchases were related to regular soda purchases as of January 1985.	Determine if diet soda purchases were statistically independent of regular soda purchases as of January 1985. $P(\text{LSODA} = 0 \cap \text{HSODA} = 0)$ $\overset{?}{=} P(\text{LSODA} = 0)$ $\times P(\text{HSODA} = 0)$

Second, a contingency table must be constructed from the sample data (see Table 4–7).

Table 4–7

Purchases of six-packs of regular soda as of January 1985 (HSODA)	Purchases of six-packs of diet soda as of January 1985 (LSODA)		
	0	*1 or more*	*Total*
0	42 (0.35)	36 (0.30)	78 (0.65)
1 or more	42 (0.35)	0 (0.00)	42 (0.35)
Total	84 (0.70)	36 (0.30)	120 (100)

Third, the relevant probabilities must be computed.

$$P(\text{LSODA} = 0 \cap \text{HSODA} = 0) = \frac{42}{120} = 0.35$$

Hence, it will be inferred that 35 percent of Stat City families did not purchase any soda as of January 1985.

$$P(\text{LSODA} > 0 \cup \text{HSODA} > 0) = P(\text{LSODA} > 0) + P(\text{HSODA} > 0)$$
$$- P(\text{HSODA} > 0 \cap \text{LSODA} > 0)$$
$$= \frac{36}{120} + \frac{42}{120} - \frac{0}{120}$$
$$= 0.30 + 0.35 - 0.00 = .65$$

Consequently, it will be inferred that 65 percent of Stat City families purchased either diet or regular soda (or both) as of January 1985.

$$P(\text{LSODA} > 0 \cap \text{HSODA} = 0) = \frac{36}{120} = 0.30$$

Hence, it will be inferred that 30 percent of Stat City families purchased only diet soda (no regular) as of January 1985.

$$P(\text{LSODA} = 0 \cap \text{HSODA} > 0) = \frac{42}{120} = 0.35$$

Consequently, it will be assumed that 35 percent of Stat City families purchased only regular soda (no diet soda) as of January 1985.

$$P(\text{LSODA} = 0 \cap \text{HSODA} = 0) \stackrel{?}{=} P(\text{LSODA} = 0) \times P(\text{HSODA} = 0)$$
$$0.35 \neq (0.70 \times 0.65 = 0.455)$$

Since the above joint probability [$P(\text{LSODA} = 0 \cap \text{HSODA} = 0)$] is not equal to the product of the above marginal probabilities [$P(\text{LSODA} = 0) \times P(\text{HSODA} = 0)$], the purchase of diet soda was statistically dependent upon the purchase of regular soda (and vice versa) as of January 1985.

The memorandum appears in Exhibit 4-4.

Exhibit 4-4

HOWARD S. GITLOW, PH.D.
STATISTICAL CONSULTANT

MEMORANDUM

TO: Mr. Marc Cooper, Manager
        A&P

FROM: Howard Gitlow

DATE: February 11, 1985

RE: Soda purchase survey as of January 1985

    In accordance with our consulting contract, I have investigated the soda-purchasing behavior of Stat City families. All statistics presented in this memorandum are based on commonly accepted statistical practices.
    The following table presents the estimated breakdown of Stat City families into nonusers and users of diet and regular soda as of January 1985.[1]

Purchases of six-packs of regular soda as of January 1985 (HSODA)	Purchases of six-packs of diet soda as of January 1985 (LSODA)		
	Nonusers (0 six-packs)	Users (1 or more six-packs)	Total
Nonusers (0 six-packs)	481 (35%)	411* (30%)	892 (65%)
Users (1 or more six-packs)	481 (35%)	0 (0%)	481 (35%)
Total	962 (70%)	411 (30%)	1,373 (100%)

  * Rounded down to force sum to 1,373.

Exhibit 4–4 (*concluded*)

The above table indicates that as of January 1985:

a. 35 percent of Stat City families did not purchase any soda.

b. 65 percent of Stat City families purchased either diet or regular soda (or both).

c. 30 percent of Stat City families purchased only diet soda (no regular soda).

d. 35 percent of Stat City families purchased only regular soda (no diet soda).

e. Diet soda purchases were related to regular soda purchases.

I hope that the above statistics satisfy your informational needs. If you have any questions, please do not hesitate to call me at 305-999-9999.

_____

Footnotes

[1] Insert standard footnote 6, using a sample size of 120.

ADDITIONAL PROBLEMS

4–5 Distribution of dwelling type by residential housing zone Ms. Sharon Vigil, chairperson of the Stat City Real Estate Board, would like information concerning the distribution of apartments and houses in each of the four residential zones in Stat City. This information is important to real estate agents when showing prospective Stat City residents the composition of dwelling units by residential zone.

You have been asked by Ms. Vigil to review the Stat City Census of Housing to determine the distribution of apartments and houses by residential zone.

An investigation of the said census (conducted in January 1985) revealed the following data:

Dwelling type	Zone				
	1	2	3	4	Total
Apartments (0)	0	0	126	748	874
Houses (1)	130	157	212	0	499
Total	130	157	338	748	1,373

Type a memorandum to Ms. Vigil reporting your findings.

4–6 Crowding and health status in Stat City The federal government sent a message to all U.S. cities indicating funds would be available to alleviate crowded housing if it was found that crowding adversely affects health. Consequently, Mr. Lee Kaplowitz, the mayor of Stat City, would like to determine if crowded living space affects health in his city.

The federal government has defined a living space to be crowded if the number of persons per room in a dwelling unit is one or more, and not crowded if the number of persons per room in a dwelling unit is less than one. Further, the federal government has defined the health status of a family to be the average number of visits per year a family makes to the hospital. The health categories are no visits, one visit, or two or more visits per year to a hospital.

You have been asked by Mr. Kaplowitz to conduct a study to investigate the above problem. Mr. Kaplowitz has provided you with sufficient funds to sample 60 Stat City dwelling units. Conduct the study and type a memorandum reporting your findings to Mr. Kaplowitz.

Please assume that you obtained the following table from your random sample of 60 Stat City dwelling units.

| | Living space | | |
| | Not crowded (less than one person per room) | Crowded (one or more persons per room) | Total |
Health status			
No visits per year	4	0	4
One visit per year on average	14	1	15
Two or more visits per year on average	29	12	41
Total	47	13	60

4–7 Beer and soda survey The Stat City Supermarket Association needs to determine, as of January 1985:

a. The percentage of Stat City families that did not purchase soda and beer.

b. The percentage of Stat City families that purchased either soda or beer (or both).

c. The percentage of Stat City families that purchased only soda (no beer).

d. The percentage of Stat City families that purchased only beer (no soda).

e. If beer purchases were related to soda purchases.

You have been asked by the Stat City Supermarket Association to conduct a survey to answer the above questions. The association has provided you with funds to draw a simple random sample of 90 Stat City families. Conduct the survey and type a memorandum reporting your findings to the association.

Please use the accompanying randomly selected sample of 90 Stat City dwelling units.

Random sample for Problem 4–7

CASE-N	ID	LSODA	HSODA	BEER	SODA	CASE-N	ID	LSODA	HSODA	BEER	SODA
1	4.	0.	6.	0.	6.	46	688.	0.	0.	0.	0.
2	10.	0.	4.	3.	4.	47	691.	1.	0.	0.	1.
3	21.	0.	0.	0.	0.	48	708.	1.	0.	4.	1.
4	27.	0.	0.	0.	0.	49	715.	0.	2.	0.	2.
5	48.	0.	0.	0.	0.	50	770.	0.	0.	4.	0.
6	87.	0.	3.	2.	3.	51	774.	0.	6.	6.	6.
7	93.	2.	0.	2.	2.	52	784.	0.	4.	3.	4.
8	95.	2.	0.	5.	2.	53	800.	2.	0.	0.	2.
9	106.	0.	5.	1.	5.	54	813.	2.	0.	5.	2.
10	114.	2.	0.	2.	2.	55	822.	0.	0.	0.	0.
11	127.	2.	0.	4.	2.	56	829.	2.	0.	2.	2.
12	142.	1.	0.	3.	1.	57	877.	0.	3.	0.	3.
13	181.	0.	0.	7.	0.	58	883.	0.	0.	0.	0.
14	212.	2.	0.	1.	2.	59	891.	0.	2.	4.	2.
15	239.	0.	3.	3.	3.	60	892.	0.	0.	0.	0.
16	254.	0.	1.	1.	1.	61	897.	0.	0.	0.	0.
17	257.	0.	0.	0.	0.	62	908.	0.	0.	0.	0.
18	268.	0.	5.	6.	5.	63	914.	0.	5.	5.	5.
19	273.	0.	0.	0.	0.	64	943.	0.	0.	0.	0.
20	276.	0.	4.	7.	4.	65	960.	2.	0.	0.	2.
21	293.	0.	0.	0.	0.	66	993.	0.	0.	0.	0.
22	305.	0.	4.	0.	4.	67	1004.	0.	4.	10.	4.
23	351.	0.	1.	0.	1.	68	1064.	0.	6.	0.	6.
24	364.	0.	4.	11.	4.	69	1066.	0.	0.	0.	0.
25	380.	2.	0.	3.	2.	70	1108.	2.	0.	4.	2.
26	392.	2.	0.	6.	2.	71	1117.	0.	0.	9.	0.
27	427.	0.	0.	0.	0.	72	1138.	0.	0.	10.	0.
28	450.	0.	4.	3.	4.	73	1139.	0.	1.	4.	1.
29	451.	2.	0.	1.	2.	74	1150.	0.	0.	0.	0.
30	497.	2.	0.	3.	2.	75	1159.	0.	0.	0.	0.
31	519.	2.	0.	1.	2.	76	1186.	0.	0.	0.	0.
32	526.	2.	0.	0.	2.	77	1199.	0.	0.	11.	0.
33	529.	0.	1.	5.	1.	78	1232.	0.	0.	0.	0.
34	546.	0.	0.	0.	0.	79	1237.	1.	0.	6.	1.
35	549.	1.	0.	4.	1.	80	1256.	2.	0.	7.	2.
36	557.	0.	0.	0.	0.	81	1266.	0.	0.	0.	0.
37	561.	0.	0.	3.	0.	82	1267.	0.	0.	0.	0.
38	562.	0.	0.	0.	0.	83	1289.	0.	1.	6.	1.
39	568.	0.	3.	3.	3.	84	1291.	2.	0.	0.	2.
40	605.	1.	0.	0.	1.	85	1297.	1.	0.	0.	1.
41	625.	1.	0.	0.	1.	86	1304.	0.	3.	9.	3.
42	646.	1.	0.	11.	1.	87	1330.	0.	5.	0.	5.
43	664.	0.	0.	0.	0.	88	1337.	2.	0.	3.	2.
44	683.	0.	0.	1.	0.	89	1342.	0.	0.	0.	0.
45	687.	0.	0.	1.	0.	90	1372.	0.	1.	4.	1.

4–8 Dwelling patterns and purchasing profiles in Stat City A survey of 50 Stat City families was conducted to collect information on the following variables: favorite gas station as of January 1985 (FAVGA), average number of trips to the gas station per month as of January 1985 (GASTR), favorite supermarket as of January 1985 (FEAT), type of dwelling unit as of January 1985 (DWELL), residential housing zone as of January 1985 (ZONE), and average weekly purchase of six-packs of diet soda (LSODA), regular soda (HSODA), and beer (BEER) as of January 1985. Compute the following probabilities:

P (FAVGA = 1 ∪ GASTR ≥ 4),   P (DWELL = 1 ∩ ZONE ≤ 2),
P (FAVGA = 2 ∪ ZONE = 1),   P (FEAT   = 1 ∩ DWELL = 1),
P (FAVGA = 1 ∪ ZONE ≥ 3),   P (LSODA = ∅ ∪ BEER = ∅),
P (GASTR ≥ 3 ∩ ZONE = 2),   P (LSODA = ∅ ∩ HSODA = ∅), and
P (GASTR ≤ 2 ∩ DWELL = ∅),   P (HSODA = ∅ ∪ BEER ≥ 1).

Use the accompanying random sample for your computations.

Random sample for Exercise 4–8

CASE-N	ID	ZONE	DWELL	GASTR	FAVGA	FEAT	LSODA	HSODA	BEER
1	6.	1.	1.	3.	1.	3.	0.	0.	1.
2	42.	1.	1.	4.	1.	2.	2.	0.	1.
3	51.	1.	1.	5.	2.	2.	0.	6.	0.
4	79.	1.	1.	9.	2.	1.	2.	0.	7.
5	80.	1.	1.	6.	2.	3.	2.	0.	1.
6	84.	1.	1.	5.	1.	3.	0.	0.	0.
7	155.	2.	1.	5.	2.	1.	2.	0.	2.
8	159.	2.	1.	4.	1.	2.	0.	6.	2.
9	196.	2.	1.	12.	1.	3.	0.	1.	3.
10	205.	2.	1.	3.	2.	3.	1.	0.	1.
11	401.	3.	0.	3.	2.	2.	2.	0.	10.
12	429.	3.	0.	7.	2.	2.	2.	0.	0.
13	498.	3.	0.	4.	1.	1.	0.	0.	2.
14	520.	3.	0.	3.	2.	1.	2.	0.	5.
15	553.	3.	1.	5.	1.	1.	0.	1.	0.
16	580.	3.	1.	6.	2.	1.	1.	0.	1.
17	586.	3.	1.	5.	2.	1.	2.	0.	0.
18	615.	3.	1.	5.	2.	2.	0.	0.	0.
19	616.	3.	1.	5.	2.	1.	0.	2.	4.
20	622.	3.	1.	2.	1.	1.	2.	0.	2.
21	636.	4.	0.	8.	1.	2.	2.	0.	8.
22	687.	4.	0.	5.	1.	2.	0.	0.	1.
23	728.	4.	0.	10.	2.	2.	0.	2.	0.
24	749.	4.	0.	5.	1.	2.	0.	0.	4.
25	852.	4.	0.	3.	1.	2.	0.	0.	0.
26	862.	4.	0.	7.	1.	2.	0.	0.	0.
27	875.	4.	0.	6.	2.	3.	0.	1.	7.
28	943.	4.	0.	11.	2.	1.	0.	0.	0.
29	967.	4.	0.	3.	2.	3.	0.	2.	0.
30	998.	4.	0.	7.	1.	2.	0.	0.	0.
31	1002.	4.	0.	8.	1.	2.	0.	0.	0.
32	1041.	4.	0.	10.	1.	2.	2.	0.	0.
33	1063.	4.	0.	10.	2.	2.	0.	0.	0.
34	1136.	4.	0.	5.	2.	2.	0.	5.	1.
35	1139.	4.	0.	4.	2.	2.	0.	1.	4.
36	1144.	4.	0.	8.	1.	2.	1.	0.	9.
37	1164.	4.	0.	5.	2.	2.	2.	0.	2.
38	1169.	4.	0.	8.	1.	2.	0.	0.	9.
39	1172.	4.	0.	3.	1.	2.	0.	0.	7.
40	1189.	4.	0.	6.	2.	2.	0.	1.	1.
41	1219.	4.	0.	5.	1.	2.	0.	5.	1.
42	1233.	4.	0.	4.	2.	2.	1.	0.	3.
43	1237.	4.	0.	5.	1.	3.	1.	0.	6.
44	1239.	4.	0.	7.	1.	2.	0.	2.	0.
45	1259.	4.	0.	7.	1.	2.	0.	5.	5.
46	1303.	4.	0.	7.	1.	2.	1.	0.	0.
47	1335.	4.	0.	5.	1.	2.	0.	1.	3.
48	1345.	4.	0.	2.	2.	2.	0.	6.	4.
49	1350.	4.	0.	4.	2.	2.	1.	0.	6.
50	1365.	4.	0.	6.	2.	2.	0.	6.	7.

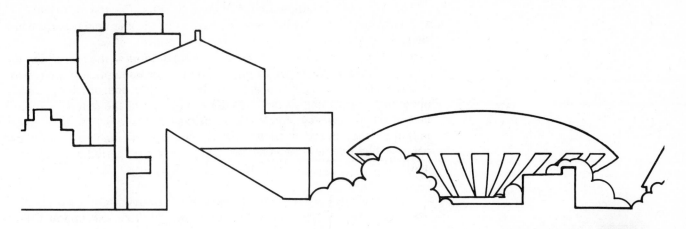

PROBABILITY DISTRIBUTIONS

Chapter 3 dealt with constructing and using frequency distributions to describe the characteristics of a population. Chapter 4 presented applications of the basic probability concepts. In this chapter you will use probability concepts to describe frequency distributions for certain Stat City characteristics (variables).

INTRODUCTION

The probability distributions presented in this chapter will help you understand more about Stat City characteristics while using less information than before; however, you will have to rely on certain critical assumptions. If the assumptions you make are reasonable, then you will make appropriate decisions. This chapter will help you become accustomed to knowing when and where these assumptions should be invoked. The topics discussed in this chapter are: the binomial distribution, the normal distribution, the normal approximation to the binomial distribution, and the Poisson distribution.

The binomial distribution is a tool used to determine the probability of x successes occurring in n trials. The assumptions required to use the binomial distribution are: (1) each trial has only two outcomes, (2) the probability of success remains constant for all n trials, and (3) each trial is independent of all other trials.[1]

---

[1] The formula used in most textbooks for computing binomial probabilities is:

$$P(x \text{ successes in } n \text{ trials}|\pi,n) = \frac{n!}{x!(n-x)!} \pi^x(1-\pi)^{n-x}$$

where

   n = the number of trials
   x = the number of successes
   π = the probability of success

Binomial probabilities for various combinations of n, x, and π can be seen in Table B–1 in Appendix B.

The normal distribution is a model used to determine the probability of a random data point falling between two values in the domain of a bell-shaped variable.[2]

The Poisson distribution is a model used to determine the probability of *x* events occurring in a specified unit of time. The assumptions required to use the Poisson distribution are: (1) there is only one of two possible outcomes in each unit of time, (2) the occurrence of the outcome in one unit of time in no way affects the occurrence of the outcome in another unit of time (independent trials), and (3) the average number of outcomes per specified time unit remains constant for the time period under examination.[3]

Example 5–1 Representativeness of the sample to be used in estimating the average number of rooms per dwelling unit in Stat City

Problem

Mr. Paul Lund, chairman of the Stat City Chamber of Commerce, selected a sample of 20 Stat City dwelling units to determine the average number of rooms per dwelling unit as of January 1985. Once the sample was drawn, it was determined that of the 20 dwelling units surveyed, all were apartments. Mr. Lund knows that 36.3 percent of Stat City dwelling units are houses. Consequently, he is concerned that the sample may not adequately reflect the actual number of rooms per dwelling unit.

Mr. Lund would like to know if the composition of dwelling unit types in his sample could be reasonably expected to occur by chance. You have been asked by Mr. Lund to resolve his dilemma. Type a memorandum to Mr. Lund reporting your findings (see Exhibit 5–1).

Solution

To determine the chances of obtaining zero houses in a sample of 20 dwelling units (if 36.3 percent of all dwelling units are houses) necessitates the use of the binomial distribution. Also, to use the binomial distribution model, one must assume the following:

---

[2] The formula in most textbooks for the normal distribution is:

$$f(x) = \frac{1}{\sqrt{2\pi}\sigma} e^{\left[-\frac{1}{2}\left(\frac{x-\mu}{\sigma}\right)^2\right]}$$

where

μ = the population mean
σ = the population standard deviation
π = 3.141592654
x = the value of the variable
e = 2.71828

Standard normal probabilities can be seen in Table B–3 in Appendix B.

[3] The formula used in most textbooks for computing Poisson probabilities is:

$$P(x) = \frac{e^{-\mu t}(\mu t)^x}{x!}$$

where

μ = the average number of occurrences per unit of time
t = the number of time units under study
e = 2.71828
x = the number of occurrences
$P(x)$ = the probability of x and occurrences in t units of time

Poisson probabilities for various combinations of μt and x can be seen in Table B–2 in Appendix B.

Assumption	Assumption met
1. Each trial has only two outcomes (house or apartment)	Yes
2. π remains constant over all n trials (the percent of houses in Stat City remains constant over the duration of the sampling process)	'Yes
3. Each trial is independent (selecting one dwelling unit in no way affects the selection of another dwelling unit)	Yes

Consequently:

$$n = 20$$
$$x = 0$$
$$\pi = 0.363$$
$$P(x|n,\pi) = \frac{20!}{(20 - 0)!0!}\,(0.363^0)(0.637^{20-0})$$
$$= 0.000\ 12$$

Approximately 12 chances in 100,000

The memorandum appears in Exhibit 5–1.

Exhibit 5–1

HOWARD S. GITLOW, PH.D.
STATISTICAL CONSULTANT
———

MEMORANDUM

TO: Mr. Paul Lund, Chairman
       Stat City Chamber of Commerce

FROM: Howard Gitlow

DATE: June 10, 1985

RE: Representativeness of the sample to be used in
       estimating the average number of rooms per dwelling
       unit in Stat City as of January 1985

   As requested and implementing commonly accepted
statistical techniques, I have determined that the chances
of obtaining zero houses in a sample of 20 Stat City dwelling
units (provided that 36.3 percent[1] of all Stat City dwelling
units are houses) is approximately 12 chances in 100,000.[2]
This result is so unlikely that if type of dwelling unit is
felt to seriously affect the statistic of concern (average
number of rooms per dwelling unit in Stat City as of January
1985), then another sample should be selected and attention
should be given to ensuring proper representation of houses
and apartments in the sample.
   If you have any further questions, please do not hesitate
to call me at 305-999-9999.

Exhibit 5–1 (*concluded*)

<u>Footnotes</u>

[1] The Stat City Census revealed that 499 out of 1,373 dwelling units (36.3 percent) as of January 1985 were houses.

[2] To determine the chances of obtaining zero houses in a sample of 20 dwelling units, if 36.3 percent of all dwelling units are houses, requires using the binomial distribution.

Binomial

Example 5–2 Composition of the Stat City Zoning Board

Problem The Stat City Zoning Board is comprised of 10 supposedly representative head-of-households. However, Mr. Pinero, a Zone 4 resident, realized that Zone 4 had no representation on the zoning board despite the fact that the Stat City census recorded 54.5 percent of all Stat City dwelling units are in Zone 4.

You have been retained by Mr. Pinero to determine the probability of Zone 4 families being excluded from the Stat City Zoning Board. Type a memorandum reporting your results to Mr. Pinero (see Exhibit 5–2).

Solution To ascertain the likelihood of obtaining zero Zone 4 families in a committee of 10 Stat City families, if 54.5 percent of all Stat City families live in Zone 4, requires using the binomial distribution. To use the binomial distribution model requires that the following assumptions have been met:

Assumption	*Assumption met*
1. Each trial has only two outcomes (family lives in Zone 4 or not)	Yes
2. π remains constant over all n trials (the percent of Zone 4 families in Stat City remains constant over the duration of the zoning board selection process)	Yes
3. Each trial is independent (selecting one family to be on the board in no way affects the selection of other families)	Yes

Consequently:

$$n = 10$$
$$x = 0$$
$$\pi = 0.545$$
$$P(x|n,\pi) = \frac{10!}{(10 - 0)!0!} (0.545)^0 (0.455)^{10-0}$$
$$= 0.00038$$
$$\approx \text{less than 4 times in 10,000}$$

The memorandum appears in Exhibit 5–2.   **Exhibit 5–2**

ROSA OPPENHEIM, PH.D.
STATISTICAL CONSULTANT
———

                        MEMORANDUM

TO: Mr. Pinero
        902 12th Street
        Zone 4
        Stat City

FROM: Rosa Oppenheim

DATE: June 12, 1985

RE: Composition of the Stat City Zoning Board

   As per your request and using commonly accepted
statistical techniques, I have determined that the chances
of obtaining "zero Zone 4 head–of–households" on the Stat
City Zoning Board are less than 4 times in 10,000.[1] In other
words, if 10,000 such zoning boards were formed, one would
expect "zero Zone 4 head–of–households" to occur only in 4 of
these 10,000 zoning boards. The chances of obtaining "zero
Zone 4 head–of–households" is so slight that it is beyond
serious discussion to assume that the Stat City Zoning Board
was formed by a chance happening.
   If you have any further questions, please do not hesitate
to call me at 305–999–9999.

                        Footnotes

   [1] This analysis required information from the Stat City Census
indicating that 54.5 percent of Stat City families live in Zone 4.
Further, to determine the chances of obtaining zero Zone 4 families
in a committee of 10 Stat City families, if 54.5 percent of all
Stat City families live in Zone 4, requires using the binomial
distribution.

Example 5–3 Distribution of the supermarket bills per person in Stat City

Problem The Stat City Supermarket Association would like to know the percentage of families that have weekly supermarket bills per person, as of January 1985, in the categories as shown in Table 5–1. The Supermarket

Table 5–1

Category of purchase	Definition
Small	Under $19.00 per week per person
Medium	Between $19.00 and $24.99 per week per person
Large	$25.00 or more per week per person

Association obtained the mean weekly supermarket bill per person ($21.86) and the standard deviation of weekly supermarket bills per person ($3.22), as of January 1985, from the Stat City Census. Unfortunately, they were unable to get any more detailed information on the distribution of weekly supermarket bills per person in Stat City.

You have been hired by the Supermarket Association to ascertain the desired information. Type a memorandum reporting your results to the Stat City Supermarket Association (see Exhibit 5–3).

Solution To compute the percentage of Stat City families in the small, medium, and large categories of weekly supermarket bills per person, you must make an assumption about the shape of the frequency distribution. It seems reasonable to assume that the distribution of weekly supermarket bills per person is normal. The normal assumption is based on the idea that most families consume $21.86 of supermarket items per person per week and that the percentage of families deviating from $21.86 trails off in both directions in a symmetric, bell-shaped pattern.

The percent of small purchasers is 18.67 percent.

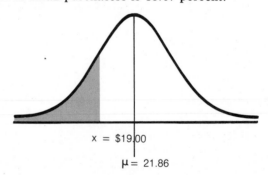

x = $19.00

μ = 21.86

$$z = \frac{x - \mu}{\sigma}$$

$$= \frac{19.00 - 21.86}{3.22}$$

$$= -0.89$$

$$P(z < -0.89) = 0.1867$$

The percent of medium purchasers is 64.98 percent.

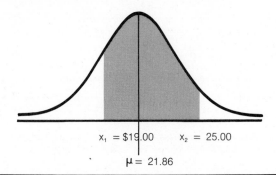

$$x_1 = \$19.00 \qquad x_2 = 25.00$$
$$\mu = 21.86$$

$$Z_1 = \frac{x_1 - \mu}{\sigma} \qquad\qquad Z_2 = \frac{x_2 - \mu}{\sigma}$$

$$= \frac{19.00 - 21.86}{3.22} \qquad\qquad = \frac{25.00 - 21.86}{3.22}$$

$$= \frac{-2.86}{3.22} = -0.89 \qquad\qquad = \frac{3.14}{3.22} = +0.98$$

$$P(-0.89 \le Z_1 \le 0) = 0.3133 \qquad P(0 \le Z_2 < 0.98) = 0.3365$$

$$P(-0.89 \le Z < 0.98) = P(\$19.00 \le X < \$25.00) = 0.6498$$

The percent of large purchasers is 16.35 percent.

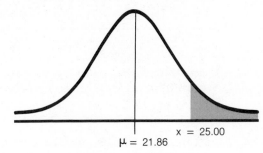

$$x = 25.00$$
$$\mu = 21.86$$

$$z = \frac{x - \mu}{\sigma}$$

$$= \frac{25.00 - 21.86}{3.22}$$

$$= 0.98$$

$$P(z \ge 0.98) = 0.1635$$

The memorandum appears in Exhibit 5–3.

Exhibit 5–3

HOWARD S. GITLOW, PH.D.
STATISTICAL CONSULTANT

MEMORANDUM

TO: Stat City Supermarket Association

FROM: Howard Gitlow

DATE: June 12, 1985

RE: Distribution of weekly supermarket bills per person
        in Stat City as of January 1985

In accordance with our consulting contract, I have investigated the supermarket shopping behavior of Stat City families. All statistics presented in this memorandum are based on commonly accepted statistical principles.

The following table presents the percent of Stat City families who have weekly supermarket bills per person, as of January 1985, in the following categories.[1]

Category of purchase	Definition of category	Percent
Small	Under $19.00 per week per person	18.67%
Medium	Between $19.00 and $24.99 per week per person	64.98
Large	$25.00 or more per week per person	16.35

I hope the above table satisfies your informational needs. If you have any further questions, please do not hesitate to call me at 305–999–9999.

Footnotes

[1] I had to make an assumption about the distribution of weekly supermarket bills per person, as of January 1985, to determine the percentage of Stat City families in the small, medium, and large categories of weekly supermarket bills per person. It seems reasonable to assume that weekly supermarket bills per person are normally distributed. The normal assumption is based on the idea that most families consume $21.86 of supermarket items per person per week and that the percent of families deviating from $21.86 trails off in both directions in a symmetric, bell–shaped pattern.

Example 5–4   HUD "home improvement" funding for Stat City

Mr. Lee Kaplowitz, the mayor of Stat City, would like to determine if his city is eligible for HUD (Housing and Urban Development) funding to improve the standard of living for homeowners. The HUD guideline for funding eligibility is that at least 15 percent of the houses in a city must have assessed values under $20,000. Consequently, Mr. Kaplowitz needs to determine the percent of Stat City houses with assessed values under $20,000 as of January 1985.

Normal

Mr. Kaplowitz obtained the average assessed value for houses ($67,980.65) and the standard deviation of assessed value for houses ($25,711.26) as of January 1985 from the Stat City Census. Unfortunately, he was unable to get any more detailed information on the distribution of assessed values of houses.

You have been retained by the mayor to determine the required information. Type a memorandum reporting your findings to Mr. Kaplowtiz (see Exhibit 5–4).

Solution

First, the relevant population must be defined. The population consists of the 499 houses in Stat City. Second, an assumption about the shape of the frequency distribution of assessed values as of January 1985 must be made. It seems reasonable to assume that assessed housing values are normally distributed. This assumption will be made because no other distributional assumption is more viable than the normal assumption. If the normal assumption is false, then the results reported to Mr. Kaplowitz will be wrong. The degree of error in the memorandum is directly related to how far the distribution of assessed values is from the normal distribution.

The percent of houses in Stat City that have assessed values under $20,000 is 3.07 percent.

x = 20,000 μ = 67,980.65

$$z = \frac{x - \mu}{\sigma}$$

$$= \frac{20,000 - 67,980.65}{25,711.26}$$

$$= -1.87$$

$$P(z < -1.87) = 0.0307$$

Consequently, Stat City is not eligible for HUD home improvement funding.

The memorandum appears in Exhibit 5–4.

ROSA OPPENHEIM, PH.D.
STATISTICAL CONSULTANT

MEMORANDUM

TO: Mr. Lee Kaplowitz, Mayor
        Stat City

FROM: Rosa Oppenheim

DATE: June 18, 1985

RE: HUD home improvement funding for Stat City

   As per your request and using commonly accepted
statistical techniques, I have determined that the
percentage of Stat City houses with assessed values under
$20,000 as of January 1985 is 3.07 percent.[1] I hope the
above statistic meets your informational needs. If you have
any further questions, please do not hesitate to call me at
305–999–9999.

_____

Footnotes

   [1] The computation of 3.07 percent was based on the assumption that
the distribution of assessed values for Stat City houses as of
January 1985 is normal. The reported 3.07 percent would change if
the distribution of assessed values was not normal. However, the
distribution of assessed values would have to depart rather
severely from the normal distribution before 15 percent or more of
Stat City houses would have assessed values under $20,000.

Example 5-5 Composition of the Stat City Planning Board

The Stat City Planning Board consists of 25 members who are purport- **Problem**
edly representative head-of-households. A Zone 4 resident (Mr. Pinero
again) realized that only three Zone 4 families were on the planning board.
It is known from the Stat City Census that 54.5 percent of all Stat City
dwelling units are in Zone 4.

You have been retained by Mr. Pinero to determine the likelihood of
acquiring three or less Zone 4 head-of-households on the planning board
by chance. Type a memorandum reporting your results to Mr. Pinero (see
Exhibit 5-5).

To determine chances of obtaining three or fewer Zone 4 head-of- **Solution**
households on the 25-member Stat City Planning Board, if 54.5 percent of
all Stat City head-of-households live in Zone 4, requires using the binomial distribution. To use the binomial distribution model requires that the
following assumptions have been met:

Assumption	Assumption met
1. Each trial has only two outcomes (Zone 4 head-of-household or not)	Yes
2. π remains constant over all n trials (the percent of Zone 4 head-of-households remains constant each time a head-of-household is selected onto the board)	Yes
3. Each trial is independent (selecting one head-of-household in no way affects the selection of any other head-of-household	Yes

Consequently:

$$n = 25$$
$$x \leq 3$$
$$\pi = 0.545$$

Unfortunately, $P(x \leq 3 | n = 25$ and $\pi = 0.545)$ is cumbersome to com-
pute. Consequently, this problem requires that the normal approximation
to the binomial distribution be used because n is large and computations
would be extremely tedious and time-consuming. However, before the
normal approximation can be used, we must check to make sure that π is
close to 0.50. Whenever $\pi = 0.50$, the binomial distribution is symmetric
and resembles the normal distribution. The further π strays from 0.50
(either higher or lower), the less symmetrical (normal looking) the binomial distribution will be. As long as π stays between 0.10 and 0.90 most
statisticians agree that the normal approximation is acceptable.

There are two commonly accepted guidelines that can be used in determining if the normal approximation can be used: $n\bar{p} > 5$ and $n(1 - \bar{p}) >$
5 if you are dealing with a sample, and $N\pi > 5$ and $N(1 - \pi) > 5$ if you
are dealing with a population.

In the problem at hand $n = 25$ and $\pi = 0.545$. Consequently,

$$n\pi = 25(0.545) = 13.625 > 5$$
$$n(1 - \pi) = 25(0.455) = 11.375 > 5$$

Hence, the normal approximation to the binomial distribution is appropriate.

The following graph and supportive formulas are used to compute the
likelihood of finding three or fewer Zone 4 head-of-households on the
25-member planning board, given that 54.5 percent of Stat City head-of-
households live in Zone 4.

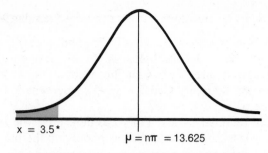

x = 3.5* $\qquad$ $\mu = n\pi = 13.625$

* 3 + 0.5 is correction for continuous approximation.

$$z = \frac{x - \mu}{\sigma}$$

$$= \frac{3.5 - 13.625}{2.49}$$

$$= -4.07$$

$$P(z < -4.07) = \text{less than 26 times in 1 million}$$

The memorandum appears in Exhibit 5–5.

Exhibit 5–5

HOWARD S. GITLOW, PH.D.
STATISTICAL CONSULTANT

MEMORANDUM

TO: Mr. Pinero
        902 12th Street
        Zone 4
        Stat City

FROM: Howard Gitlow

DATE: June 24, 1985

RE: Composition of the Stat City Planning Board

As per your request and using commonly accepted statistical techniques, I have determined that the likelihood of finding three or fewer Zone 4 head–of–households on the 25–member Stat City Planning Board is less than 26 times in 1 million.[1] This result is so unlikely that it precludes serious discussion of the board having been randomly selected. In other words, it is not reasonable to assume that the planning board is a fair representation of Stat City with respect to zonal representation.

If you have any further questions, please do not hesitate to call me at 305–999–9999.

Exhibit 5–5 (*concluded*)

Footnotes

[1] To determine the chances of obtaining three or fewer Zone 4 head-of-households being on a 25-member planning board, given 54.5 percent of all Stat City head-of-households reside in Zone 4, requires using the normal approximation to the binomial distribution. (Note: The Stat City Census revealed that 54.5 percent of all dwelling units in Stat City are in Zone 4.)

Example 5–6   Staffing the Stat City Chamber of Commerce switchboard

Problem

The Stat City Chamber of Commerce receives an average of five telephone calls per minute. An average switchboard operator can handle three telephone calls per minute. The Chamber of Commerce is trying to determine the number of switchboard operators they need so that no more than 1 percent of all incoming telephone calls go unanswered. The chamber can hire part-time employees.

You have been hired by Mr. Paul Lund, chairman of the Chamber of Commerce, to determine how many switchboard operators the chamber must hire. Type a memorandum to Mr. Lund reporting your findings (see Exhibit 5–6).

Solution

To compute the number of switchboard operators required to answer at least 99 percent of all incoming telephone calls to the Stat City Chamber of Commerce requires using the Poisson distribution. Use of the Poisson distribution is predicated upon the following assumptions:

	Assumption	*Assumption met*
1.	Number of events occurring in one time interval is independent of what happened in previous time intervals	Yes
2.	The average number of occurrences per unit of time remains constant for the time period under examination	Yes
3.	It is extremely rare for more than one event to occur in a small interval of time	Yes

Consequently, $\mu t = (5)(1) = 5$, and

$$\left\{ P(x > X) = 1 - \sum_{i=0}^{X} \frac{e^{-(\mu t)}(\mu t)^i}{i!} \right\} \leq 0.01$$

	i	$P(x = i)$	$P(x \leq i)$	$P(x > i)$
	0	.006738	.006738	
	1	.033690	.040428	
	2	.084224	.124652	
	3	.140374	.265026	
	4	.175467	.440493	
	5	.175467	.615960	
	6	.146223	.762183	
	7	.104445	.866628	
	8	.065278	.931906	
	9	.036266	.968172	
$X = i \rightarrow$	10	.018133	.986305	.013695
	11	.008242	.994570	.005430 $= P(x > 10) = P(x \geq 11)$

As you can see, the chamber switchboard receives 11 or more telephone calls per minute approximately 0.50 percent of the time. Hence, the chamber must have the capability of answering at least 10 calls per minute if they are to miss only 1 percent or less of incoming telephone calls. To answer 10 calls per minute, given an average operator can answer three calls per minute, the chamber must hire three and one-third operators (10 calls per minute divided by 3 calls per minute equals 3⅓ operators).

The memorandum appears in Exhibit 5–6.

Exhibit 5–6

ROSA OPPENHEIM, PH.D.
STATISTICAL CONSULTANT
———

MEMORANDUM

TO: Mr. Paul Lund, Chairman
        Stat City Chamber of Commerce

FROM: Rosa Oppenheim

DATE: February 21, 1985

RE: Staffing the Stat City Chamber of Commerce
        switchboard

The Stat City Chamber of Commerce must employ three and one-third switchboard operators if you are to have the capability of answering at least 99 percent of all incoming telephone calls.[1] The above calculations presume that an average of five telephone calls come into the chamber per minute and that a typical switchboard operator can handle three calls per minute.

The above computations are based on commonly accepted statistical techniques. If you have any further questions, please do not hesitate to call me at 305–999–9999.

———

Footnote

[1] The computations presented in this memorandum are based upon the Poisson distribution and consequently are subject to its assumptions.

5–7 Composition of the Stat City task force on problem drinking   Mr. Lee Kaplowitz, the mayor of Stat City, convened a five-person task force to discuss problem drinking in Stat City. Mr. Upton (identification number 115), the local tavern owner, claimed that the task force had not been randomly chosen but had been personally selected to come up with anti-alcohol recommendations. Mr. Upton claimed that not one person on the task force came from a family that purchased one or more six-packs of beer on a weekly basis. This determination was made by canvassing each member of the task force. Mr. Upton decided to use beer drinking as an extremely conservative measure of attitude toward alcohol; that is, non-beer drinkers have negative attitudes toward alcohol, and beer drinkers have positive attitudes.

Mr. Upton checked the Stat City Census and found that 46.76 percent of Stat City families (642) purchase one or more six-packs of beer weekly. You have been hired by Mr. Upton to determine the chances of randomly selecting a five-person task force in which not one member comes from a family that purchase beer on a weekly basis. Type a memorandum reporting your findings to Mr. Upton.

5–8 Federal request for information on housing costs in Stat City   The mayor of Stat City, Mr. Lee Kaplowitz, must determine if more than 30 percent of Stat City families spent less than $200 per month on housing (rent or mortgage) as of January 1985. This information is required by the federal Department of Health and Human Services (HHS).

Mr. Kaplowitz obtained the average housing cost ($462.80) and the standard deviation of housing costs ($171) as of January 1985 from the Stat City Census. Unfortunately, he was unable to obtain more detailed information on the distribution of housing costs in Stat City.

You have been retained by Mr. Kaplowitz to determine the required information. Type a memorandum reporting your findings to the mayor. (Assume housing costs are normally distributed.)

5–9 Staffing the maternity ward in Park View Hospital   Ms. Arlene Davis, director of the Park View Hospital, knows from experience that approximately three women enter the hospital per day to give birth. Further, she also knows that the staff obstetrician, Dr. Deborah Jones, can deliver a maximum of five babies per day.

Ms. Davis has hired you to determine the average number of days per year that Dr. Jones will be unable to handle the inflow of expectant mothers. Type a memorandum reporting your findings to Ms. Davis.

5–10 Aiding library patrons   Ms. Elissa Gitlow, chief librarian (the only librarian) of the Stat City Library, knows that an average of five people per hour enter the library seeking aid from a librarian. She also knows that she can help an average of seven people per hour.

She has retained you to determine the percentage of the time that she will not be able to help patrons seeking the advices of a librarian. Type a memorandum reporting your findings to Ms. Gitlow.

5–11 Demographic probabilities in Stat City The Stat City Census revealed the following parameters:

$$P(\text{HSODA} = 0) = .673$$
$$P(\text{LSODA} = 0) = .689$$
$$P(\text{BEER} > 3) = .238$$
$$P(\text{FEAT} = 2) = .606$$
$$P(\text{FAVGA} = 1) = .404$$
$$P(\text{PEPLE} = 1) = .046$$
$$P(\text{CARS} < 2) = .731$$
$$P(3 \le \text{ROOMS} \le 8) = .829$$

Compute the following binomial probabilities:

1. The probability of randomly selecting 3 families that do not purchase any six-packs of regular soda weekly out of 10 randomly selected families.
2. The probability of randomly selecting more than 8 families that do not purchase any six-packs of diet soda weekly out of 12 randomly selected families.
3. The probability of randomly selecting less than three families that purchase more than three six-packs of beer weekly out of five randomly selected families.
4. The probability of randomly selecting one or more families that prefer the Grand Union out of seven randomly selected families.
5. The probability of randomly selecting between three and five families (inclusive) that favor Paul's Texaco Station out of nine randomly selected families.
6. The probability of randomly selecting between two and four single-person families (inclusive) out of nine families.
7. The probability of randomly selecting 3 families that own less than two cars out of 100 randomly selected families.
8. The probability of randomly selecting 50 or more families that own homes with between three and eight rooms, inclusive, out of 75 randomly selected families.

5–12 Advertising claims by the A&P Mr. Marc Cooper, the manager of the A&P, recently formed a consumer panel consisting of 50 Stat City families (each family is represented by the person most responsible for the family's supermarket shopping). He asked the panel which supermarket in Stat City they prefer. Thirty-one of the 50 family representatives said they prefer the A&P. Consequently, Mr. Cooper claimed in a TV commercial that the A&P is Stat City's favorite supermarket.

The management of the Food Fair and the Grand Union (especially the Grand Union) were extremely upset at this "false" claim. Grand Union management knew from the Stat City Census that only 19.5 percent of all Stat City families prefer the A&P. Food Fair and Grand Union management can accept that Mr. Cooper had not known about the census figures and had conducted a study of his own to arrive at the 62 percent (31/50) statistic. However, Food Fair and Grand Union management are concerned that Mr. Cooper unfairly selected his consumer panel; for example, he may have selected the first 50 people who came into the A&P on a given day.

Consequently, Food Fair and Grand Union management have hired you to determine the chances of finding 31 or more families that favor the A&P out of 50 families, given that 19.5 percent of all families favor the A&P. Using the normal curve below, indicate the information and region of interest. Report your findings to Food Fair and Grand Union management.

Normal Approx to Binomial

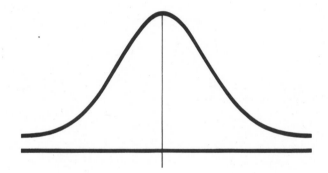

5–13 Distribution of utility and food costs in Stat City   Assume that HEAT, ELEC, PHONE, GAS, and EAT are all normally distributed random variables. The Stat City Census revealed the following parameters for the aforementioned random variables:

	Mean (μ)	Standard deviation (σ)
HEAT	$648.89	$271.57
ELEC	55.76	19.73
PHONE	36.70	26.50
GAS	134.37	44.88
EAT	90.19	33.14

Sketch the appropriate areas under the normal curve and compute the following quantities:

$P(\text{HEAT} > \$700) =$
$P(\text{HEAT} < \$600) =$
$P(190 \leq \text{HEAT} \leq \$600) =$
$P(\text{ELEC} > \$40) =$
$P(\$60 \leq \text{ELEC} \leq \$70) =$
$P(\$25 \leq \text{PHONE} \leq \$35) =$
$P(\text{PHONE} \leq \$10) =$
$P(\text{GAS} < \$100 \text{ OR GAS} > \$200) =$
$P(\text{EAT} < \$70 \text{ OR EAT} > \$100) =$

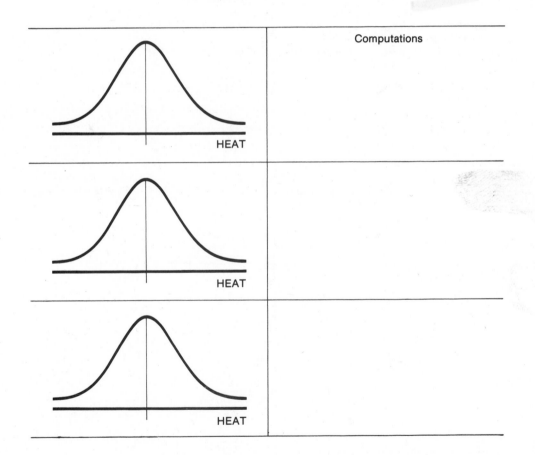

Computations

HEAT

HEAT

HEAT

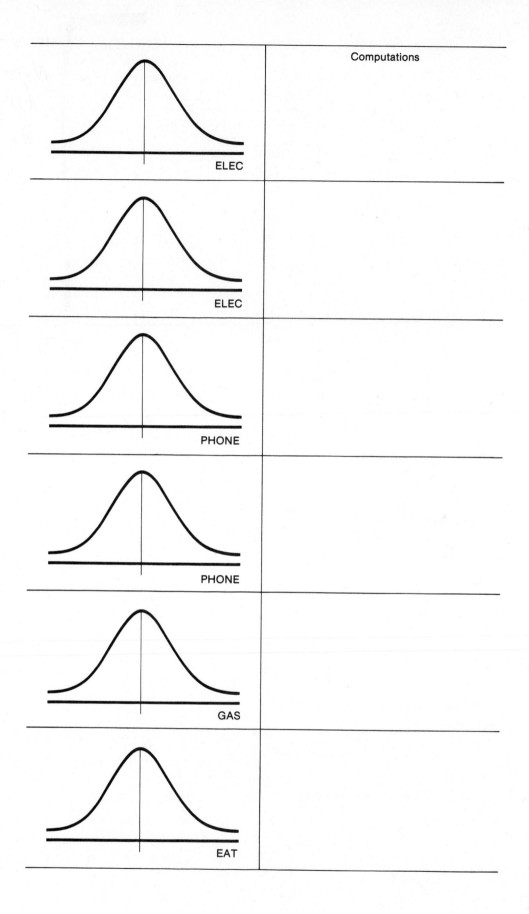

Computations

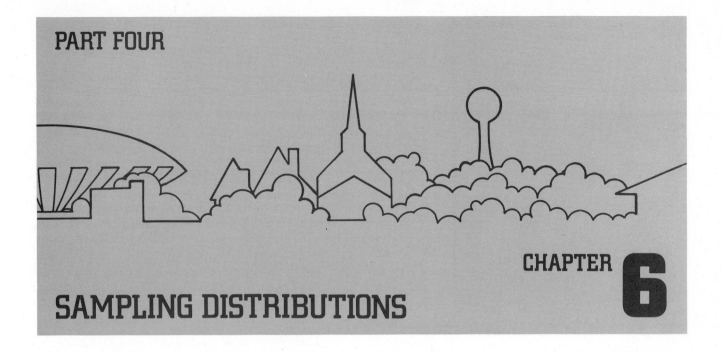

PART FOUR

SAMPLING DISTRIBUTIONS

CHAPTER 6

INTRODUCTION

In this chapter we are going to formalize something we have been doing since Chapter 1—drawing conclusions about populations from samples. The process of making a generalization from a sample statistic to a population parameter is called statistical inference. In Chapter 1, we drew a simple random sample of 10 Stat City dwelling units and computed the average number of rooms per dwelling unit to be 6.5. We then inferred that the average number of rooms for all Stat City dwelling units is 6.5. This generalization is a statistical inference.

A major advantage of random sampling is that it allows a researcher to assign a probability that a sample statistic accurately estimates an unknown population parameter. Learning how to assign probability statements to sample statistics will be a major focus of this chapter.

To use statistical inference you must understand the behavior of sample statistics. For example, logic would dictate that if two random samples of size 10 were drawn from Stat City to estimate the average number of rooms per dwelling unit, each sample would have different sample means. This would happen because different samples include different elements of the population. Consequently, you must learn to deal with errors which occur when estimating population parameters due to sample-to-sample variation. Sample-to-sample variation is called sampling error. You will use the probability models you studied in Chapter 5 to better understand sampling error and the problems it causes when estimating population parameters.

The frequency distribution of all possible sample statistics that can be drawn from a population is called a sampling distribution. An understanding of sampling distributions will allow you to make proper statistical inferences about a population from a sample by accounting for sampling error.

Understanding a statistical inference is more complex than a person untrained in statistics would suspect. Consequently, it is imperative that the footnotes in Exhibit 6–1 (or some rendition thereof) appear whenever a statistical inference is being reported in a memorandum. These footnotes serve as a partial disclaimer for the errors which can occur when

Exhibit 6–1 Standard footnotes 1 and 2

[1] The summary statistics reported are estimates derived from a sample survey of _____ (_____) elementary units; drawn via a _____ random sampling plan.

"There are two types of errors possible in an estimate based on a sample survey—sampling and nonsampling. Sampling errors occur because observations are made only on a sample, not on the entire population. Nonsampling errors can be attributed to many sources; for example, inability to obtain information about all cases in the sample, definitional difficulties, differences in the interpretation of questions, inability or unwillingness to provide correct information on the part of respondents, mistakes in recording or coding the data obtained and other errors of collection, response, processing, coverage, and estimation for missing data. Nonsampling errors also occur in complete censuses. The 'accuracy' of a survey result is determined by the joint effects of sampling and nonsampling errors." [Journal of the American Statistical Association 70, no. 351, part II (September 1975), p. 6.]

[2] "The particular sample used in this survey is one of a large number of all possible samples of the same size that could have been selected using the same sample design. Estimates derived from the different samples would differ from each other. The difference between a sample estimate and the average of all possible samples is called the sampling deviation. The standard or sampling error of a survey estimate is a measure of the variation among the estimates from all possible samples and thus is a measure of the precision with which an estimate from a particular sample approximates the average result of all possible samples. The relative standard error is defined as the standard error of the estimate divided by the value being estimated.

"As calculated for this report, the standard error also partially measures the effect of certain nonsampling errors but does not measure any systematic biases in the data. Bias is the difference, averaged over all possible samples, between the estimate and the desired value. Obviously, the accuracy of a survey result depends on both the sampling and nonsampling errors measured by the standard error and the bias and other types of nonsampling error not measured by the standard error." [Journal of the American Statistical Association 70, no. 351, part II (September 1975), p. 6.]

computing statistics from a particular sample; remember, a sample statistic is subject to sampling and nonsampling error.

This chapter will give you insights into the sampling distributions of the sample mean and proportion. Hopefully, these insights, in conjunction with the theory your professor is giving you, will make it clear how to make proper statistical inferences.

There will not be any problems in this chapter; there will just be illustrations of how the sampling distributions of the mean and proportion behave under various sets of "real-world circumstances."

GENERAL DISCUSSION OF THE SAMPLING DISTRIUBTION OF THE SAMPLE MEAN ($\bar{x}$)

If all possible random samples of size n are drawn from a population of size N and the mean for each sample is computed, the frequency distribution of all the sample means is called the sampling distribution of $\bar{x}$. The shape of the sampling distribution of $\bar{x}$ changes in accordance with certain real-world conditions. Table 6–1 indicates that there are three basic shapes for the sampling distribution of $\bar{x}$: normal, t, and unknown.

Each of these basic shapes must be thoroughly understood to make proper statistical inferences about μ from $\bar{x}$. Table 6–1 also delineates the relationship between real-world conditions and the shape of the sampling distribution of $\bar{x}$. Consequently, once you can isolate which real-world conditions are present in your problem, you can determine the basic shape of the sampling distribution of $\bar{x}$ and can make appropriate statisti-

Table 6–1

Case	Sample size n (n ≥ 30 is large) (n < 30 is small)	Population standard deviation (σ)	Distribution of original population	Shape of the sampling distribution of x̄		
				Shape 1	*Shape 2*	*Shape 3*
1............	Large	Known	Normal	Normal		
2............	Large	Known	Unknown	Normal		
3............	Large	Unknown	Normal	Normal		
4............	Large	Unknown	Unknown	Normal		
5............	Small	Known	Normal	Normal		
6............	Small	Unknown	Normal		*t*	
7............	Small	Known	Unknown			Unknown
8............	Small	Unknown	Unknown			Unknown

cal inferences about μ from your $\bar{x}$. It is important to realize that sampling distributions are never constructed in practice; they are theoretical abstractions which allow researchers to make appropriate statistical inferences.

Sampling distribution of $\bar{x}$ when the sample size is large, the population standard deviation is known, and the distribution of the original population is normal (Case 1)

To understand the sampling distribution of $\bar{x}$ for Case 1, 100 random samples of sizes 2, 5, 10, and 30 were drawn concerning family incomes of Stat City families. For each sample size, all 100 sample means were computed.

Table 6–2 shows the relationship between the sample size and the standard error of the mean for total family income in Stat City, given the population standard deviation is known ($\sigma = \$22,847.12$).

Table 6–2

Case	Sample size (n)	Population standard deviation (σ)	Population size (N)	Finite population correction factor $\sqrt{\frac{N-n}{N-1}}$	Standard error of x̄ (σ_x̄)	
					Infinite population $\sigma/\sqrt{n}$	Finite population $(\sigma/\sqrt{n})\sqrt{\frac{N-n}{N-1}}$
1.........	1	$22,847.12	1,373	1.0	$22,847.12	$22,847.12
2.........	2	22,847.12	1,373	0.99964	16,155.35	16,149.46
3.........	5	22,847.12	1,373	0.99854	10,271.54	10,202.64
4.........	10	22,847.12	1,373	0.99671	7,224.89	7,201.16
5.........	30	22,874.12	1,373	0.98938	4,171.29	4,126.97

As you can see, the standard error of $\bar{x}$ decreases as the sample size increases. This simply demonstrates that as n is increased, the chances of any given $\bar{x}$ deviating from μ decreases, because the sample mean is based on a larger proportion of the population. The above rule holds for infinite and finite populations.[1]

Frequency distributions of the 100 sample means for each sample size are shown in Exhibit 6–2. This table graphically illustrates that the larger the sample size, the more closely the relative frequency distribution of the sample means (sampling distributions of $\bar{x}$) approaches the normal distribution. Further, the mean of the sampling distribution of

---

[1] For all five cases in Table 6–1, the finite population correction factor (fpc) is negligible; $\frac{\sigma}{\sqrt{n}}\sqrt{\frac{N-n}{N-1}} \cong \frac{\sigma}{\sqrt{n}}[1] \cong \frac{\sigma}{\sqrt{n}}$. The fpc is unimportant when $[n/N]$ is less than 10 percent. If $[n/N]$ is greater than 10 percent (as n approaches N), the fpc becomes a significant modifier of the standard error of $\bar{x}$.

$\bar{x}$ is μ, and the standard deviation (called the standard error of $\bar{x}$) is $\sigma/\sqrt{n}$ if the population is infinite and $\dfrac{\sigma}{\sqrt{n}} \sqrt{\dfrac{N-n}{N-1}}$ if the population is finite.

The above phenomenon can be demonstrated mathematically and is called the central limit theorem for means.

Exhibit 6–2

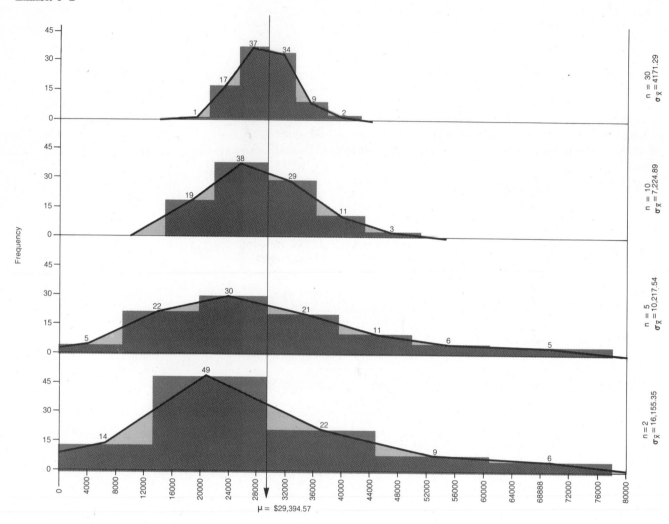

Central limit theorem for means

If random samples of n observations are drawn from a population with a finite mean (μ) and a finite standard deviation (σ), as n becomes large (greater than 30), the distribution of the sample means ($\bar{x}$) will approximate a normal distribution with mean μ and standard error $\sigma/\sqrt{n}$ for infinite populations, and $\dfrac{\sigma}{\sqrt{n}} \sqrt{\dfrac{N-n}{N-1}}$ for finite populations.

Now that it has been established that the frequency distribution of the sample means is approximately normal for large n, you can use what you know about the normal distribution to interpret the ONE SAMPLE MEAN you have in practice.

Remember, you know that approximately 99.73 percent of all data in a normal population lies within three standard deviations (σ) of the population mean (μ).

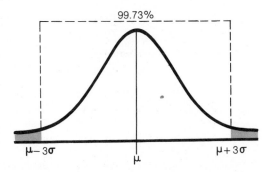

99.73%

$\mu - 3\sigma$ μ $\mu + 3\sigma$

Hence, if the sampling distribution of means is normal, then approximately 99.73 percent of all sample means will fall within three standard errors of the unknown population mean. Consequently, if the standard error is large, 99.73 percent of all sample means will fall within a WIDE range around μ (three standard errors either way of μ) and you cannot have "confidence" that your sample mean will provide a good estimate of μ. However, if the standard error is small, 99.73 percent of all sample means will fall within a SMALL range around μ (three standard errors either way of μ) and you can have "confidence" that your sample mean will provide a good estimate of μ.

Remember, even if the standard error of the mean is small, you never know for certain if your $\bar{x}$ is an accurate estimate of μ. As the standard error decreases, the chances of your $\bar{x}$ deviating from μ decrease, but there is always a finite probability that even with a small standard error, your $\bar{x}$ may be one of the $\bar{x}$s which will not provide a good estimate of μ. However, since you never know if your $\bar{x}$ is a good estimator of μ or not, you will base your evaluation of $\bar{x}$ on the following criteria: assume that $\bar{x}$ is a good estimator of μ if the standard error is small or assume that $\bar{x}$ may be a poor estimator of μ if the standard error is large.

Finally, a way of increasing your confidence in $\bar{x}$ as an estimator of μ is to increase the sample size upon which $\bar{x}$ is based. As n gets larger, the standard error of the mean gets smaller[2] and the $\bar{x}$s are more closely grouped around μ. In other words, the interval in which 99.73 percent of all sample means would fall (three standard errors on either side of μ) is reduced as n is increased. Hence, each $\bar{x}$ has a better chance of being a reasonable estimator of μ. Do not forget that n is usually restricted by budgetary considerations. This precludes researchers from taking a complete census and makes sampling all the more important.

Sampling distribution of $\bar{x}$ when the sample size is large, regardless of whether the population standard deviation is known or the distribution of the original population is normal (Cases 2, 3, and 4).

The sampling distribution of the $\bar{x}$s is normal when n is large (n is greater than or equal to 30), regardless of whether the population standard deviation is known or the original population is normal. Upon reflection, you can logically accept the foregoing statement so another example is unnecessary.

---

[2] The standard error of $\bar{x}$ is:

$$\frac{\sigma}{\sqrt{n}} \text{ for infinite populations, and}$$

$$\frac{\sigma}{\sqrt{n}} \sqrt{\frac{N-n}{N-1}} \text{ for finite populations.}$$

Sampling distribution of $\bar{x}$ when the sample size is small, the population standard deviation is known, and the distribution of the original population is normal (Case 5).

The sampling distribution of the $\bar{x}$s is normal when n is small if σ is known and the original population is normal. Upon reflection, you can logically accept the foregoing statement so another example is unnecessary.

Sampling distribution of $\bar{x}$ when the sample size is small, the population standard deviation is unknown, and the distribution of the original population is normal (Case 6).

The sampling distribution of the $\bar{x}$s is distributed according to student's t-distribution (this probability model has yet to be discussed) when: n is small ($n < 30$), σ is unknown, and the original population is normal.

> **Central limit theorem
> for means based on
> small sample sizes**
>
> If random samples of n observations are drawn from a normal population with finite mean (μ) and unknown and finite standard deviation (σ), the distribution of the sample means ($\bar{x}$) will be t-distributed with mean μ and standard error $s/\sqrt{n}$ for infinite populations, and $[s/\sqrt{n}]\sqrt{(N - n)/(N - 1)}$ for finite populations.

The manner in which you will interpret a sample mean is changed only in the percent of sample means that lies within a specified number of standard errors either way of μ. To ascertain the percent of sample means that lies within a specified number of standard errors of μ, you must consult a t-table. (See Table B–4 in Appendix B.)

Sampling distribution of $\bar{x}$ when the sample size is small, the original population distribution is unknown, and the population standard deviation is known or unknown (Cases 7 and 8).

Statistical inferences about μ cannot be made from $\bar{x}$ if the shape of the sampling distribution of $\bar{x}$ is unknown. Since the shape of the sampling distribution of $\bar{x}$ is unknown, you cannot know with any specificity what percent of the $\bar{x}$s lie within a given range around μ (you could use Chebychev's inequality). Since this percent is unknown, you will be unable to determine if $\bar{x}$ is a good estimator of μ.

GENERAL DISCUSSION OF THE SAMPLING DISTRIBUTION OF THE SAMPLE PROPORTION ($\bar{p}$)

The shape of the sampling distribution of $\bar{p}$ changes in accordance with certain "real-world conditions" as does the sampling distribution of $\bar{x}$. Table 6–3 indicates that there are four basic shapes for the sampling distribution of $\bar{p}$: normal, binomial, Poisson, and hypergeometric (the latter being a probability model we have not yet discussed in this text).

The hypergeometric distribution and the binomial distribution are both concerned with the probability of obtaining x successes in n trials. However, the binomial model assumes that the population under study is infinite while the hypergeometric model assumes that the population under study is finite. The effect of the population being finite is that the probability of success changes from trial to trial; it is not constant.

Table 6–3

Case	Distribution of the original population	Sample size (n)	Proportion between .1 & .9 ($.1 \leq \pi \leq .9$)	Sampling fraction ($n/N \leq .1$)	Comments	Shape of the sampling distribution of $\bar{p}$
1		Large	Yes		$n\pi > 5$ $n(1 - \pi) > 5$	Normal
2	Infinite Bernoulli process (binomial distribution) or sampling with replacement	Large	No	Not relevant for infinite population	$n\pi$ and/or $n(1 - \pi) < 5$	Poisson
3		Small	Yes		$n\pi$ and/or $n(1 - \pi) < 5$	Binomial
4		Small	No		$n\pi$ and/or $n(1 - \pi) < 5$	Binomial
5		Large	Yes	No	$n\pi > 5$ $n(1 - \pi) > 5$	Normal with fpc
6		Large	Yes	Yes	$n\pi > 5$ $n(1 - \pi) > 5$	Normal
7		Large	No	No	$n\pi$ and/or $n(1 - \pi) < 5$	Binomial with fpc
8		Large	No	Yes	$n\pi$ and/or $n(1 - \pi) < 5$	Binomial
9	Finite Bernoulli process (hypergeometric distribution) or sampling without replacement	Small	Yes	No	$n\pi$ and/or $n(1 - \pi) < 5$ and $n/N > .1$	Hypergeometric
10		Small	Yes	Yes	$n\pi$ and/or $n(1 - \pi) < 5$ and $n/N \leq .1$	Binomial
11		Small	No	No	$n\pi$ and/or $n(1 - \pi) < 5$ and $n/N > .1$	Hypergeometric
12		Small	No	Yes	$n\pi$ and/or $n(1 - \pi) < 5$ and $n/N \leq .1$	Binomial

The formula for the hypergeometric distribution is:

$$P(X|n,N,R) = \frac{\begin{bmatrix} R \\ X \end{bmatrix}\begin{bmatrix} N - R \\ n - X \end{bmatrix}}{\begin{bmatrix} N \\ n \end{bmatrix}}$$

$$= \frac{\left[\dfrac{R!}{X!\,(R - X)!}\right]\left[\dfrac{(N - R)!}{(\{N - R\} - \{n - X\})!\,(n - X)!}\right]}{\left[\dfrac{N!}{n!\,(N - n)!}\right]}$$

where

n = the sample size
N = the population size
R = the number of successes in the population
$N - R$ = the number of failures in the population
X = the number of successes in the sample
$P(X|n,N,R)$ = the probability of obtaining X successes in n trials, given a finite population of N units containing R successes

Each of the shapes must be understood to make proper statistical inferences about π from $\bar{p}$. Table 6–3 delineates the relationship between real-world conditions and the shape of the sampling distribution of $\bar{p}$. Consequently, once you can isolate which real-world conditions are present in your problem, you can determine the basic shape of the sampling distribution of $\bar{p}$ and can make proper statistical inferences about π from your $\bar{p}$. Remember, sampling distributions are never constructed in practice. They are theoretical abstractions which allow people to make sound statistical inferences.

Sampling distribution of $\bar{p}$ is normal (Cases 1, 5, and 6).

If $n\pi$ and $n(1 - \pi)$ are both greater than 5, the distribution of sample proportions is approximately normal. To demonstrate, 100 samples of size 2, 5, 10, and 30 were drawn concerning types of dwelling units for Stat City families. The sample proportions were computed for all 100 samples of each size.

Table 6–4 shows the relationship between sample size and the standard error of the proportion of houses in Stat City, given that the population

Table 6–4

Sample size (n)	Population standard deviation (σ)	Population size (N)	Finite population correction factor (fpc)	Standard error of $\bar{p}$ ($\sigma_{\bar{p}}$) Infinite pop.	Standard error of $\bar{p}$ ($\sigma_{\bar{p}}$) Finite pop.
2	.481	1373	.99964	.340	.340
5	.481	1373	.99854	.215	.215
10	.481	1373	.99671	.152	.151
30	.481	1373	.98938	.088	.087

standard deviation is known ($\sigma = .481 = \sqrt{\pi(1 - \pi)} = \sqrt{.363 \times .637}$). Table 6–4 clearly indicates that the standard error of $\bar{p}$ decreases as the sample size increases. The table demonstrates that as n is increased, the chances of $\bar{p}$ deviating from π decreases because $\bar{p}$ is based on a larger percentage of the population. The above rule works for both infinite and finite populations.[3]

Frequency distributions of the 100 sample proportions for each sample size are shown in Exhibit 6–3. The larger the sample size, the more closely the relative frequency distribution of the sample proportion ($\bar{p}$) approaches the normal distribution. The above phenomena is called the central limit theorem for proportions and only works when $n\pi$ and $n(1 - \pi)$ are both greater than 5 (in this example $n\pi = 10.89$ and $n[1 - \pi] = 19.11$).

**Central limit theorem
for proportions**

If random samples of n observations are drawn from an infinite population, which is a Bernoulli process, as n enlarges (such that $n\pi$ and $n[1 - \pi]$ are greater than 5), the distribution of the sample proportions ($\bar{p}$) will approximate a normal distribution with mean π and standard error $\sqrt{\pi(1 - \pi)/n}$ for infinite populations, and $\sqrt{\pi(1 - \pi)/n}\sqrt{(N - n)/(N - 1)}$ for finite populations.

---

[3] The fpc is inconsequential in this problem because $[n/N]$ is less than 10 percent, or $\frac{\sigma}{\sqrt{n}} \sqrt{\frac{N - n}{N - 1}} \cong \frac{\sigma}{\sqrt{n}} \sqrt{1} \cong \frac{\sigma}{\sqrt{n}}$.

Exhibit 6–3

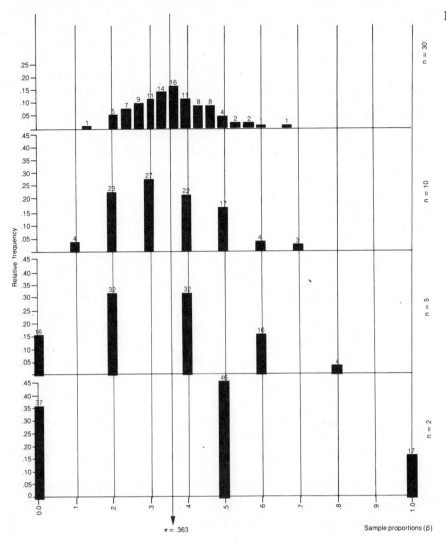

Relative frequency

Sample proportions ($\bar{p}$)

$\pi = .363$

Now that it has been established that the frequency distribution of $\bar{p}$s (sampling distribution of the proportions) is approximately normal (if $n\pi$ and $n[1 - \pi]$ are both greater than 5), you can use what you know about the normal distribution to interpret the ONE SAMPLE PROPORTION you have in practice.

If the sampling distribution of proportions is normal, then approximately 99.73 percent of all sample proportions will fall within three standard errors of the population proportion. Consequently, if the standard error is large, 99.73 percent of all sample proportions will fall within a WIDE range around π (three standard errors either way of π) and you cannot have "confidence" that your sample proportion will provide a good estimate of π. However, if the standard error is small, 99.73 percent of all sample proportions will fall within a SMALL range around π (three standard errors either way of π) and you can have "confidence" that your sample proportion will provide a good estimate of π.

Remember, even if the standard error of the proportion is small, you can never be certain that your $\bar{p}$ is an accurate estimator of π. As the standard error decreases, the chance of your $\bar{p}$ deviating from π decreases, but there is always a finite probability that even with a small standard error, your $\bar{p}$ may be one of the $\bar{p}$s which will not provide a good estimator of π. However, since you never know if your $\bar{p}$ is a good estimator of π or not, you will base your evaluation of $\bar{p}$ on the following criteria: assume that $\bar{p}$ is a good estimator of π if the standard error is small or assume that $\bar{p}$ may be a poor estimator of π if the standard error is large.

Finally, a way of increasing your confidence in $\bar{p}$ as an estimator of π is to increase the sample size upon which $\bar{p}$ is based. As n increases, the standard error becomes smaller and the $\bar{p}$s are more closely grouped around π. To elaborate, the interval in which 99.73 percent of all sample proportions would fall (three standard errors either way of π) is reduced as n is increased. Hence, each $\bar{p}$ has a better chance of being a reasonable estimator of π.

If π is unknown and $n\bar{p}$ and $n(1 - \bar{p})$ are both greater than 5, the distributions of sample proportions is also approximately normal. However, the standard error of $\bar{p}$ is:

$$\sqrt{\frac{\bar{p}(1 - \bar{p})}{n}} \text{ assuming an infinite population, and}$$

$$\sqrt{\frac{\bar{p}(1 - p)}{n}} \sqrt{\frac{N - n}{N - 1}} \text{ assuming a finite population.}$$

Everything else in the prior discussion holds true when π is unknown.

Note: The following three sections are directed toward advanced statistics classes or for the more ambitious introductory statistics student.

Sampling distribution of $\bar{p}$ is binomial (Cases 3, 4, 7, 8, 10, 12).

Given the frequency distribution of ps is binomial, it is possible to compute the percentage of $\bar{p}$s that fall within a specified number of standard errors of the unknown population proportion. Consequently, probability statements concerning the sample proportion's accuracy can be made.

Finally, if π is unknown, $\bar{p}$ can be substituted in its place without any loss of generality.

Sampling distribution of $\bar{p}$ is Poisson (Case 2).

If n is large and π is small (or large), such that $n\pi$ and/or $n(1 - \pi)$ are less than 5, given an infinite population, the distribution of sample proportions is distributed Poisson. Given the frequency distribution of sample proportions is Poisson, it is possible to compute the percentage of $\bar{p}$s that fall within a specified number of standard errors of the unknown population proportion. Hence, a probability statement concerning the sample proportion's accuracy can be made.

Finally, if π is unknown, $\bar{p}$ can be substituted in its place without any loss of generality.

Sampling distribution of $\bar{p}$ is hypergeometric (Cases 9 and 11).

If $n\pi$ and/or $n(1 - \pi)$ is less than 5, and the sampling fraction is greater than .1 $(n/N > .1)$, the sampling distribution of the proportion is hypergeometric. Since the frequency distribution of $\bar{p}$s is hypergeometric, probability statements concerning the sample proportion's accuracy can be made. Unfortunately, the hypergeometric distribution is often computationally burdensome.

Finally, if π is unknown, $\bar{p}$ can be substituted in its place without any loss of generality.

SUMMARY

Chapter 6 is an exercise in abstract thinking. In this chapter you have been asked to conceptualize the shape of the distribution of all possible sample statistics drawn from a population. This is abstract because sampling distributions are never constructed in practice. Understanding the shapes of sampling distributions (a theoretical abstraction) allows you to state the percentage of possible sample statistics that fall within a specified number of standard errors from the true, but unknown, population parameter (another theoretical abstraction). Being able to state the probability that your sample statistic is within a specified distance of the unknown population parameter lends "confidence" to the use of your sample statistic.

CHAPTER 7

ESTIMATION OF μ AND π

Estimating parameters, such as μ and π, from simple random samples of size n was presented in Chapter 3. In that chapter, you noted that different samples of size n yield different estimates, such as $\bar{x}$ and $\bar{p}$, of parameters. This sample-to-sample variability is called sampling error.

Chapter 6 dealt with the sampling distributions of $\bar{x}$ and $\bar{p}$. You learned that if the sampling distribution of $\bar{x}$ (or $\bar{p}$) is known, it is possible to state the percentage of sample means (or proportions) that fall within a specified number of standard errors from the true, but unknown, population mean (or proportion). Consequently, it is possible to state the probability that the ONE SAMPLE MEAN (or proportion) you have computed is within a specified distance of the unknown population mean (or proportion). Being able to make the above probability statement lends "confidence" to the use of your sample mean (or proportion). Remember, you never know for certain if your sample mean (or proportion) is one of the many sample means (or proportions) that is within the specified distance from the unknown population mean (or proportion).

Now you have all the information necessary to estimate μ (or π) while considering sampling error. In this chapter you will take the concepts you learned in Chapters 3 and 6 and combine them so you can make probability statements about μ (or π), called confidence intervals, from your sample mean (or proportion). You will also be instructed how to compute the sample size required to estimate μ or π.

Understanding and interpreting confidence intervals is not a trivial task, especially to a person without formal training in statistics. Consequently, the footnote seen in Exhibit 7–1 (or some version thereof) should be attached to every memorandum in which a confidence interval is reported.

INTRODUCTION

Example 7–1   Continuation of the survey for the Stat City Tax Assessor's Office (see Example 3–2)

Rewrite the report you prepared for the Stat City Tax Assessor's Office concerning the distribution of rooms per dwelling unit in Zone 2 of Stat

Problem

Exhibit 7–1 Standard footnote 3

[3] "The sample estimate and an estimate of its standard error permit us to construct interval estimates with prescribed confidence that the interval includes the average result of all possible samples (for a given sampling rate).

"To illustrate, if all possible samples were selected, if each of these were surveyed under essentially the same conditions, and if an estimate and its estimated standard error were calculated from each sample, then:

"a. Approximately two–thirds of the intervals from one standard error below the estimate to one standard error above the estimate would include the average value of all possible samples. We call an interval from one standard error below the estimate to one standard error above the estimate a two–thirds confidence interval.

"b. Approximately nine–tenths of the intervals from 1.6 standard errors below the estimate to 1.6 standard errors above the estimate would include the average value of all possible samples. We call an interval from 1.6 standard errors below the estimate to 1.6 standard errors above the estimate a 90 percent confidence interval.

"c. Approximately 19/20 of the intervals from two standard errors below the estimate to two standard errors above the estimate would include the average value of all possible samples. We call an interval from two standard errors below the estimate to two standard errors above the estimate a 95 percent confidence interval.

"d. Almost all intervals from three standard errors below the sample estimate to three standard errors above the sample estimate would include the average value of all possible samples.

"The average value of all possible samples may or may not be contained in any particular computed interval. But for a particular sample, one can say with specified confidence that the average of all possible samples is included in the constructed interval." [Journal of the American Statistical Association 70, no. 351, part II (September 1975), p. 6.]

Table 7–1

Random identification number	Number of rooms
198	6
272	9
287	9
131	6
134	7
281	9
280	8
274	7
278	10
230	8
149	8
270	5
255	7
139	9
233	8
168	7
282	9
152	8
204	8
266	7
262	5
132	7
188	8
210	8
171	10
222	8
269	7
283	8
218	7
161	8

City (see Exhibit 7–2). Recall, the tax assessor needed this information to determine new tax assessment rates so that Stat City can obtain a target total tax.

Include information in the new report on the mean, minimum, maximum, range, and standard deviation of the number of rooms per dwelling unit in Zone 2. Be sure to discuss the sampling error of the mean by using a 95 percent confidence interval for the mean in your report.

Use the same sample that you used in writing your original report to the Stat City Tax Assessor's Office. This simple random sample is repeated in Table 7–1 for your convenience.

The sample statistics, calculated in Example 3–2, are shown below: **Solution**

1. Mean $= \bar{x} = \Sigma x/n = 231/30$ $= 7.7$
2. Minimum $=$ 5.0
3. Maximum $=$ 10.0
4. Range $=$ maximum $-$ minimum $= 10 - 5 = 5$
5. Standard deviation $= s$ $= 1.236$
6. Standard error of $\bar{x} = [s/\sqrt{n}] \sqrt{\dfrac{N-n}{N-1}} = \left[\dfrac{1.236}{\sqrt{30}}\right] \sqrt{\dfrac{157-30}{157-1}}$

$$= .2257\,(.9023) = .2036$$
$$\approx \text{approximately } 1/5 \text{ of a room}$$

7. Confidence interval for the mean $= \bar{x} \pm Z \left[\dfrac{s}{\sqrt{n}} \sqrt{\dfrac{N-n}{N-1}}\right]$

$$= 7.7 \pm 1.96\,(.2036)$$
$$= 7.7 \pm .3991$$
$$\approx 7.7 \pm .4$$
$$\approx 7.3 \text{ to } 8.1$$

The memorandum appears in Exhibit 7–2.

Exhibit 7–2

ROSA OPPENHEIM, PH.D.
STATISTICAL CONSULTANT

MEMORANDUM

TO: Stat City Tax Assessor's Office

FROM: Rosa Oppenheim

DATE: July 22, 1985

RE: Revision of the report on the distribution of rooms
        per dwelling unit in Zone 2 of Stat City as of January
        1985

    As per your request and using commonly accepted
statistical techniques, I have computed several summary
measures concerning the number of rooms per dwelling unit in
Zone 2.[1] My research indicates that the typical Zone 2
dwelling unit has approximately 7.7 rooms ($\pm$.2 rooms).[2]
Allowing for the errors in the survey estimates, you can have
"95 percent confidence" that the average number of rooms per
dwelling unit in Zone 2 falls in the range from 7.3 rooms to
8.1 rooms.[3] Further, the number of rooms per dwelling unit in
Zone 2 ranges between 5 and 10,[4] indicating some degree of
variability in dwelling unit size.[5]
    If you have any further questions, please do not hesitate
to call me at 305–999–9999.

Exhibit 7–2 (*concluded*)

<u>Footnotes</u>

[1] Insert standard footnote 1, using 30 Zone 2 dwelling units and a simple random sampling plan.

[2] Insert standard footnote 2.

[3] "The sample estimate and an estimate of its standard error permit us to construct interval estimates with prescribed confidence that the interval includes the average result of all possible samples (for a given sampling rate).

"To illustrate, if all possible samples were selected, if each of these were surveyed under essentially the same conditions, and if an estimate and its estimated standard error were calculated from each sample, then:

"1. Approximately two–thirds of the intervals from one standard error below the estimate to one standard error above the estimate would include the average value of all possible samples. We call an interval from one standard error below the estimate to one standard error above the estimate a two–thirds confidence interval.

"2. Approximately nine–tenths of the intervals from 1.6 standard errors below the estimate to 1.6 standard errors above the estimate would include the average value of all possible samples. We call an interval from 1.6 standard errors below the estimate to 1.6 standard errors above the estimate a 90 percent confidence interval.

"3. Approximately 19/20 of the intervals from two standard errors below the estimate to two standard errors above the estimate would include the average value of all possible samples. We call an interval from two standard errors below the estimate to two standard errors above the estimate a 95 percent confidence interval.

"4. Almost all intervals from three standard errors below the sample estimate to three standard errors above the sample estimate would include the average value of all possible samples.

"The average value of all possible samples may or may not be contained in any particular computed interval. But for a particular sample, one can say with specified confidence that the average of all possible samples is included in the constructed interval." [<u>Journal of the American Standard Association</u> 70, no. 351, part II (September 1975), p. 6.]

[4] Dwelling units range in size from two bedrooms, one bathroom, one kitchen, one living room up to four bedrooms, three bathrooms, one kitchen, one den, one living room.

[5] The degree of variability in the number of rooms per dwelling unit in Zone 2 of Stat City is also evidenced by the fact that the standard deviation of the number of rooms per dwelling unit in Zone 2 is 1.236 rooms. This number indicates that approximately 95 percent of Zone 2 dwelling units have between 5.28 and 10.12 rooms. The above statement is true if the sample mean and standard deviation accurately reflect the true, but unknown, population mean and standard deviation and the distribution of rooms in Zone 2 is bell–shaped. I believe that this assumption is tenable; consequently, the standard deviation can be interpreted as stated above.

Do not be confused by the discrepancy between 95 percent of the dwelling units having between 5.28 rooms and 10.12 rooms and the number of rooms per dwelling unit in Zone 2 ranging between 5 and 10 rooms. The discrepancy can be easily explained by realizing that all the statistics presented are only sample estimates. If the true, but unknown, minimum and maximum number of rooms per dwelling unit in Zone 2 were available, this discrepancy would no longer exist. In other words, the minimum and maximum number of rooms per dwelling unit would be larger than the interval created in which 95 percent of Zone 2 dwelling units would fall.

Point 1 The lengthy footnotes appended to Exhibit 7–2 are necessary if the reader is to be properly informed about the potential errors arising from sample surveys. It is imperative that the footnotes discuss sampling and nonsampling errors, the sampling plan used, the interpretation of standard errors, and the interpretation of confidence intervals.

Point 2 The finite population correction factor (fpc) was used in calculating the standard error because the sampling fraction (n/N) was greater than 0.10 (30/157 = 0.1911). A large sampling fraction indicates that the fpc will be significantly less than 1.0 and must be used to modify the standard error. For Zone 2 of Stat City, the relationship between the sampling fraction and the fpc is shown in Table 7–2.

As you can see, the fpc becomes significant between a sample size of 10 and 30.

Point 3 Reporting only the sample mean of the number of rooms per dwelling unit in Zone 2 of Stat City (7.7 rooms) would be inadequate to properly explain the size of the typical Zone 2 dwelling unit. The above inadequacy stems from the unqualified use of a point estimate ($\bar{x}$) which could impart a false sense of exactness (accuracy) to a reader. Consequently, point estimates (like 7.7 rooms) should always be qualified by their standard errors (like 0.2036 rooms).

Table 7–2

	$N = 157$	
n	n/N	$\sqrt{(N - n)/(N - 1)}$
10	0.0637	0.9707
30	0.1911	0.9023
100	0.6369	0.6045
157	1.000	0.0000

Example 7–2 Survey to estimate the average monthly electric bill in Stat City

Problem

Mr. Saul Reisman, chief financial officer of the Stat City Electric Company, needs to estimate the average monthly electric bill for Stat City families as of January 1985. Mr. Reisman has retained you to conduct a survey of 25 Stat City families. Prepare a typed memorandum for Mr. Reisman reporting your findings (see Exhibit 7–4). Be sure to report all appropriate statistics: mean, minimum, maximum, range, standard deviation, standard error, and 95 percent confidence interval.

Use the simple random sample of 25 Stat City dwelling units listed in Table 7–3.

Solution

The statistics are shown below:

1. Mean = $\bar{x} = \Sigma x/n = 1,476/25 = \$59.04$
2. Minimum = \$21
3. Maximum = \$101
4. Range = \$101 − \$21 = \$80
5. Standard deviation = $s = \$19.58$
6. Standard error = $s/\sqrt{n} = \$3.92$
7. 95 percent confidence interval = $\bar{x} \pm t\,[s/\sqrt{n}]$
$$59.04 \pm 2.064\,(3.92)$$
$$59.04 \pm 8.09$$
$$\$50.95 \text{ to } \$67.13$$

The above confidence interval was constructed using t (not Z) standard errors. Consequently, the footnote which interprets confidence intervals (standard footnote 3) must be altered to account for the use of t (not Z) standard errors. Exhibit 7–3 displays the format of a standard footnote (3A) that explains how to interpret confidence intervals based on t standard errors.

Table 7–3

ID	ELEC
14.	101.
42.	65.
45.	91.
151.	72.
174.	80.
196.	78.
204.	74.
236.	79.
257.	59.
289.	21.
302.	65.
433.	85.
458.	47.
495.	47.
515.	44.
648.	43.
699.	67.
793.	43.
804.	45.
886.	49.
988.	50.
1077.	38.
1079.	39.
1227.	51.
1323.	43.

Exhibit 7–3   Standard footnote 3A

[3A] The sample estimate and an estimate of its standard error permit us to construct interval estimates with prescribed confidence that the interval includes the average result of all possible samples for a sample of size _____.

To illustrate, if all possible samples of size _____ were selected, if each of these were surveyed under essentially the same conditions, and if an estimate and its estimated standard error were calculated from each sample, then:

a. Approximately two–thirds of the intervals from _____ standard errors below the estimate to _____ standard errors above the estimate would include the average value of all possible samples. We call an interval from _____ standard errors below the estimate to _____ standard error(s) above the estimate a two–thirds confidence interval.

b. Approximately 19/20 of the intervals from _____ standard errors below the estimate to _____ standard errors above the estimate would include the average value of all possible samples. We call an interval from _____ standard errors below the estimate to _____ standard errors above the estimate a 95 percent confidence interval.

The average value of all possible samples may or may not be contained in any particular computed interval. But for a particular sample, one can say with specified confidence that the average of all possible samples is included in the constructed interval.

The memorandum appears in Exhibit 7–4.

Exhibit 7–4

HOWARD S. GITLOW, PH.D.
STATISTICAL CONSULTANT
_____

MEMORANDUM

TO: Mr. Saul Reisman, Chief Financial Officer
        Stat City Electric Company

FROM: Howard Gitlow

DATE: July 23, 1985

RE: Survey to estimate the average monthly electric bill
        in Stat City as of January 1985

I have computed several summary measures concerning the monthly electric bills in Stat City using commonly accepted statistical techniques.[1] My research indicates that the average monthly electric bill for a Stat City family, as of January 1985, is $59.04 (±$3.92).[2] Allowing for errors in survey estimates, you can have "95 percent confidence" that the average monthly electric bill, as of January 1985, is contained within the range from $50.95 to $67.13.[3] Further, the monthly electric bills range between $21.00 and $101.00, indicating some degree of variability.[4]

If you have any questions, please call me at 305–999–9999.

< no>
Exhibit 7–4 (*concluded*)

Footnotes

[1] Insert standard footnote 1, using a sample size of 25 drawn via simple random sampling.

[2] Insert standard footnote 2.

[3] The sample estimate and an estimate of its standard error permit us to construct interval estimates with prescribed confidence that the interval includes the average result of all possible samples for a sample of size 25.

To illustrate, if all possible samples of size 25 were selected, if each of these were surveyed under essentially the same conditions, and if an estimate and its estimated standard error were calculated from each sample, then:

a. Approximately two–thirds of the intervals from 1 standard error below the estimate to 1 standard error above the estimate would include the average value of all possible samples. We call an interval from 1 standard error below the estimate to 1 standard error above the estimate a two–thirds confidence interval.

b. Approximately 19/20 of the intervals from 2.064 standard errors below the estimate to 2.064 standard errors above the estimate would include the average value of all possible samples. We call an interval from 2.064 standard errors below the estimate to 2.064 standard errors above the estimate a 95 percent confidence interval.

The average value of all possible samples may or may not be contained in any particular computed interval. But for a particular sample, one can say with specified confidence that the average of all possible samples is included in the constructed interval.

[4] The degree of variability in monthly electric bills in Stat City is also evidenced by the fact that the standard deviation is $19.58. This number indicates that approximately 95 percent of Stat City dwelling units have monthly electric bills between $20.66 and $97.42. The above statement is true if the sample mean and standard deviation accurately reflect the true, but unknown, population mean and standard deviation and the distribution of electric bills in Stat City is bell–shaped. I believe that this assumption is tenable; consequently, the standard deviation can be interpreted as stated above.

Example 7–3 Survey to estimate the average monthly telephone bill in Stat City

Mr. Jack Davis, chairman of the Stat City Telephone Company, needs to estimate the average monthly Stat City telephone bill as of January 1985. Mr. Davis wants to have 95 percent confidence in the estimate. Further, he wants the estimate to be accurate to within $5 of the true average telephone bill.

You have been retained by Mr. Davis to perform the desired survey. The pilot study shown in Table 7–4 should be helpful in your endeavor. Type a memorandum to Mr. Davis reporting your findings (see Exhibit 7–5). Note: This problem requires you to compute the proper sample size!

C.I. μ

Problem

Table 7–4

ID	PHONE
258.	24.
295.	65.
386.	23.
415.	22.
416.	17.
736.	14.
858.	21.
1016.	19.
1149.	27.
1242.	22.

Solution   The first step is to compute the proper sample size (n).[1]

$$e = \text{the tolerable error} = (\bar{x} - \mu) = \$5$$

Consequently,

$$n_0 = \frac{(1.96)^2(\$206.93)}{(\$5)^2} = 31.8$$

$$n = \frac{31.8}{\left[\frac{(31.8 + 1{,}372)}{1{,}373}\right]} = 31.1 = 32$$

The sample size of 32 implies that 32 Stat City families must be randomly sampled to estimate the population mean telephone bill with 95 percent confidence, allowing for a tolerable error of $5.00.

Once the sample size has been determined, the random sample can be drawn. Table 7–5 lists the random sample selected in this study.

Finally, the data can be analyzed and conclusions can be drawn. The chart below shows the results of the survey.

$$n = 32$$
$$\bar{x} = \$45.13$$
$$s = \$37.76$$
$$s_{\bar{x}} = \$6.68$$
$$\text{95 percent confidence interval} = \$45.13 \pm 1.96(6.68)$$
$$= \$45.13 \pm \$13.09$$
$$= \$32.04 \text{ to } \$58.22$$

The sample indicates that the average monthly telephone bill in Stat City, as of January 1985, is $45.13. The 95 percent confidence interval ranges from $32.04 to $58.22.

Table 7–5

CASE-N	ID	PHONE
1	13.	14.
2	44.	16.
3	60.	15.
4	121.	16.
5	266.	20.
6	285.	21.
7	337.	118.
8	341.	68.
9	358.	98.
10	505.	21.
11	533.	14.
12	536.	55.
13	687.	97.
14	695.	17.
15	698.	19.
16	782.	10.
17	809.	35.
18	825.	14.
19	846.	15.
20	884.	139.
21	887.	80.
22	943.	65.
23	1000.	107.
24	1035.	92.
25	1057.	10.
26	1118.	72.
27	1152.	40.
28	1211.	15.
29	1216.	74.
30	1257.	21.
31	1274.	20.
32	1346.	26.

---

[1] Since the population is finite, the following formulas are used:

$$n_0 = \frac{Z^2 \hat{s}^2}{e^2}$$

$$n = \frac{n_0}{\left(\frac{n_0 + (N - 1)}{N}\right)}$$

where

n = the sample size modified by the finite population correction factor
n_0 = the sample size unmodified by the finite population correction factor
N = the population size (1,373 dwelling units)
z = the number of standard errors required to set the desired level of confidence (z = 1.96 for 95 percent confidence)
$\hat{s}^2$ = the pilot study variance. $\hat{s}^2$ is computed by drawing a random sample of, say, 10 Stat City dwelling units and computing the variance from the pilot study:

ID	PHONE
258.	24.
295.	65.
386.	23.
415.	22.
416.	17.
736.	14.
858.	21.
1016.	19.
1149.	27.
1242.	22.

Pilot study $\bar{x} = \$\ 25.40$
Pilot study $\hat{s}^2 = \$206.93$

The memorandum is shown in Exhibit 7–5.

Exhibit 7–5

ROSA OPPENHEIM, PH.D.
STATISTICAL CONSULTANT
—————

MEMORANDUM

TO: Mr. Jack Davis, Chairman
        Stat City Telephone Company

FROM: Rosa Oppenheim

DATE: September 3, 1985

RE: Estimating the average monthly telephone bill in
        Stat City as of January 1985

Per your request, I have conducted a survey of Stat City
families to estimate the average monthly telephone bills as
of January 1985.[1] My research indicates that the typical
household's average monthly telephone bill is $45.13
(±$6.68).[2] Allowing for the errors in survey estimates, you
can have "95 percent confidence" that the average monthly
telephone bill per household in Stat City falls in the range
from $32.04 to $58.22.[3] Furthermore, average monthly
telephone bills range between $10 and $139, indicating some
degree of variability in telephone bill size.
    If you have any further questions, please call me at
305–999–9999.

—————————

Footnotes

[1] Insert standard footnote 1 (insert a sample size of 32 drawn via
simple random sampling).
[2] Insert standard footnote 2.
[3] Insert standard footnote 3.

Example 7–4   Estimating the percentage of childless Zone 4 families

Problem

Mr. Hector Chavez, manager of the Grand Union supermarket, believes that the presence of children in a household strongly affects supermarket shopping behavior. Consequently, Mr. Chavez would like to determine the percentage of Zone 4 families that have no children.

Mr. Chavez has agreed that the instance of a single-parent family is rare in Stat City; hence, if a household has two or more people, it is safe to assume that two are adults and the remaining people are children.

You have been retained by Mr. Chavez to determine the percent of childless Zone 4 families. Type a memorandum reporting your findings to Mr. Chavez (see Exhibit 7–6).

Mr. Chavez has provided you with funds to draw a sample of 131 Zone 4 families. Use the simple random sample shown in Table 7–6 in conducting your analysis.

Table 7–6

CASE-N	ID	PEPLE	CASE-N	ID	PEPLE
1	627.	6.	66	952.	4.
2	632.	2.	67	956.	6.
3	643.	6.	68	966.	3.
4	645.	4.	69	972.	1.
5	649.	5.	70	973.	4.
6	653.	6.	71	974.	2.
7	656.	7.	72	980.	4.
8	661.	8.	73	991.	4.
9	662.	1.	74	1008.	4.
10	665.	3.	75	1010.	7.
11	670.	7.	76	1016.	5.
12	672.	5.	77	1018.	4.
13	673.	6.	78	1028.	1.
14	676.	5.	79	1040.	7.
15	682.	4.	80	1044.	5.
16	684.	5.	81	1046.	3.
17	686.	7.	82	1078.	6.
18	691.	1.	83	1082.	6.
19	695.	5.	84	1090.	6.
20	698.	5.	85	1098.	7.
21	700.	5.	86	1102.	4.
22	702.	4.	87	1104.	4.
23	703.	3.	88	1124.	2.
24	705.	4.	89	1127.	4.
25	715.	2.	90	1130.	6.
26	726.	4.	91	1131.	2.
27	737.	6.	92	1132.	5.
28	739.	2.	93	1138.	5.
29	743.	4.	94	1140.	5.
30	758.	8.	95	1143.	4.
31	759.	4.	96	1144.	5.
32	769.	6.	97	1148.	4.
33	770.	4.	98	1149.	4.
34	774.	3.	99	1152.	7.
35	778.	6.	100	1156.	6.
36	780.	4.	101	1161.	6.
37	799.	2.	102	1165.	2.
38	806.	4.	103	1167.	2.
39	819.	3.	104	1171.	5.
40	820.	5.	105	1172.	7.
41	822.	5.	106	1177.	3.
42	825.	4.	107	1188.	7.
43	830.	7.	108	1190.	3.
44	836.	4.	109	1213.	6.
45	847.	7.	110	1218.	1.
46	864.	4.	111	1261.	5.
47	870.	7.	112	1266.	3.
48	877.	5.	113	1272.	5.
49	878.	3.	114	1277.	6.
50	884.	6.	115	1286.	6.
51	891.	3.	116	1289.	5.
52	897.	7.	117	1290.	6.
53	900.	6.	118	1295.	4.
54	903.	7.	119	1302.	2.
55	905.	7.	120	1303.	6.
56	911.	6.	121	1316.	1.
57	917.	4.	122	1323.	6.
58	927.	4.	123	1336.	4.
59	936.	2.	124	1340.	4.
60	937.	8.	125	1348.	4.
61	938.	5.	126	1350.	4.
62	940.	4.	127	1357.	6.
63	943.	7.	128	1362.	8.
64	945.	2.	129	1370.	1.
65	950.	4.	130	1371.	5.
			131	1373.	6.

Solution The statistics are shown below and in Table 7–7.

Table 7–7

Number of people in household as of January 1985	Number	Percentage
1 or 2 (single or married—no children).................	19	14.5
3 or more (married with children)	112	85.5
Total	131	100.0

1. Sample proportion $= \bar{p} = .145$
2. Estimated standard error of the proportion

$$= \sqrt{\bar{p}(1 - \bar{p})/n}\sqrt{(N - n)/(N - 1)}$$
$$= .031(.909)$$
$$= .028$$

3. 95 percent confidence interval for the population proportion

$$= \bar{p} \pm Z[\sqrt{\bar{p}(1 - \bar{p})/n}\sqrt{(N - n)/(N - 1)}]$$
$$= .145 \pm 1.96[.028]$$
$$= .145 \pm .055$$
$$= .090 \text{ to } .200$$

The memorandum appears in Exhibit 7–6.

Exhibit 7–6

HOWARD S. GITLOW, PH.D.
STATISTICAL CONSULTANT

MEMORANDUM

TO: Mr. Hector Chavez, Manager
        Grand Union

FROM: Howard Gitlow

DATE: July 24, 1985

RE: Estimating the percentage of childless Zone 4
        families as of January 1985

Using commonly accepted statistical techniques, I have
computed several summary measures concerning the
percentage of childless[1] Zone 4 families.[2] My research
indicates that approximately 14.5 percent (±2.8 percent)[3]
of Zone 4 families are childless. Allowing for the
errors in survey estimates, you can have "95 percent
confidence" that the percentage of Zone 4 families
that are childless is contained within the range from
9.0 percent to 20.0 percent.[4]

If you have any further questions, please do not hesitate
to call me at 305-999-9999.

Footnotes

[1] A dwelling unit containing one or two persons is considered
childless. This determination is reasonable because the instance
of single-parent families is rare in Stat City. In other words, it
is safe to assume that if a household has two or more people, two
are adults and the remainder are children.
[2] Insert standard footnote 1 (insert a sample size of 131 drawn via
simple random sampling).
[3] Insert standard footnote 2.
[4] Insert standard footnote 3.

Example 7–5 Estimating the percentage of Stat City families that have their automotive repairs performed by a gas station

Problem

Mr. Paul Sugrue, the owner of Paul's Texaco Station, wants to determine the percentage of Stat City families that have their automotive repairs performed in a gas station, as opposed to "dealer" or "do-it-yourself" repairs. Mr. Sugrue wants to have 95 percent confidence in his estimate, with a tolerable error in estimation of 10 percentage points.

You have been retained by Mr. Sugrue to ascertain the desired information. Type a memorandum reporting your results to Mr. Sugrue (see Exhibit 7–7).

Note: This problem requires that you compute the appropriate sample size to meet Mr. Sugrue's specifications.

Solution   The first step is to compute the proper sample size (n).[2]

$$e = \text{the tolerable error (.10)}$$

Hence,

$$n_0 = \left[\frac{z^2(\hat{\pi})(1 - \hat{\pi})}{e^2} \right] = \frac{(1.96)^2(.5)(.5)}{(.1)^2} = 96.04$$

$$n = \frac{n_0}{\left[\dfrac{n_0 + (N - 1)}{N} \right]} = 89.8 = 90$$

The sample size of 90 implies that 90 Stat City families must be randomly sampled to estimate the population percentage of families that have their automotive repairs performed by a gas station with 95 percent confidence, allowing for a tolerable error of 10 percentage points.

Once the sample size has been computed, the random sample can be drawn. Table 7–8 lists the random sample selected in this study.

Finally, the data can be analyzed and conclusions can be drawn. The chart below shows the results of the survey.

$$n = 90$$
$$\bar{p} = .422$$
$$s_{\bar{p}} = .052 = \sqrt{\frac{.422 \times .578}{90}}$$

95 percent confidence interval =
.422 ± 1.96(.052) = .422 ± .102
= .320 to .524

The sample indicates that the proportion of Stat City families that have their automotive repairs performed by a gas station, as of January 1985, is .422. The 95 percent confidence interval ranges from .320 to .524.

The memorandum appears in Exhibit 7–7.

---

[2] Since the population is finite, the following formulas are used:

$$n_0 = \frac{z^2(\hat{\pi})(1 - \hat{\pi})}{e^2}$$

$$n = \frac{n_0}{\left[\dfrac{n_0 + (N - 1)}{N} \right]}$$

where

n = the sample size modified by the finite population correction factor
n_0 = the sample size unmodified by the finite population correction factor
N = the population size (1,373)
$\hat{\pi}$ = the "guesstimated" value of the population parameter ($\hat{\pi} = .5$ because no better information is available; hence, maximum variation is assumed). $\hat{\pi}$ could also be estimated from a prior or pilot study.
Z = the number of standard errors required to set the desired level of confidence ($Z = 1.96$ for 95 percent confidence)

CASE-N	ID	REPAR	CASE-N	ID	REPAR
1	4.	0.	46	568.	0.
2	9.	1.	47	570.	0.
3	12.	1.	48	572.	0.
4	27.	0.	49	599.	0.
5	40.	0.	50	602.	1.
6	41.	1.	51	615.	0.
7	49.	0.	52	622.	0.
8	58.	1.	53	641.	1.
9	81.	1.	54	664.	1.
10	90.	0.	55	673.	0.
11	126.	0.	56	688.	0.
12	133.	0.	57	771.	1.
13	180.	0.	58	776.	1.
14	184.	0.	59	785.	1.
15	211.	0.	60	817.	0.
16	215.	0.	61	819.	1.
17	230.	1.	62	836.	0.
18	235.	1.	63	897.	1.
19	263.	0.	64	910.	1.
20	268.	1.	65	955.	1.
21	274.	0.	66	990.	1.
22	278.	0.	67	995.	0.
23	282.	0.	68	1001.	0.
24	284.	0.	69	1036.	0.
25	300.	0.	70	1045.	0.
26	302.	0.	71	1058.	1.
27	354.	1.	72	1073.	0.
28	386.	0.	73	1087.	0.
29	388.	0.	74	1108.	1.
30	414.	0.	75	1114.	1.
31	424.	0.	76	1130.	0.
32	434.	1.	77	1159.	1.
33	460.	1.	78	1165.	1.
34	465.	0.	79	1173.	1.
35	472.	1.	80	1175.	1.
36	489.	1.	81	1247.	0.
37	490.	1.	82	1248.	1.
38	498.	0.	83	1253.	0.
39	509.	0.	84	1278.	0.
40	514.	0.	85	1279.	1.
41	521.	0.	86	1327.	1.
42	530.	0.	87	1337.	1.
43	531.	1.	88	1339.	0.
44	545.	0.	89	1358.	1.
45	566.	0.	90	1371.	0.

Table 7–8

Note: Code for repairs: 1 = Done by service station (originally 1).
2 = Done somewhere else (originally 0 or 2).

Exhibit 7–7

```
┌─────────────────────────────────────────────────────────┐
│                                                         │
│   ROSA OPPENHEIM, PH.D.                                 │
│   STATISTICAL CONSULTANT                                │
│   ─────────                                             │
│                                                         │
│                      MEMORANDUM                         │
│                                                         │
│   TO:    Mr. Paul Sugrue, Owner                        │
│          Paul's Texaco Station                         │
│                                                         │
│   FROM:  Rosa Oppenheim                                │
│                                                         │
│   DATE:  March 8, 1985                                 │
│                                                         │
│   RE:    Estimating the percentage of Stat City        │
│          families that have their automotive repairs    │
│          performed by a gas station as of January 1985  │
│                                                         │
│     In accordance with our consulting contract, I have  │
│   conducted a survey to estimate the percentage of Stat │
│   City families that have their automotive repairs      │
│   performed by a gas station as of January 1985.[1] My  │
│   research, which was performed using commonly accepted  │
│   statistical techniques, indicates that the percentage │
│   of Stat City families that have their automotive      │
│   repairs performed by a gas station is 42.2 percent    │
│   (±5.2 percent).[2] Allowing for the errors in survey  │
│   estimates, you can have "95 percent confidence" that  │
│   the percentage of Stat City families that have their  │
│   automotive repairs performed by a gas station is      │
│   contained within the range from 32.0 percent to 52.4  │
│   percent.[3]                                           │
│     If you have any further questions, call me at       │
│   305-999-9999.                                         │
│                                                         │
│   ─────────────                                         │
│                                                         │
│                      Footnotes                          │
│                                                         │
│    [1] Insert standard footnote 1, using a sample size  │
│   of 90 drawn via simple random sampling.               │
│    [2] Insert standard footnote 2.                      │
│    [3] Insert standard footnote 3.                      │
│                                                         │
└─────────────────────────────────────────────────────────┘
```

ADDITIONAL PROBLEMS

7–6 Need for a promotional campaign to increase supermarket patronage in Stat City The Stat City Supermarket Association would like to estimate the average weekly supermarket bill per family in Stat City as of January 1985. The association requires 95 percent confidence in the estimate. They have provided you with sufficient funds to draw a random sample of 50 Stat City families. Type a memorandum to the association reporting your findings.

Use the following simple random sample for computing the required statistics.

CI

ID	EAT	ID	EAT	ID	EAT
2.	81.	585.	76.	1043.	69.
14.	155.	604.	85.	1045.	101.
120.	45.	624.	87.	1089.	127.
158.	182.	640.	67.	1104.	80.
159.	99.	670.	161.	1113.	112.
164.	71.	674.	70.	1123.	29.
177.	93.	682.	76.	1146.	70.
223.	60.	688.	86.	1150.	61.
232.	97.	752.	98.	1160.	81.
271.	72.	757.	142.	1220.	53.
314.	153.	813.	101.	1292.	130.
316.	122.	816.	97.	1305.	107.
399.	86.	841.	26.	1313.	90.
474.	144.	875.	87.	1320.	142.
533.	109.	915.	87.	1332.	42.
541.	93.	959.	70.	1361.	72.
579.	112.	1025.	115.		

7–7 Another study to estimate the average monthly electric bill in Stat City Mrs. Jessica Mara, assistant financial officer of the Stat City Electric Company, must estimate the average monthly electric bill in Stat City as of January 1985. She wants to have 95 percent confidence in the estimate and be accurate to within $10 of the true average electric bill. Unfortunately, she is not aware of the study performed for Mr. Reisman (Example 7–2).

She has retained you to perform the required study. She has provided you with the following data from a pilot survey of eight Stat City families.

ID	ELEC
22.	59.
83.	86.
147.	81.
298.	127.
393.	32.
448.	47.
531.	61.
776.	35.

Type a memorandum to her reporting your findings.

Note: After you compute the proper sample size, your professor will hand out the random sample you should use to compute the required statistics. This is in order that everyone in the class will have the same sample.

7–8 Estimating the percentage of Stat City families who do not use the hospital Ms. Arlene Davis, director of the Stat City Hospital, wants to ascertain the percentage of Stat City families that do not use the Stat City hospital. Ms. Davis wants 90 percent confidence in the estimate. Further, she wants the estimate to be accurate to within 4 percentage points.

You have been retained by Ms. Davis to conduct a survey to compute the desired percentage. Ms. Davis told you that a similar study performed several years ago indicated that 6.7 percent of Stat City families do not use the hospital. Type a memorandum to Ms. Davis reporting your findings.

COMPUTATIONAL EXERCISES

7–9 A&P study of buying habits Mr. Marc Cooper, manager of the A&P supermarket, is interested in studying several characteristics of Stat City dwelling units. In particular, Mr. Cooper is interested in: the average weekly supermarket bill as of January 1985 (EAT) and the average weekly purchase of six-packs of diet soda (LSODA), regular soda (HSODA), and beer (BEER) as of January 1985. Mr. Cooper has supplied you with enough funds to survey a random sample of 49 Stat City families. Compute the 99 percent confidence interval for each variable.

Use the random sample and summary statistics which accompany this problem for your computations.

Random sample for Exercise 7–9

CASE-N	ID	EAT	LSODA	HSODA	BEER
1	1.	128.	2.	0.	5.
2	30.	79.	1.	0.	2.
3	131.	53.	1.	0.	0.
4	136.	89.	2.	0.	9.
5	202.	156.	1.	0.	0.
6	216.	94.	0.	6.	0.
7	226.	68.	2.	0.	6.
8	261.	68.	0.	5.	7.
9	267.	34.	0.	0.	0.
10	354.	120.	0.	0.	7.
11	373.	66.	0.	4.	0.
12	420.	108.	0.	0.	0.
13	480.	101.	2.	0.	5.
14	498.	117.	0.	0.	2.
15	534.	63.	0.	0.	0.
16	563.	144.	0.	5.	0.
17	591.	72.	1.	0.	3.
18	666.	51.	2.	0.	0.
19	674.	70.	0.	0.	0.
20	703.	65.	0.	0.	0.
21	737.	135.	2.	0.	0.
22	743.	85.	2.	0.	0.
23	786.	134.	0.	4.	0.
24	807.	71.	2.	0.	10.
25	820.	89.	2.	0.	0.
26	847.	138.	0.	5.	10.
27	895.	161.	0.	0.	0.
28	901.	75.	2.	0.	0.
29	909.	59.	0.	0.	6.
30	935.	136.	0.	1.	11.
31	936.	43.	1.	0.	0.
32	937.	126.	2.	0.	0.
33	953.	121.	0.	0.	0.
34	984.	110.	0.	0.	0.
35	987.	89.	0.	0.	0.
36	988.	129.	0.	4.	3.
37	1035.	80.	0.	3.	3.
38	1098.	132.	2.	0.	5.
39	1126.	87.	0.	1.	0.
40	1170.	84.	0.	5.	0.
41	1174.	132.	0.	0.	4.
42	1193.	114.	0.	0.	0.
43	1218.	26.	0.	5.	6.
44	1226.	91.	0.	4.	0.
45	1232.	79.	0.	0.	0.
46	1272.	104.	0.	0.	0.
47	1280.	110.	0.	1.	0.
48	1281.	93.	2.	0.	2.
49	1367.	110.	0.	1.	3.

```
VARIABLE  EAT        AVERAGE WEEKLY FOOD BILL AS OF 1-80

MEAN          95.694              STD ERROR      4.603         STD DEV       32.224
VARIANCE    1038.384              KURTOSIS       -.666         SKEWNESS       -.013
RANGE        135.000              MINIMUM       26.000         MAXIMUM      161.000
SUM         4689.000

VALID OBSERVATIONS -     49                    MISSING OBSERVATIONS -       0
```

- -

```
VARIABLE  LSODA      AVE WKLY DIET SODA PURCHASE AS OF 1-80

MEAN            .633              STD ERROR       .126         STD DEV         .883
VARIANCE        .779              KURTOSIS      -1.232         SKEWNESS        .810
RANGE          2.000              MINIMUM        .000          MAXIMUM        2.000
SUM           31.000

VALID OBSERVATIONS -     49                    MISSING OBSERVATIONS -       0
```

- -

```
VARIABLE  HSODA      AVE WKLY REG. SODA PURCHASE AS OF 1-80

MEAN           1.102              STD ERROR       .275         STD DEV        1.928
VARIANCE       3.719              KURTOSIS        .346         SKEWNESS       1.431
RANGE          6.000              MINIMUM        .000          MAXIMUM        6.000
SUM           54.000

VALID OBSERVATIONS -     49                    MISSING OBSERVATIONS -       0
```

- -

```
VARIABLE  BEER       AVE WKLY BEER PURCHASE AS OF 1-80

MEAN           2.163              STD ERROR       .465         STD DEV        3.255
VARIANCE      10.598              KURTOSIS        .678         SKEWNESS       1.348
RANGE         11.000              MINIMUM        .000          MAXIMUM       11.000
SUM          106.000

VALID OBSERVATIONS -     49                    MISSING OBSERVATIONS -       0
```

7-10 Survey of dwelling unit characteristics for the *Stat City Beacon* Donna Nelson, a reporter for the *Stat City Beacon,* is writing an article about certain characteristics of Stat City dwelling units. In particular, Ms. Nelson is reporting on: the average yearly heating bill as of January 1985 (HEAT), the average monthly telephone bill as of January 1985 (PHONE), the 1984 total family income (INCOM), the number of automobiles in a household as of January 1985 (CARS), the average bimonthly automobile gasoline bill as of January 1985 (GAS), and the average number of trips to a gas station per month as of January 1985 (GASTR). Ms. Nelson has supplied you with funds to survey a random sample of 10 Stat City dwelling units. Compute the mean, standard deviation, standard error, and 90 percent confidence interval for each of the above variables. Assume that all of the above variables are normally distributed.

Use the accompanying simple random sample.

Random sample for Exercise 7-10

ID	HEAT	ELEC	PHONE	INCOM	CARS	GAS	GASTR
258.	1136.	73.	24.	116203.	2.	127.	3.
295.	538.	42.	65.	25328.	1.	99.	6.
386.	792.	72.	23.	32667.	2.	160.	6.
415.	413.	37.	22.	26972.	1.	65.	6.
416.	457.	37.	17.	41549.	3.	174.	7.
736.	392.	38.	14.	18404.	2.	120.	12.
858.	256.	21.	21.	17052.	2.	120.	5.
1016.	512.	46.	19.	24800.	3.	185.	5.
1149.	651.	69.	27.	19670.	2.	120.	5.
1242.	471.	47.	22.	24743.	3.	168.	8.

7-11 Survey of attributes for Stat City Chamber of Commerce Paul Lund, chairman of the Stat City Chamber of Commerce, is interested in investigating the following attributes of Stat City families as of January 1985, the percentage of Stat City families that did not purchase any six-packs of diet soda (LSODA = 0), the percentage of Stat City families that

Random sample for Exercise 7–11	ID	LSODA	HSODA	PEPLE	GASTR	REPAR
	6.	0.	0.	2.	3.	1.
	47.	0.	0.	4.	5.	2.
	53.	2.	0.	1.	6.	1.
	68.	1.	0.	1.	4.	1.
	88.	1.	0.	2.	2.	2.
	92.	0.	1.	5.	8.	1.
	164.	0.	1.	3.	3.	1.
	167.	0.	5.	5.	5.	2.
	291.	2.	0.	4.	8.	2.
	459.	0.	0.	6.	6.	0.
	469.	0.	0.	5.	6.	1.
	477.	0.	0.	6.	11.	1.
	490.	0.	0.	4.	6.	1.
	498.	0.	0.	5.	4.	0.
	528.	2.	0.	5.	9.	1.
	538.	2.	0.	7.	5.	2.
	539.	2.	0.	3.	6.	0.
	588.	0.	1.	4.	7.	1.
	597.	0.	4.	4.	2.	1.
	599.	2.	0.	4.	6.	2.
	616.	0.	2.	6.	5.	2.
	657.	0.	0.	4.	2.	0.
	684.	0.	4.	5.	4.	1.
	705.	0.	0.	4.	6.	1.
	724.	0.	0.	2.	6.	1.
	764.	0.	0.	2.	3.	0.
	788.	2.	0.	7.	3.	1.
	795.	2.	0.	5.	4.	1.
	797.	2.	0.	4.	3.	1.
	816.	0.	5.	4.	8.	1.
	835.	0.	1.	3.	7.	0.
	854.	0.	0.	7.	5.	0.
	873.	2.	0.	6.	7.	1.
	897.	0.	0.	7.	4.	1.
	962.	2.	0.	6.	3.	0.
	970.	0.	0.	6.	12.	1.
	996.	0.	0.	4.	4.	1.
	1073.	2.	0.	6.	7.	0.
	1086.	0.	0.	5.	5.	1.
	1092.	0.	0.	4.	6.	2.
	1134.	2.	0.	8.	4.	1.
	1136.	0.	5.	8.	5.	2.
	1192.	0.	0.	7.	2.	1.
	1215.	2.	0.	5.	7.	0.
	1244.	2.	0.	3.	2.	1.
	1279.	0.	1.	2.	6.	1.
	1297.	1.	0.	5.	6.	0.
	1299.	0.	6.	4.	7.	2.
	1320.	0.	0.	7.	4.	1.
	1348.	2.	0.	4.	3.	0.

did not purchase any six-packs of regular soda (HSODA = 0), the percentage of childless Stat City families (PEPLE ≤ 2), the percentage of Stat City families making an average of nine or more trips to the gas station per month (GASTR ≥ 9), and the percentage of Stat City families that have their automotive repairs performed at a dealer (REPAR = 2). Mr. Lund has supplied you with funds to survey 50 randomly selected Stat City dwelling units and to compute percentages, standard errors, and 95 percent confidence intervals for each variable under study.

Use the accompanying random sample for your computations.

HYPOTHESIS TESTING

CHAPTER 8

INTRODUCTION

In Chapter 7, you learned about the branch of statistical inference concerned with making point and interval estimates of parameters. In this chapter, you will study another branch of statistical inference: the branch concerned with testing preconceived ideas (hypotheses) about parameters. Generally, if a preconceived idea about a parameter remains tenable after it has been tested, the idea gains credibility; if not, it loses credibility. For example, the Stat City Supermarket Association believes that the typical Stat City family purchases at least one six-pack of diet soda per week, on average. If the idea does not remain tenable after a hypothesis test, it loses credibility; this may cause the association to mount a promotional campaign to increase diet soda sales.

The steps involved in testing a hypothesis are as follows:

1. List the states of nature.
2. List the alternative courses of action.
3. Construct a decision matrix (see Table 8–1).

Table 8–1

States of nature

	State 1	State 2
Act 1		
Act 2		

Alternative courses of action

4. State the null hypothesis (H_0).
5. State the alternative hypothesis (H_A).
6. State the level of significance (α).
7. State the proper test statistic.
8. Set up the rejection region(s) for the null hypothesis.
9. Compute the sample size (n).
10. Draw the sample and compute the test statistic.
11. Ascertain if the test statistic falls within the rejection region(s).

12. State the statistical decision (reject the H_0 or not).
13. Relate the statistical decision to the appropriate person(s).

Understanding hypothesis testing is more difficult than a person without formal training in statistics would suspect. Hence, it is important that the footnote in Exhibit 8–1 (or some version thereof) appears whenever a hypothesis test is reported in a memorandum. The footnote in Exhibit 8–1 serves as a partial disclaimer for the errors which can occur when performing a hypothesis test from a sample survey; remember, a sample statistic is subject to sampling error.

Exhibit 8–1   Standard footnote 4

> [4] The hypothesized population parameter, the sample estimate, and an estimate of the standard error permit a hypothesis test to be performed for a given sample size.
>
> To illustrate, if all possible samples were selected, if each of these were surveyed under essentially the same conditions, and if a sample estimate and its standard error were calculated, then:
>
> a. Approximately nine-tenths of the hypothesis tests conducted at the 90 percent level of confidence (or 10 percent level of significance) would fail to reject the null hypothesis when it was tenable (or 10 percent would reject the null hypothesis when it was tenable).
>
> b. Approximately 19/20 of the hypothesis tests conducted at the 95 percent level of confidence (or 5 percent level of significance) would fail to reject the null hypothesis when it was tenable (or 5 percent would reject the null hypothesis when it was tenable).
>
> Once the level of confidence (or significance) has been set, the chances of failing to reject the null hypothesis when it is false (type II error) must be computed. If these probabilities are unacceptably high, then corrective action must be taken.

The formulas presented in this chapter for computing sample sizes for hypothesis testing problems are limited to one sample and one-tailed tests for means and proportions.[1]

The purpose of this chapter is to develop your hypothesis testing skills. The following examples should be helpful in your gaining a firm grasp of hypothesis testing.

Example 8–1   Applying for HHS funds for Zone 4 families

The U.S. Department of Health and Human Services (HHS) has issued **Problem** a statement that any urban area with an average 1984 family income below $20,000 is entitled to federal subsidies. Mr. Kaplowitz, the mayor of Stat City, would like to receive federal aid for families living in Zone 4 of Stat City.

The courses of action open to the mayor are twofold: to apply or not to apply for the federal aid. If the mayor applies for the federal aid, when in fact Stat City is not eligible, he runs the risk of incurring HHS's disfavor. Consequently, Stat City may have an extremely difficult time obtaining other HHS funds in the future. The mayor is far more concerned about applying for the funds when he should not than he is with refusing to apply for the funds when Stat City may be eligible.

Mr. Kaplowitz has hired you to determine if the average family income in 1984 in Zone 4 was less than $20,000; that is, if he should apply for the funds. Conduct the necessary survey and type of memorandum to the mayor reporting your findings (see Exhibit 8–3).

---

[1] The formula for computing the sample size for a hypothesis test concerning a mean is:

$$n = \frac{\hat{s}^2(Z_\alpha + Z_\beta)^2}{(\mu_0 - \mu_1)^2}$$

where

n = the sample size
$\hat{s}^2$ = an estimate of the variance of the variable under study (obtained from a pilot or prior study)
μ_0 = value of the population mean under the null hypothesis
μ_1 = value of the population mean under the alternative hypothesis
Z_α = Z value for a given α level of significance (omit the sign of Z_α)
Z_β = Z value for a given β risk of a type II error (omit the sign of Z_β)

The formula for computing the sample size for a hypothesis test concerning a proportion is:

$$n = \frac{\{Z_\alpha\sqrt{\pi_0(1 - \pi_0)} + Z_\beta\sqrt{\pi_1(1 - \pi_1)}\}^2}{(\pi_1 - \pi_0)^2}$$

where

n = the sample size
π_0 = value of the population proportion under the null hypothesis
π_1 = value of the population proportion under the alternative hypothesis
Z_α = Z value for a given α level of significance (omit sign of Z_α)
Z_β = Z value for a given β risk of a type II error (omit sign of Z_β)

Use of the finite population correction factor in conjunction with the above formulas can cause logical inconsistencies.

Solution

BEFORE the survey is performed, it is important to specify the hypothesis being tested. The alternative courses of action open to the mayor are: applying for the HHS funds or not applying for the HHS funds. The relevant states of nature are whether or not the average 1984 family income in Zone 4 of Stat City is less than $20,000. The mayor should apply for the funds if the average 1984 family income in Stat City is less than $20,000.

Table 8–2 illustrates the structure of the decision to be made. The box marked A indicates a correct decision—applying for the HHS funds if the average 1984 family income in Zone 4 is less than $20,000 per year. The box marked B also indicates a correct decision—not applying for HHS funds if the average family income in Zone 4 is greater than or equal to $20,000 per year. The box marked C indicates an incorrect decision—not applying for HHS funds if the average 1984 family income in Zone 4 is less than $20,000 per year. Finally, the box marked D also indicates an error—applying for HHS funds if the average 1984 family income in Zone 4 is greater than or equal to $20,000 per year. Recall that this is the error of utmost concern to the mayor.

Table 8–2

	States of nature	
Alternative courses of action	Average 1984 family income in Zone 4 is less than $20,000 per year ($\mu < 20,000$)	Average 1984 family income in Zone 4 is greater than or equal to $20,000 per year ($\mu \geq 20,000$)
Apply for HHS funds	A	D
Do not apply for HHS funds.....................	C	B

It is customary when working with one-tail tests to designate as the null hypothesis that hypothesis which, if rejected when tenable, leads to the type I error. In this case the null hypothesis is:

$$H_0: \mu \geq \$20,000$$

Note, $\mu \geq \$20,000$ is designated as the null hypothesis because if it is rejected when it is tenable, the type I error is committed; the mayor applies for HHS funds which Stat City does not deserve and consequently incurs HHS's displeasure. The alternative hypothesis is:

$$H_A: \mu < \$20,000$$

If the null hypothesis is rejected and the alternative hypothesis is deemed tenable, then action is taken; application is made for the HHS monies.

The problem would be set up as follows:

State of nature	Alternative courses of action
$H_0: \mu \geq \$20,000$	Don't apply for HHS funds
$H_A: \mu < \$20,000$	Apply for HHS funds

The level of significance for the problem (α) is set to control the more severe error. If $\alpha = 0.05$, then the mayor is willing to reject the null hypothesis when it is tenable 5 percent of the time. In other words, if the mayor were making this decision 100 times, he would tolerate being wrong (and incur HHS's displeasure) 5 times. In fact, the mayor only makes this decision once.

The appropriate statistical test is:

$$H_0: \mu \geq \$20,000$$
$$H_A: \mu < \$20,000$$
$$\alpha = 0.05$$
$$z = \frac{\bar{X} - \mu}{\frac{s}{\sqrt{n}}}$$

The next step is to determine the proper sample size (n) for a hypothesis test concerning a mean, using a random sample of, say, five Zone 4 dwelling units for a pilot study to estimate the variance:

ID	INCOM
0631.	12,037.
0712.	13,527.
0891.	10,024.
0918.	6,305.
1262.	11,557.

$$\bar{x} = \$10,690$$
$$\hat{s} = \$2,751.66$$
$$\hat{s}^2 = \$7,571,632.76$$

Z_β can be computed after the decision maker states that he/she wishes to have, say, a 90 percent chance (power) of rejecting the null hypothesis (of $20,000) when the population mean is really equal to $19,000 and he/she is willing to permit α to be 5 percent. Hence, $Z_\beta = 1.28$, as shown in Exhibit 8–2.

Exhibit 8–2

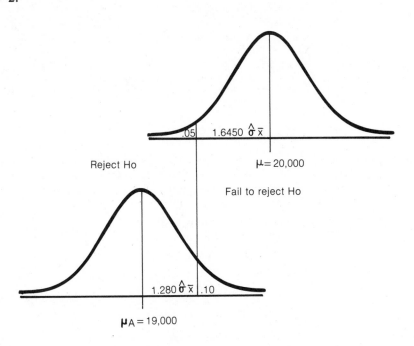

.05 1.6450 $\hat{\sigma}\bar{x}$

Reject Ho

$\mu = 20,000$

Fail to reject Ho

1.280 $\hat{\sigma}\bar{x}$ .10

$\mu_A = 19,000$

Consequently,

$$n = \frac{(\$7,571,632.76)\,(+1.645 + 1.28)^2}{(20,000 - 19,000)^2}$$

$$n = \frac{(7,571,632.76)\,(+2.925)^2}{(1,000)^2} = 64.78 = 65^*$$

* Since $n/N \le .1$, it is not necessary to use the finite population correction factor.

The sample size of 65 implies that 65 Zone 4 families must be randomly sampled if Mr. Kaplowitz is willing to have a 5 percent risk of comitting a type I error and a 90 percent chance of rejecting the null hypothesis (of \$20,000) and detecting that the population mean is really \$19,000.

Once the sample size has been determined, the random sample can be drawn. Table 8–3 enumerates the random sample selected in this study.

Finally, the data can be analyzed and conclusions can be drawn. The chart below shows the results of the survey.

Table 8–3

CASE-N	ID	INCOM
1	626.	14081.
2	634.	17987.
3	641.	30195.
4	648.	21242.
5	649.	18425.
6	651.	27781.
7	655.	23675.
8	666.	27030.
9	672.	11480.
10	692.	18489.
11	704.	13755.
12	711.	10961.
13	716.	12364.
14	720.	16777.
15	752.	11187.
16	772.	4322.
17	774.	4886.
18	799.	19071.
19	802.	16371.
20	827.	23872.
21	831.	14573.
22	844.	16144.
23	847.	15130.
24	854.	12097.
25	861.	15055.
26	863.	16916.
27	880.	11445.
28	896.	15904.
29	899.	15823.
30	946.	4087.
31	949.	11867.
32	973.	22127.
33	978.	24772.
34	1000.	20070.
35	1003.	19710.
36	1011.	16007.
37	1013.	20760.
38	1038.	15340.
39	1050.	8664.
40	1060.	7288.
41	1073.	7306.
42	1078.	6894.
43	1102.	19518.
44	1129.	18771.
45	1135.	21870.
46	1136.	23923.
47	1143.	12770.
48	1155.	14018.
49	1161.	11075.
50	1210.	16986.
51	1212.	22910.
52	1221.	24703.
53	1244.	16095.
54	1259.	16406.
55	1272.	13933.
56	1275.	16418.
57	1297.	18922.
58	1303.	18216.
59	1308.	28058.
60	1324.	20395.
61	1325.	13694.
62	1336.	10960.
63	1337.	14338.
64	1338.	11897.
65	1365.	9715.

$$n = 65$$
$$\bar{x} = \$16,269.55$$
$$s = \$5,851.37$$
$$s_{\bar{x}} = \$725.77$$

The sample indicates that the average income in Zone 4 is \$16,269.55. The question arises: Is \$16,269.55 below \$20,000.00 by a statistically significant amount so that the mayor can decide to apply for the HHS funds and only be wrong 5 percent of the time ($\alpha = 0.05$). The hypothesis test is shown below.

$$H_0: \mu \ge \$20,000$$
$$H_A: \mu < \$20,000$$
$$\alpha = 0.05$$
$$Z = \frac{\bar{x} - \mu}{\frac{s}{\sqrt{n}}} = \frac{16,269.55 - 20,000.00}{725.77}$$
$$= \frac{-3,730.45}{(725.77)} = -5.14$$

Since -5.14 is less than -1.645, you can conclude that the null hypothesis, $\mu \ge \$20,000$, is not tenable. This conclusion would lead the mayor to apply for the HHS funds.

The memorandum appears in Exhibit 8–3.

Exhibit 8–3

HOWARD S. GITLOW, PH.D.
STATISTICAL CONSULTANT

MEMORANDUM

TO: Mr. Lee Kaplowitz
        Mayor, Stat City

FROM: Howard Gitlow

DATE: September 17, 1985

RE: Applying for HHS funds for Zone 4 families

In accordance with our consulting contract and commonly accepted statistical principles, I have computed several summary measures concerning the average 1984 family income in Zone 4.[1] My research indicates that the average 1984 family income in Zone 4 was $16,269.55 per year (±$725.77).[2] Allowing for the errors in survey estimates, you can have "95 percent confidence" that the average 1984 family income in Zone 4 was below the $20,000 guideline.[3]

If you have any further questions, please do not hesitate to call me at 305-999-9999.

Footnotes

[1] Insert standard footnote 1, using a sample size of 65 drawn via simple random sampling.
[2] Insert standard footnote 2.
[3] Insert standard footnote 4.

Example 8–2   Instituting a two-pronged promotional campaign for the Stat City Electric Company

Problem

Mr. Saul Reisman, chief financial officer of the Stat City Electric Company, believes there is a significant difference in average monthly electric bills between homeowners and apartment dwellers. If a difference does exist, Mr. Reisman is considering having two marketing campaigns to promote the idea of saving energy—one directed toward apartment dwellers and the other toward homeowners.

Mr. Reisman has retained you to determine if there is a statistically significant difference in average monthly electric bills between homeowners and apartment dwellers; that is, if he should spend his stockholders' money to buy the two-pronged campaign. Conduct the necessary survey and type a memorandum to the Stat City Electric Company reporting your findings at the 5 percent level of significance (see Exhibit 8–4).

Solution   It is important to state the statistical hypothesis to be tested before any data are collected. The alternative courses of action open to the Electric Company are twofold: purchasing the two-pronged marketing campaign or continuing with the present marketing campaign. The relevant states of nature are whether or not the average monthly electric bills of apartment dwellers and homeowners are equal.

Remember that when working with two-tail tests the equality always goes in the null hypothesis. In this case the null hypothesis is:

$$H_0: \mu_{\text{house}} = \mu_{\text{apartment}}$$

The alternative hypothesis is:

$$H_A: \mu_{\text{house}} \neq \mu_{\text{apartment}}$$

If the null hypothesis is rejected and the alternative hypothesis is deemed tenable, then the two-pronged campaign will be instituted.

The problem would be set up as follows:

State of nature	Alternative course of action
$H_0: \mu_{\text{house}} = \mu_{\text{apartment}}$	Retain the present marketing campaign
$H_A: \mu_{\text{house}} \neq \mu_{\text{apartment}}$	Institute the new two-pronged marketing campaign
Set $\alpha = 0.05$	If $\alpha = 0.05$, the management of the Stat City Electric Company is willing to reject the null hypothesis when it is tenable 5 percent of the time

The next steps are to determine the proper sample size and to draw the sample. For purposes of ease, use the simple random sample shown in Table 8–4.

Finally, the data can be analyzed and conclusions can be drawn. Table 8–5 shows the results of the survey.

Please note the average monthly electric bill for homeowners ($70.90) is quite different from the average monthly electric bill for apartment dwellers ($46.39). The question arises: Is $70.90 different from $46.39 by a statistically significant amount? If there is a statistically significant difference, then the two-pronged promotional campaign should be instituted.

The test statistic is $t = 8.39$.[2]

Table 8–5   Average monthly electric bill as of January 1985

	House	Apartment
$\bar{x}$	$70.90	$46.39
s	$18.75	$16.79
n	62	88

---

[2]
$$t(n_{\text{house}} + n_{\text{apartment}} - 2) = \frac{\bar{x}_{\text{house}} - \bar{x}_{\text{apartment}}}{\sqrt{s_p^2 \left(\frac{1}{n_{\text{house}}} + \frac{1}{n_{\text{apartment}}} \right)}}$$

where

$$s_p^2 = \frac{(n_{\text{house}} - 1)s^2_{\text{house}} + (n_{\text{apartment}} - 1)s^2_{\text{apartment}}}{n_{\text{house}} + n_{\text{apartment}} - 2}$$

Table 8–4

CASE-N	ID	ELEC	CASE-N	ID	ELEC	CASE-N	ID	ELEC
1	3.	66.	1	397.	83.	45	890.	39.
2	5.	54.	2	403.	44.	46	905.	35.
3	11.	62.	3	404.	41.	47	922.	95.
4	14.	101.	4	409.	31.	48	940.	38.
5	39.	74.	5	416.	37.	49	966.	38.
6	53.	51.	6	419.	37.	50	967.	47.
7	56.	37.	7	444.	45.	51	970.	47.
8	73.	82.	8	451.	48.	52	982.	26.
9	87.	73.	9	480.	35.	53	1007.	40.
10	88.	42.	10	482.	73.	54	1062.	27.
11	91.	70.	11	502.	63.	55	1069.	60.
12	100.	85.	12	521.	80.	56	1075.	44.
13	102.	66.	13	625.	51.	57	1079.	39.
14	117.	62.	14	632.	24.	58	1086.	20.
15	120.	54.	15	645.	43.	59	1094.	40.
16	132.	66.	16	646.	82.	60	1095.	45.
17	140.	50.	17	647.	71.	61	1104.	62.
18	142.	75.	18	648.	43.	62	1115.	52.
19	149.	72.	19	656.	50.	63	1135.	26.
20	151.	72.	20	660.	68.	64	1140.	51.
21	165.	91.	21	665.	51.	65	1150.	68.
22	172.	72.	22	670.	46.	66	1155.	61.
23	178.	82.	23	679.	54.	67	1165.	24.
24	181.	86.	24	688.	62.	68	1166.	46.
25	205.	77.	25	698.	35.	69	1169.	30.
26	206.	81.	26	722.	40.	70	1172.	38.
27	227.	67.	27	729.	49.	71	1174.	41.
28	230.	67.	28	733.	37.	72	1185.	33.
29	235.	44.	29	745.	39.	73	1192.	52.
30	238.	80.	30	753.	87.	74	1214.	21.
31	279.	87.	31	757.	42.	75	1234.	49.
32	280.	71.	32	759.	73.	76	1241.	58.
33	295.	42.	33	786.	42.	77	1263.	35.
34	301.	96.	34	787.	28.	78	1265.	84.
35	308.	58.	35	799.	21.	79	1266.	53.
36	313.	92.	36	814.	22.	80	1293.	21.
37	326.	65.	37	815.	57.	81	1312.	31.
38	329.	82.	38	825.	44.	82	1335.	27.
39	335.	106.	39	827.	45.	83	1337.	64.
40	340.	33.	40	836.	68.	84	1344.	61.
41	347.	42.	41	856.	44.	85	1348.	65.
42	355.	78.	42	870.	47.	86	1360.	30.
43	360.	67.	43	884.	33.	87	1370.	30.
44	361.	98.	44	889.	36.	88	1373.	38.
45	367.	82.						
46	369.	78.						
47	372.	59.						
48	381.	60.						
49	523.	60.						
50	538.	125.						
51	543.	98.						
52	553.	34.						
53	557.	81.						
54	566.	53.						
55	568.	42.						
56	579.	71.						
57	587.	84.						
58	612.	73.						
59	613.	72.						
60	617.	75.						
61	624.	69.						
62	625.	100.						

$$H_0: \mu_{\text{house}} = \mu_{\text{apartment}}$$
$$H_A: \mu_{\text{house}} \neq \mu_{\text{apartment}}$$
$$\alpha = 0.05$$

$$t(148) = \frac{70.90 - 46.39}{\sqrt{310.61 \left(\frac{1}{62} + \frac{1}{88}\right)}}$$

where

$$s_p^2 = \frac{61(351.56) + 87(281.90)}{148}$$

$$= \frac{21,445.16 + 24,525.30}{148}$$

$$= \frac{45,970.46}{148} = 310.61$$

$$t(148) = \frac{70.90 - 46.39}{2.922} = \frac{24.51}{2.92} = 8.39$$

Since 8.39 is greater than $|1.98|$ $[t(1 - \alpha/2 = .975, 148) \approx t(1 - \alpha/2 = .975, 120) = 1.98]$ $\mu_{\text{house}} = \mu_{\text{apartment}}$ is not tenable. This conclusion would lead to purchasing the new two-pronged promotional campaign.

The memorandum appears in Exhibit 8–4.

Exhibit 8–4

ROSA OPPENHEIM, PH.D.
STATISTICAL CONSULTANT

MEMORANDUM

TO: Mr. Saul Reisman, Chief Financial Officer
        Stat City Electric Company

FROM: Rosa Oppenheim

DATE: September 26, 1985

RE: Instituting a two–pronged promotional campaign for
        the Stat City Electric Company

As per your request and using commonly accepted statistical techniques, I have computed several summary measures concerning the average monthly electric bills of Stat City apartment dwellers and homeowners.[1] My research indicates that the average monthly electric bill is $70.90 for a homeowner and $46.39 for an apartment dweller. In other words, the difference in the average monthly electric bills in Stat City between homeowners and apartment dwellers is $24.51 ($\pm$$2.92).[2] Allowing for the errors in survey estimates, you can have "95 percent confidence" that there is a statistically significant difference in average monthly electric bills between apartment dwellers and homeowners.[3]

If you have any further questions, call me at 305–999–9999.

Footnotes

[1] Insert standard footnote 1, using a sample size of 150 drawn via simple random sampling.
[2] Insert standard footnote 2.
[3] Insert standard footnote 4.

Example 8–3   Opening an automotive parts outlet in Stat City

Problem

The Pep Boys, Manny, Mo, and Jack are considering opening an automotive parts outlet in Stat City. They will open the outlet if more than 20 percent of Stat City families perform their own automotive repairs; 20 percent is the break-even percentage to open the automotive parts outlet. The Pep Boys are far more concerned about opening the outlet if there is insufficient demand than they are about not opening the outlet if there is sufficient demand.

The Pep Boys have retained you to determine if more than 20 percent of Stat City dwelling units perform their own automotive repairs as of January 1985; that is, if they should open the outlet in Stat City.

Conduct the necessary survey and type a memorandum to Manny, Mo, and Jack reporting your findings at the 5 percent level of significance (see Exhibit 8–6).

Solution

BEFORE the survey is performed it is important to specify the hypothesis being tested. The alternative courses of action open to the Pep Boys are: opening the outlet or not opening the outlet. The revelant states of nature are whether or not more than 20 percent of the Stat City families perform their own repairs. The outlet is deemed financially feasible if more than 20 percent of the Stat City families perform their own automotive repairs. Table 8–6 shows the structure of the decision to be made.

Table 8–6

	States of nature	
Alternative courses of action	Less than (or equal to) 20 percent of Stat City families perform their own repairs ($\pi \leq .20$)	More than 20 percent of Stat City families perform their own repairs ($\pi > .20$)
Open outlet	C	B
Do not open outlet	A	D

The box marked A indicates a correct decision—not opening the outlet if less than (or equal to) 20 percent of Stat City families perform their own automotive repairs. The box marked B also indicates a correct decision—opening the outlet if more than 20 percent of the Stat City families perform their own repairs. The cell marked C indicates an incorrect decision—opening the outlet if less than (or equal to) 20 percent of Stat City families perform their own repairs. This error would result in opening an unsuccessful store (an extremely expensive error). The cell marked D also indicates an incorrect decision—not opening the outlet if more than 20 percent of Stat City families perform their own repairs. This error is far less serious to the Pep Boys than the previously stated error. This error (D) represents the opportunity loss of a potentially good business. However, Manny, Mo, and Jack would still have their capital for some other business venture.

Recall, it is customary when working with one-tail tests to designate as the null hypothesis that hypothesis which, if rejected when tenable, leads to the type I error. In this case the null hypothesis is:

$$H_0: \pi \leq .20$$

Note, $\pi \leq .20$ is designated as the null hypothesis because if it is rejected when it is tenable, the type I error is committed; the Pep Boys open the store when demand is not sufficient to warrant opening the store. Remember, opening a store when there is insufficient demand means that the Pep Boys will lose their capital. The alternatives hypothesis is:

$$H_A: \pi > .20$$

If the null hypothesis is rejected and the alternative hypothesis is deemed tenable, action is then taken; the outlet would be opened because more than 20 percent of Stat City families perform their own repairs (there is sufficient demand to support an automotive parts outlet).

The problem would be set up as follows:

State of nature	Alternative course of action
$H_0: \pi \le .20$	Do not open outlet
$H_A: \pi > .20$	Open outlet

The level of significance for the problem (α) is set to control the more severe error. If $\alpha = 0.05$, then the Pep Boys are willing to reject the null hypothesis when, in fact, it is tenable 5 percent of the time. In lay language, if the Pep Boys were making this decision 100 times, they would tolerate being wrong (and losing their capital) 5 times. The fact remains—the Pep Boys only make this decision once. The appropriate statistical test is:

$$H_0: \pi \le .20$$
$$H_A: \pi > .20$$
$$\alpha = 0.05$$
$$Z = \frac{\bar{p} - \pi}{\hat{\sigma}_{\bar{p}}}$$

where

$$\hat{\sigma}_{\bar{p}} = \sqrt{\frac{\pi(1 - \pi)}{n}}$$

The next step is to determine the proper sample size (n) for a hypothesis test concerning a proportion. Z_β can be computed after the decision maker states that he/she wishes to have, say, an 80 percent chance (power) of rejecting the null hypothesis (of $\pi = 0.20$) when the population proportion is actually, say, 30 percent and he/she is willing to permit α to be 5 percent. Hence, $Z_\beta = 0.84$, as shown in Exhibit 8–5.

Consequently,

$$n = \frac{\{1.645\sqrt{.2 \times .8} + .84\sqrt{.3 \times .7}\}^2}{(.30 - .20)^2}$$
$$= \frac{(.658 + .385)^2}{.1^2} = \frac{1.043^2}{.1^2} = 108.8 \cong 109$$

Exhibit 8–5

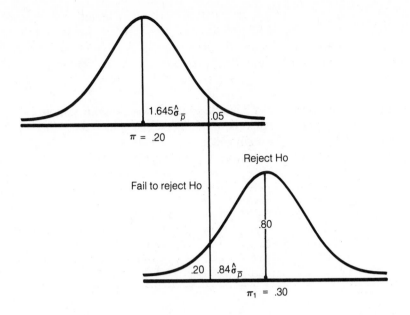

Table 8–7

CASE-N	ID	REPAR	CASE-N	ID	REPAR
1	45.	2.	55	789.	1.
2	79.	2.	56	790.	0.
3	80.	1.	57	792.	0.
4	111.	2.	58	804.	0.
5	124.	0.	59	810.	0.
6	126.	0.	60	822.	0.
7	130.	2.	61	880.	0.
8	145.	2.	62	901.	1.
9	150.	1.	63	904.	1.
10	179.	2.	64	924.	1.
11	186.	2.	65	937.	1.
12	189.	0.	66	951.	1.
13	208.	1.	67	954.	1.
14	212.	2.	68	957.	2.
15	221.	1.	69	967.	1.
16	238.	1.	70	971.	1.
17	276.	0.	71	980.	2.
18	304.	2.	72	983.	0.
19	309.	1.	73	984.	1.
20	331.	2.	74	989.	0.
21	332.	1.	75	994.	2.
22	354.	1.	76	1040.	1.
23	366.	0.	77	1044.	1.
24	391.	2.	78	1051.	1.
25	413.	1.	79	1061.	0.
26	420.	2.	80	1105.	1.
27	424.	2.	81	1119.	2.
28	425.	0.	82	1120.	0.
29	431.	2.	83	1146.	1.
30	435.	0.	84	1155.	1.
31	459.	0.	85	1160.	0.
32	463.	1.	86	1170.	1.
33	465.	2.	87	1180.	1.
34	481.	2.	88	1193.	1.
35	486.	2.	89	1210.	0.
36	496.	2.	90	1211.	2.
37	507.	0.	91	1231.	0.
38	532.	1.	92	1235.	0.
39	559.	1.	93	1240.	2.
40	569.	1.	94	1247.	2.
41	580.	2.	95	1288.	0.
42	608.	2.	96	1291.	1.
43	619.	2.	97	1298.	0.
44	627.	0.	98	1299.	2.
45	628.	0.	99	1305.	2.
46	633.	1.	100	1316.	0.
47	689.	0.	101	1317.	1.
48	704.	1.	102	1318.	0.
49	705.	1.	103	1320.	1.
50	727.	0.	104	1326.	2.
51	753.	1.	105	1335.	1.
52	764.	0.	106	1344.	0.
53	772.	0.	107	1356.	1.
54	779.	1.	108	1362.	0.
			109	1363.	1.

(Since $n/N < .10$, the finite population correction factor is not necessary.)[3] The sample size of 109 implies that 109 Stat City families must be randomly sampled in order for the Pep Boys to have a 5 percent risk of committing a type I error and an 80 percent chance of rejecting the null hypothesis (of 20 percent) while detecting that the population proportion is really 30 percent.

Once the sample size has been determined the random sample can be drawn. Table 8–7 enumerates the random sample selected for this study.

Finally, the data can be analyzed and conclusions can be drawn. Table 8–8 shows the results of the survey.

Table 8–8

Favorite place for automotive repairs as of January 1985	*Sample frequency*	*Estimated percent*
Repairs by owner = 0	35	32.1
Repairs performed by gas station = 1	43	39.4
Repairs performed by dealer = 2	31	28.4
Total	109	100.0

The above sample indicates that 32.1 percent of Stat City families perform their own automotive repairs. The question arises: Is 32.1 percent above 20 percent by a statistically significant amount so that the Pep Boys can open the outlet with the understanding that there is only a 5 percent chance of opening the outlet when it should not be opened ($\alpha = 0.05$)? The hypotheses test follows:

$$H_0: \pi \leq 0.20$$
$$H_A: \pi > 0.20$$
$$\alpha = 0.05$$
$$Z = \frac{.321 - .200}{\sqrt{\frac{.2 \times .8}{109}}} = \frac{.121}{.0383}$$
$$= 3.16$$

Since 3.16 is greater than 1.645, you can conclude that $\pi \leq 0.20$ is untenable. This conclusion would lead to opening the automotive parts outlet.

---

[3] Use of the finite population correction factor may cause logical inconsistencies in problems of this type.

The memorandum appears in Exhibit 8–6.

Exhibit 8–6

HOWARD S. GITLOW, PH.D.
STATISTICAL CONSULTANT
___

MEMORANDUM

TO: Manny Pep, Mo Pep, and Jack Pep

FROM: Howard Gitlow

DATE: September 4, 1985

RE: Opening an automotive parts outlet in Stat City

As per your request and using commonly accepted
statistical techniques, I have computed several summary
measures concerning the proportion of Stat City families
that perform their own automotive repairs.[1] My research
indicates that approximately 32.1 percent (±3.83 percent)[2]
of Stat City families perform their own automotive repairs.
Allowing for the errors in survey estimates, you can have "95
percent confidence" that the proportion of Stat City
families that perform their own repairs exceeds the
break-even percentage of 20 percent.[3]

If you have any further questions, please do not hesitate
to call me at 305-999-9999.

___

Footnotes

[1] Insert standard footnote 1, using a sample size of 109 drawn
via simple random sampling.
[2] Insert standard footnote 2.
[3] Insert standard footnote 4.

Example 8–4 Instituting a two-pronged promotional campaign for Howie's Gulf Station

Problem   Mr. Howie, the owner of Howie's Gulf Station, is considering having two promotional campaigns—one directed toward apartment dwellers and one for homeowners. Mr. Howie will pay for the two campaigns if there is a significant difference between apartment dwellers and homeowners in respect to their favorite gas station.

Mr. Howie has retained you to determine if there is a significant difference between apartment dwellers and homeowners with respect to their choice of gas station; that is, should he consider an expensive promotional campaign. Conduct the necessary survey and type a memorandum to Mr. Howie reporting your findings at the 1 percent level of significance (see Exhibit 8–7).

Solution   Remember, it is important to state the statistical hypothesis being tested before any data are collected. The alternative courses of action open to Mr. Howie are: instituting a new two-pronged promotional campaign or retaining the present one-pronged promotional campaign. The relevant states of nature are whether or not the percentage of apartment dwellers favoring Howie's Gulf is equal to the percentage of homeowners favoring Howie's Gulf.

Recall that when working with two-tail tests the equality always goes in the null hypothesis. In this case the null hypothesis is:

$$H_0: \pi_{\text{apartment}} = \pi_{\text{house}}$$

The alternative hypothesis is:

$$H_A: \pi_{\text{apartment}} \neq \pi_{\text{house}}$$

If the null hypothesis is rejected and the alternative hypothesis is deemed tenable, then the two-pronged campaign will be instituted.

The problem would be set up as follows:

State of nature	Alternative course of action
$H_0: \pi_{\text{apartment}} = \pi_{\text{house}}$	Continue present promotional campaign
$H_A: \pi_{\text{apartment}} \neq \pi_{\text{house}}$	Institute the new two-pronged promotional campaign
Set $\alpha = 0.01$	If $\alpha = 0.01$, Mr. Howie is willing to reject the null hypothesis when, in fact, it is tenable 1 percent of the time

The next steps are to determine the proper sample size and to draw the sample. For purposes of ease, use the simple random sample given in Table 8–9.

Table 8–9

CASE-N	ID	DWELL	FAVGA	CASE-N	ID	DWELL	FAVGA
1	22.	1.	1.	51	785.	0.	1.
2	41.	1.	2.	52	799.	0.	1.
3	48.	1.	2.	53	807.	0.	2.
4	99.	1.	2.	54	833.	0.	1.
5	108.	1.	1.	55	837.	0.	2.
6	125.	1.	1.	56	839.	0.	2.
7	128.	1.	1.	57	841.	0.	1.
8	132.	1.	1.	58	870.	0.	1.
9	133.	1.	2.	59	885.	0.	1.
10	134.	1.	1.	60	888.	0.	2.
11	137.	1.	2.	61	910.	0.	2.
12	142.	1.	2.	62	912.	0.	1.
13	160.	1.	2.	63	913.	0.	2.
14	165.	1.	2.	64	914.	0.	1.
15	181.	1.	1.	65	916.	0.	2.
16	182.	1.	2.	66	934.	0.	2.
17	199.	1.	2.	67	940.	0.	1.
18	205.	1.	2.	68	945.	0.	2.
19	230.	1.	2.	69	977.	0.	2.
20	231.	1.	1.	70	983.	0.	2.
21	249.	1.	2.	71	996.	0.	2.
22	263.	1.	2.	72	1011.	0.	2.
23	266.	1.	2.	73	1031.	0.	2.
24	298.	1.	1.	74	1039.	0.	1.
25	302.	1.	2.	75	1041.	0.	1.
26	335.	1.	1.	76	1045.	0.	2.
27	341.	1.	2.	77	1052.	0.	2.
28	345.	1.	2.	78	1089.	0.	2.
29	349.	1.	1.	79	1095.	0.	2.
30	350.	1.	2.	80	1113.	0.	1.
31	352.	1.	1.	81	1122.	0.	2.
32	372.	1.	2.	82	1129.	0.	2.
33	383.	1.	1.	83	1136.	0.	2.
34	461.	0.	1.	84	1143.	0.	1.
35	482.	0.	1.	85	1148.	0.	1.
36	487.	0.	2.	86	1157.	0.	2.
37	491.	0.	1.	87	1158.	0.	2.
38	515.	0.	2.	88	1159.	0.	1.
39	536.	1.	2.	89	1163.	0.	2.
40	545.	1.	2.	90	1171.	0.	1.
41	565.	1.	1.	91	1176.	0.	2.
42	567.	1.	2.	92	1180.	0.	2.
43	622.	1.	1.	93	1246.	0.	2.
44	644.	0.	2.	94	1247.	0.	2.
45	668.	0.	2.	95	1275.	0.	1.
46	672.	0.	2.	96	1307.	0.	2.
47	691.	0.	2.	97	1321.	0.	2.
48	716.	0.	1.	98	1328.	0.	2.
49	721.	0.	2.	99	1346.	0.	1.
50	732.	0.	2.	100	1356.	0.	2.

Finally, the data can be analyzed and conclusions can be drawn. Table 8–10 shows the results of the survey.

Table 8–10

Type of dwelling	Favorite gas station		
	Paul's Texaco	Howie's Gulf	Total
Apartment	22	40	62
House	15	23	38
Total	37	63	100

The above sample indicates that 64.5 percent (40/62) of apartment dwellers favor Howie's Gulf while 60.5 percent (23/38) of homeowners favor Howie's Gulf. The question arises: Is 64.5 percent different from 60.5 percent by a statistically significant amount? If there is a statistically significant difference, then the two-pronged promotional campaign will be instituted.

The hypothesis test is shown below:

$$H_0: \pi_{\text{apartment}} = \pi_{\text{house}}$$
$$H_A: \pi_{\text{apartment}} \neq \pi_{\text{house}}$$
$$\alpha = 0.05$$

The assumptions required to perform the above test are: (1) independent random samples, (2) large sample sizes ($n_{\text{apartment}}$ and n_{house} are both greater than 30), and (3) the equality of the population standard deviations.

The test statistic is:[4]

$$Z = \frac{0.645 - 0.605}{\sqrt{0.63 \times 0.37 \left(\frac{1}{62} + \frac{1}{38} \right)}} = \frac{0.040}{0.099} = 0.404$$

where

$$\bar{p} = \frac{40 + 23}{62 + 38} = \frac{63}{100} = 0.63$$

Since $Z = 0.404$ is less than $|1.96|$, you can conclude that $\pi_{\text{apartment}} = \pi_{\text{house}}$ is tenable. This conclusion favors retention of the present promotional campaign.

---

[4]

$$z \cong \frac{\bar{p}_{\text{apartment}} - \bar{p}_{\text{house}}}{\sqrt{\bar{p}(1 - \bar{p}) \left(\frac{1}{n_{\text{apartment}}} + \frac{1}{n_{\text{house}}} \right)}}$$

where

$$\bar{p} = \frac{X_{\text{apartment}} + X_{\text{house}}}{n_{\text{apartment}} + n_{\text{house}}}$$

$X_{\text{apartment}}$ = number of families favoring Howie's Gulf out of apartment dwellers

X_{house} = number of families favoring Howie's Gulf out of homeowners

The memorandum appears in Exhibit 8–7.

Exhibit 8–7

```
ROSA OPPENHEIM, PH.D.
STATISTICAL CONSULTANT

                    MEMORANDUM

TO:    Mr. Howie, Owner
       Howie's Gulf Station

FROM:  Rosa Oppenheim

DATE:  September 4, 1985

RE:    Promotional campaign for Howie's Gulf Station

   In accordance with our consulting contract and using
commonly accepted statistical techniques, I have computed
several summary measures concerning the proportion of Stat
City apartment dwellers and homeowners who prefer Howie's
Gulf Station over Paul's Texaco Station.¹ My research
indicates that approximately 64.5 percent of apartment
dwellers favor Howie's Gulf while 60.5 percent of
homeowners favor Howie's Gulf. In other words, the
difference in the percent of Stat City families that favor
Howie's Gulf between apartment dwellers and homeowners is 4
percent (±9.9 percent).² Allowing for the errors in survey
estimates, you can have "99 percent confidence" that there
is no statistically significant difference between the
proportion of apartment dwellers and homeowners who favor
Howie's Gulf.³
   If you have any further questions, please do not hesitate
to call me at 305-999-9999.
```

```
                    Footnotes

  ¹ Insert standard footnote 1, using a sample size of 100 drawn
via simple random sampling.
  ² Insert standard footnote 2.
  ³ Insert standard footnote 4.
```

8–5 Decision to purchase a new promotional campaign to boost diet soda sales The Stat City Supermarket Association believes that the typical Stat City family purchases less than one six-pack of diet soda per week on an average. The association would like to test this hypothesis because if it is tenable, they will mount a promotional campaign to increase diet soda sales. If the hypothesis is not tenable, the association will continue with their current promotional format. The association is more concerned with spending funds for the new promotional campaign when it is not necessary than they are about retaining the present promotional campaign when they should have purchased the new promotional campaign.

You have been retained by the association to test their hypothesis and help them determine if they should purchase the new promotional campaign. The association is willing to have a 10 percent chance of purchasing

ID
0083
1211
1296
0595
0486

the new campaign when it is unnecessary and wants a 90 percent chance of rejecting the notion that the average purchase is one six-pack per week when, in fact, the average weekly purchase is four cans (two-thirds of a six-pack) per week.

Use the pilot sample in the margin of five Stat City dwelling units to obtain an estimate of the variance of average weekly purchases of six-packs of diet soda as of January 1985.

Once you have computed the appropriate sample size, use the accompanying random number table to draw your simple random sample (start in the upper left-hand corner).

Random number table for Problem 8–5

2668	7422	4354	4569	9446	8212	3737	2396	6892	3766
6067	7516	2451	1510	0201	1437	6518	1063	6442	6674
4541	9863	8312	9855	0995	6025	4207	4093	9799	9308
6987	4802	8975	2847	4413	5997	9106	2876	8596	7717
0376	8636	9953	4418	2388	8997	1196	5158	1803	5623
8468	5763	3232	1986	7134	4200	9699	8437	2799	2145
9151	4967	3255	8518	2802	8815	6289	9549	2942	3813
1073	4930	1830	2224	2246	1000	9315	6698	4491	3046
5487	1967	5836	2090	3832	0002	9844	3742	2289	3763
4896	4957	6536	7430	6208	3929	1030	2317	7421	3227
9143	7911	0368	0541	2302	5473	9155	0625	1870	1890
9256	2956	4747	6280	7342	0453	8639	1216	5964	9772
4173	1219	7744	9241	6354	4211	8497	1245	3313	4846
2525	7811	5417	7824	0922	8752	3537	9069	5417	0856
9165	1156	6603	2852	8370	0995	7661	8811	7835	5087
0014	8474	6322	5053	5015	6043	0482	4957	8904	1616
5325	7320	8406	5962	6100	3854	0575	0617	8019	2646
2558	1748	5671	4974	7073	3273	6036	1410	5257	3939
0117	1218	0688	2756	7545	5426	3856	8905	9691	8890
8353	1554	4083	2029	8857	4781	9654	7946	7866	2535
1990	9886	3280	6109	9158	3034	8490	6404	6775	8763
9651	7870	2555	3518	2906	4900	2984	6894	5050	4586
9941	5617	1984	2435	5184	0379	7212	5795	0836	4319
7769	5785	9321	2734	2890	3105	6581	2163	4938	7540
3224	8379	9952	0515	2724	4826	6215	6246	9704	1651
1287	7275	6646	1378	6433	0005	7332	0392	1319	1946
6389	4191	4548	5546	6651	8248	7469	0786	0972	7649
1625	4327	2654	4129	3509	3217	7062	6640	0105	4422
7555	3020	4181	7498	4022	9122	6423	7301	8310	9204
4177	1844	3468	1389	3884	6900	1036	8412	0881	6678
0927	0124	8176	0680	1056	1008	1748	0547	8227	0690
8505	1781	7155	3635	9751	5414	5113	8316	2737	6860
8022	8757	6275	1485	3635	2330	7045	2106	6381	2986
8390	8802	5674	2559	7934	4788	7791	5202	8430	0289
3630	5783	7762	0223	5328	7731	4010	3845	9221	5427
9154	6388	6053	9633	2080	7269	0894	0287	7489	2259
1441	3381	7823	8767	9647	4445	2509	2929	5067	0779
8246	0778	0993	6687	7212	9968	8432	1453	0841	4595
2730	3984	0563	9636	7202	0127	9283	4009	3177	4182
9196	8276	0233	0879	3385	2184	1739	5375	5807	4849
5928	9610	9161	0748	3794	9683	1544	1209	3669	5831
1042	9600	7122	2135	7868	5596	3551	9480	2342	0449
6552	4103	7957	0510	5958	0211	3344	5678	1840	3627
5968	4307	9327	3197	0876	8480	5066	1852	8323	5060
4445	1018	4356	4653	9302	0761	1291	6093	5340	1840
8727	8201	5980	7859	6055	1403	1209	9547	4273	0857
9415	9311	4996	2775	8509	7767	6930	6632	7781	2279
2648	7639	9128	0341	6875	8957	6646	9783	6668	0317
3707	3454	8829	6863	1297	5089	1002	2722	0578	7753
8383	8957	5595	9395	3036	4767	8300	3505	0710	6307

Type a memorandum to the Stat City Supermarket Association reporting your findings.

8–6 Establishment of a policy for showing houses to wealthy prospective Stat City families Ms. Sharon Vigil, chairperson of the Stat City Real Estate Board, believes that Zone 2 dwelling units have more rooms on an average than Zone 1 dwelling units. She would like to test this hypothesis because if a difference exists, the board will direct wealthier prospective Stat City families toward homes in Zone 2, as opposed to Zone 1. If little difference exists, the board will direct wealthier prospective Stat City families toward homes in both Zones 1 and 2. Ms. Vigil is far more concerned about directing wealthy families exclusively to Zone 2 when it would be inappropriate than directing wealthy families to both zones when only Zone 2 should be considered.

Ms. Vigil has hired you to determine if the homes in Zone 2 have significantly more rooms than the homes in Zone 1; that is, if the board should direct wealthier prospective Stat City families exclusively to homes in Zone 2. Ms. Vigil wants "95 percent confidence" in the statistics.

Use the following random sample to estimate the required statistics.

ID	ROOMS	ID	ROOMS	ID	ROOMS
8.	9.	113.	8.	181.	10.
42.	8.	118.	8.	187.	10.
45.	9.	141.	7.	191.	5.
58.	11.	150.	7.	202.	11.
70.	6.	161.	8.	207.	6.
71.	8.	168.	8.	217.	5.
80.	9.	170.	7.	223.	9.
111.	8.	176.	8.	233.	8.
				266.	7.

Type a memorandum to Ms. Vigil reporting your findings.

8–7 Percentage of families with children in Zones 1 and 2 of Stat City Ms. Sharon Vigil, chairperson of the Stat City Real Estate Board, has a policy of indiscriminately directing wealthy prospective Stat City families who want to live in an area with children into Zones 1 and 2. Ms. Vigil would like to determine if the above policy is reasonable. Consequently, she needs to determine if the proportion of Stat City families with children in Zones 1 and 2 are equal (three or more people in a Stat City family indicate the presence of children). If the proportion of families with children in Zones 1 and 2 are equal, the board's policy will be maintained. However, if the proportion of families with children in Zones 1 and 2 are different, the board will have to formulate a new policy for directing wealthy prospective Stat City families into residential housing zones.

The board has retained you to determine if there is a significant difference (at the 5 percent level of significance) between Zone 1 and Zone 2 in respect to the percent of families with children. Conduct the necessary survey and type a memorandum to the Stat City Real Estate Board reporting your findings.

Use the accompanying simple random sample of Stat City dwelling units.

Random sample for Problem 8–7

CASE-N	ID	ZONE	PEPLE	CASE-N	ID	ZONE	PEPLE
1	7.	1.	4.	1	134.	2.	3.
2	16.	1.	4.	2	147.	2.	4.
3	22.	1.	3.	3	154.	2.	4.
4	30.	1.	4.	4	157.	2.	4.
5	48.	1.	3.	5	160.	2.	2.
6	51.	1.	1.	6	163.	2.	4.
7	57.	1.	4.	7	165.	2.	6.
8	59.	1.	4.	8	166.	2.	3.
9	66.	1.	5.	9	169.	2.	5.
10	68.	1.	1.	10	178.	2.	3.
11	74.	1.	3.	11	179.	2.	2.
12	78.	1.	2.	12	183.	2.	3.
13	80.	1.	4.	13	185.	2.	2.
14	84.	1.	3.	14	188.	2.	3.
15	89.	1.	6.	15	189.	2.	2.
16	91.	1.	4.	16	190.	2.	2.
17	94.	1.	3.	17	195.	2.	4.
18	95.	1.	1.	18	196.	2.	4.
19	96.	1.	2.	19	200.	2.	4.
20	99.	1.	4.	20	210.	2.	3.
21	103.	1.	4.	21	213.	2.	3.
22	105.	1.	3.	22	218.	2.	2.
23	108.	1.	4.	23	220.	2.	6.
24	111.	1.	5.	24	222.	2.	2.
25	114.	1.	4.	25	224.	2.	3.
26	117.	1.	3.	26	229.	2.	5.
27	119.	1.	6.	27	230.	2.	3.
28	124.	1.	1.	28	231.	2.	4.
29	126.	1.	2.	29	232.	2.	4.
30	129.	1.	3.	30	233.	2.	3.
				31	234.	2.	2.
				32	238.	2.	4.
				33	239.	2.	6.
				34	250.	2.	5.
				35	252.	2.	4.
				36	255.	2.	2.
				37	259.	2.	4.
				38	260.	2.	5.
				39	264.	2.	6.
				40	282.	2.	5.

8–8 Federal funding for the Stat City Hospital Mr. Marc Wurgaft, **ASSIGNMENT PROBLEM**
subdirector of the Department of Health and Human Services (HHS), has
issued a statement to the Stat City Hospital stating that:

1. If 25 percent or more of Stat City families use the Stat City Hospital at
   least once per year, federal funding for the hospital will not be
   changed.
2. If less than 25 percent of Stat City families use the Stat City Hospital
   at least once per year, federal funding for the hospital will be restricted.

Mr. Wurgaft is more concerned about unnecessary reductions in the Stat
City Hospital's federal funding than he is about missing an opportunity to
change the Stat City Hospital's federal funding when it is appropriate.

   You have been hired by Mr. Wurgaft to determine if 25 percent or more
of Stat City families use the hospital; that is, if HHS should leave the Stat
City Hospital's funding intact. Mr. Wurgaft is willing to have a 10 percent
chance of reducing the hospital's funding when it is not called for and
wants a 70 percent chance of rejecting the notion that the percentage of
families that use the hospital is 25 percent when, in fact, it is 15 percent.
Conduct the necessary survey using the accompanying random number
table to draw your sample; begin at the upper left-hand corner of the table.
On the chart below, indicate the appropriate alternative courses and
states of nature and explain A, B, C, and D.

Alternative courses	States of nature	
	B	D
	C	A

   On the curves below, label the means and indicate the positions and
values of Z_α and Z_β

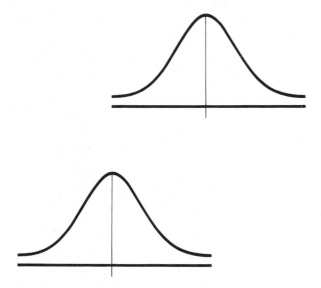

Type a memorandum to Mr. Wurgaft reporting your findings.

9263	7824	1926	9545	5349	2389	3770	7986	7647	6641
7944	7873	7154	4484	2610	6731	0070	3498	6675	9972
5965	7196	2738	5000	0535	9403	2928	1854	5242	0608
3152	4958	7661	3978	1353	4808	5948	6068	8467	5301
0634	7693	9037	5139	5588	7101	0920	7915	2444	3024
2870	5170	9445	4839	7378	0643	8664	6923	5766	8018
6810	8926	9473	9576	7502	4846	6554	9658	1891	1639
9993	9070	9362	6633	3339	9526	9534	5176	9161	3323
9154	7319	3444	6351	8383	9941	5882	4045	6926	4856
4210	0278	7392	5629	7267	1224	2527	3667	2131	7576
1713	2758	2529	2838	5135	6166	3789	0536	4414	4267
2829	1428	5452	2161	9532	3817	6057	0808	9499	7846
0933	5671	5133	0628	7534	0881	8271	5739	2525	3033
3129	0420	9371	5128	0575	7939	8739	5177	3307	9706
3614	1556	2759	4208	9928	5964	1522	9607	0996	0537
2955	1843	1363	0552	0279	8101	4902	7903	5091	0939
2350	2264	6308	0819	8942	6780	5513	5470	3294	6452
5788	8584	6796	0783	1131	0154	4853	1714	0855	6745
5533	7126	8847	0433	6391	3639	1119	9247	7054	2977
1008	1007	5598	6468	6823	2046	8938	9380	0079	9594
3410	8127	6609	8887	3781	7214	6714	5078	2138	1670
5336	4494	6043	2283	1413	9659	2329	5620	9267	1592
8297	6615	8473	1943	5579	6922	2866	1367	9931	7687
5482	8467	2289	0809	1432	8703	4289	2112	3071	4848
2546	5909	2743	8942	8075	8992	1909	6773	8036	0879
6760	6021	4147	8495	4013	0254	0957	4568	5016	1560
4492	7092	6129	5113	4759	8673	3556	7664	1821	6344
3317	3097	9813	9582	4978	1330	3608	8076	3398	6862
8468	8544	0620	1765	5133	0287	3501	6757	6157	2074
7188	5645	3656	0939	9695	3550	1755	3521	6910	0167
0047	0222	7472	1472	4021	2135	0859	4562	8398	6374
2599	3888	6836	5956	4127	6974	4070	3799	0343	1887
9288	5317	9919	9380	5698	5308	1530	5052	5590	4302
2513	2681	0709	1567	6068	0441	2450	3789	6718	6282
8463	7188	1299	8302	8248	9033	9195	7457	0353	9012
3400	9232	1279	6145	4812	7427	2836	6656	7522	3590
5377	4574	0573	8616	4276	7017	9731	7389	8860	1999
5931	9788	7280	5496	6085	1193	3526	7160	5557	6771
2047	6655	5070	2699	0985	5259	1406	3021	1989	1929
8618	8493	2545	2604	0222	5201	2182	5059	5167	6541
2145	6800	7271	4026	6128	1317	6381	4897	5173	5411
9806	6837	8008	2413	7235	9542	1180	2974	8164	8661
0178	6442	1443	9457	7515	9457	6139	9619	0322	3225
6246	0484	4327	6870	0127	0543	2295	1894	9905	4169
9432	3108	8415	9293	9998	8950	9158	0280	6947	6827
0579	4398	2157	0990	7022	1979	5157	3643	3349	7988
1039	1428	5218	0972	2578	3856	5479	0489	5901	8925
3517	5698	2554	5973	6471	5263	3110	6238	4948	1140
2563	8961	7588	9825	0212	7209	5718	5588	0932	7346
1646	4828	9425	4577	4515	6886	1138	1178	2269	4198

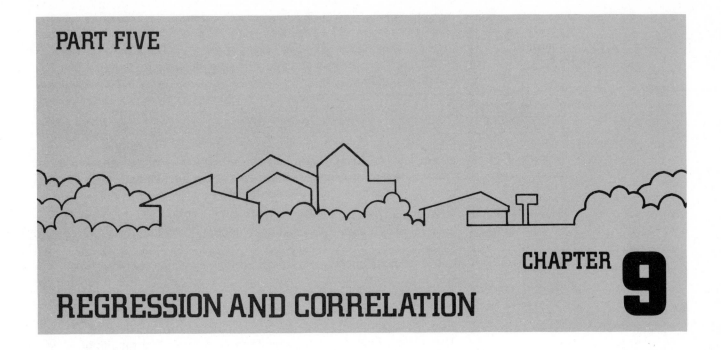

PART FIVE

REGRESSION AND CORRELATION

This chapter focuses on understanding the relationship between variables. There are essentially two types of relationship measurement problems: (1) problems in which the variables under investigation are categorical (race, religion, region of the county, and so on) and where there is no basis for estimating the value of one variable from knowledge of the other variable and (2) problems in which the variables under investigation are numerical (age, height, price, number of units purchased, and so forth) where there is a basis for estimating the value of one variable (called the dependent, criterion, or response variable) from knowledge of the other variable(s) (called the independent or predictor variable(s)).

There are two types of numerical relationship problems: deterministic (functional) and statistical. A deterministic (functional) relationship exists between variables if the value of the dependent variable (Y) is exactly and uniquely defined by specifying the value of the independent variable(s) (Xs). For example, if the rental fee for a chain saw (Y, in dollars) is $8 per day ($X$) plus a fixed charge of $10, then the deterministic relationship is as follows:

$$Y = \$10 + \$8\,(X)$$

A statistical relationship exists between variables if the value of the dependent variable (Y) is not exactly or uniquely defined by specifying the value of the independent variable(s) (Xs). For example, if the weekly dollar sales volume in a store (Y) is related to the store's weekly advertising budget (X), then a statistical relationship can be formed! Please note that this is a statistical relationship because there can be several weekly sales volumes for each weekly advertising budget; that is, the sales volume is not uniquely defined for an advertising budget.

The analysis of categorical relationships is called contingency table analysis or cross-tabulation analysis. The analysis of statistical-numeric relationships is called regression and correlation analysis. In Chapter 4, you saw how contingency tables illustrated the relationship (or lack of one) between categorical variables. In this chapter, you will see how

INTRODUCTION

regression and correlation analysis can be used to determine the relationship between numerically related variables and how the statistical importance of the relationships can be measured.

All problems in this chapter will rely on sample statistics for their solution, not on population parameters. Formulas will not be presented in this chapter because there are no standard notations for computing regression-related statistics. Consequently, computations will be presented which you can adapt to the formulas presented in your textbook.

Example 9–1 Relationship between average monthly telephone bill and 1984 family income as it pertains to the telephone company's marketing plan

Problem   Mr. Jack Davis, chairman of the Stat City Telephone Company, wants to determine if there is a relationship between the average monthly telephone bill of a Stat City family as of January 1985 and the family's total 1984 income. If there is a relationship, the telephone company will diversify their commercials to include different income groups to encourage increased telephone usage.

You have been retained by Mr. Davis to help the telephone company decide if they should have specialized income-based commercials. Mr. Davis has provided you with funds to draw a random sample of 30 Stat City families. Conduct the study and type a memorandum reporting your findings (see Exhibit 9–3). Present a scatter diagram, coefficient of determination, and regression line in your memorandum.

Please use the following randomly selected identification numbers (variable 4) to draw your sample:

72, 86, 180, 195, 271, 285, 322, 340, 426, 432, 468,
520, 611, 616, 620, 759, 807, 853, 855, 941, 1061,
1078, 1104, 1180, 1184, 1217, 1222, 1240, 1246, 1311.

Solution   The random sample is shown in Table 9–1.

Table 9–1

ID	PHONE (Y)	INCOM (X)	ID	PHONE (Y)	INCOM (X)
72.	39.	29910.	759.	77.	12717.
86.	17.	42818.	807.	78.	24995.
180.	12.	65255.	853.	41.	15113.
195.	16.	60714.	855.	38.	14460.
271.	14.	116346.	941.	33.	9726.
285.	21.	50449.	1061.	26.	1399.
322.	19.	13452.	1078.	91.	6894.
340.	15.	18485.	1104.	20.	24259.
426.	87.	26316.	1180.	19.	11156.
432.	93.	32862.	1184.	13.	6549.
468.	21.	47583.	1217.	27.	20175.
520.	19.	39430.	1222.	15.	20390.
611.	14.	18591.	1240.	76.	25241.
616.	48.	27071.	1246.	58.	24855.
620.	21.	24156.	1331.	73.	19738.

Some of the following quantities should be helpful in computing the desired statistics; see Exhibit 9-1 for details.

$$\Sigma x = 851,105 \qquad \Sigma(xy) = 27,547,299$$
$$\Sigma(x^2) = 39,120,966,527 \qquad (\Sigma x)^2 = 724,379,721,025$$
$$\Sigma y = 1,141 \qquad (\Sigma y)^2 = 1,301,881$$
$$\Sigma(y^2) = 64,931 \qquad n = 30$$

$$\Sigma(x - \bar{x})^2 = 14,974,975,825.5$$
$$\Sigma(y - \bar{y})^2 = 21,534.965$$
$$\Sigma[(x - \bar{x})(y - \bar{y})] = -4,823,065.8$$

Exhibit 9-1

COBOL Coding Form

SYSTEM				PUNCHING INSTRUCTIONS			PAGE 1 OF 4
PROGRAM Problem 9-1				GRAPHIC		CARD FORM #	*
PROGRAMMER		DATE		PUNCH			

COBOL STATEMENT — $(x-\bar{x})^2$

X	X²	(x − x̄)	(x − x̄)²	y	y²	(y − ȳ)	(y − ȳ)² IDENTIFICATION
29910	894608100	1539.833	2371085.7	39	1521	0.97	0.934
42818	1833381124	14447.833	208739878.4	17	289	-21.03	442.401
65255	4258215025	36884.833	1360490905	12	144	-26.03	677.734
60714	3686189796	32343.833	1046123533	16	256	-22.03	485.468
116346	13536391716	87975.833	7739747192	14	196	-24.03	577.601
50449	2545101601	22078.833	487474866.6	21	441	-17.03	290.134
13452	180956304	-14918.167	222551706.6	19	361	-19.03	362.268
18485	341695225	-9885.167	97716526.6	15	225	-23.03	530.534
26316	692531856	-2054.167	4219602.1	87	7569	48.97	2397.735
32862	1079911044	4491.833	20176563.7	93	8649	54.97	3021.335
47583	2264141889	19212.833	369132951.9	21	441	-17.03	290.134
39430	1554724900	11059.833	122319906	19	361	-19.03	362.268
18591	345625281	-9779.167	95632107.2	14	196	-24.03	577.601
27071	732839041	-1299.167	1687834.9	48	2304	9.97	99.334
24156	583512336	-4214.167	17759203.5	21	441	-17.03	290.134
12712	161722089	-15653.167	245021637.1	77	5929	38.97	1518.401
24995	624750025	-3375.167	11391752.3	78	6084	39.97	1597.335
15113	228402769	-13257.167	175752476.9	41	1681	2.97	8.801
14460	209091600	-13910.167	193492746	38	1444	-0.03	0.001
9726	94595076	-18644.167	347604963.1	33	1089	-5.03	25.334
1399	1957201	-26971.167	727443849.3	26	676	-12.03	144.801
6894	47527236	-21476.167	461235749	91	8281	52.97	2805.468
24259	588499081	-4111.167	16901694.1	20	400	-18.03	325.201
11156	124456336	-17214.167	296327545.5	19	361	-19.03	362.268

COBOL Coding Form

SYSTEM				PUNCHING INSTRUCTIONS			PAGE 2 OF 4
PROGRAM Problem 9-1				GRAPHIC		CARD FORM #	*
PROGRAMMER		DATE		PUNCH			

COBOL STATEMENT — $(x-\bar{x})^2$

X	X²	(x − x̄)	(x − x̄)²	y	y²	(y − ȳ)	(y − ȳ)² IDENTIFICATION
6549	42889401	-21821.167	476163329.2	13	169	-25.03	626.668
20175	407030625	-8195.167	67160762.2	27	729	-11.03	121.734
20390	415752100	-7980.167	63683065.4	15	225	-23.03	530.534
25241	637108081	-3129.167	9791686.1	76	5776	37.97	1441.468
24855	617771025	-3515.167	12356399	58	3364	19.97	398.668
19738	389588644	-8632.167	74514307.1	73	5329	34.97	1222.668
851105	39120966527	-0.01		1141	64931	0.00	21534.965
			14974975825.5				

Exhibit 9–1 (concluded)

COBOL Coding Form

SYSTEM | PUNCHING INSTRUCTIONS
PROGRAM: *Problem 9-1*
PROGRAMMER | DATE
GRAPHIC | PUNCH | CARD FORM #

$(x)(y)$	$(x-\bar{x})(y-\bar{y})$
1166490	1489.0185
7217906	-303837.93
783060	-960112.2
971424	-712534.64
1628844	-2114059.3
1059429	-376002.53
2555588	283892.72
2172275	227655.4
2289492	-100592.56
3056166	246916.06
999243	-327194.55
749170	-210468.62
2602274	2349993.38
1299408	-12952.695
507276	71767.264
979209	-610003.92
1949610	-134905.42
619633	-39373.786
549480	417.305
320958	93780.16
36374	324463.14
627354	-1137592.6
485180	74124.341
2111964	327585.6

COBOL Coding Form

SYSTEM | PUNCHING INSTRUCTIONS
PROGRAM: *Problem 9-1*
PROGRAMMER | DATE
GRAPHIC | PUNCH | CARD FORM #

$(x)(y)$	$(x-\bar{x})(y-\bar{y})$
85137	546183.81
5447725	90392.692
305850	183783.25
1918316	-118814.47
1441590	-70197.885
1440874	-301866.88
2 27547299	-4823065.8

Exhibit 9–2 lists various statistics of interest.

Exhibit 9–2

Mean income ($\bar{x}$)	\$28,370.17
Standard deviation of income (s_x)	\$22,723.96
Mean telephone bill ($\bar{y}$)	\$ 38.03
Standard deviation of telephone bill (s_y)	\$ 27.25
Correlation (r)	−.26858
Coefficient of determination (r^2)	.07213
Y-intercept (b_0 or a or $\hat{\beta}_0$)	\$ 47.17
Slope (b_1 or b or $\hat{\beta}_1$)	−0.00032

The memorandum appears in Exhibit 9–3.

Exhibit 9–3

HOWARD S. GITLOW, PH.D.
STATISTICAL CONSULTANT

MEMORANDUM

TO: Mr. Jack Davis, Chairman
        Stat City Telephone Company

FROM: Howard Gitlow

DATE: May 13, 1985

RE: Relationship between average monthly telephone
        bill and 1984 family income as it pertains to the
        telephone company's marketing plan (advertising
        copy).

As per your instructions and using commonly accepted
statistical techniques, I have investigated the
relationship between "average monthly telephone bill as
of January 1985 (PHONE)" and "family income in 1984
(INCOM)." A plot of PHONE versus INCOM is shown below:[1]

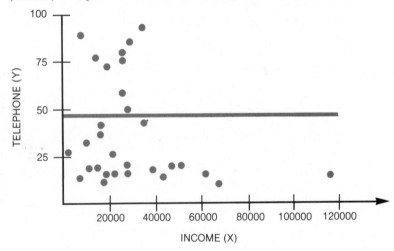

INCOME (X)

Only 7.2 percent of the variation in PHONE is explained
by INCOM.[2] The weak relationship between PHONE and INCOM
is expressed in the following equation:[3]

$$PHONE = \$47.17 - .00032\,(INCOM)$$

Consequently, one standard commercial stressing
increased telephone usage or multiple commercials
stressing increased telephone usage by variables other
than income should be employed according to the telephone
company's marketing strategy.

If you have any further questions, please do not
hesitate to call me at 305–999–9999.

Footnotes

[1] Insert standard footnote 6, using a sample size of 30.
[2] The coefficient of determination (r^2) is 7.2 percent. This
indicates that the variability in PHONE is reduced by 7.2 percent
when INCOM is considered.
[3] The small proportionate reduction in the variability of
PHONE (7.2 percent) when considering INCOM suggests that the
regression relation may not be very useful in forecasting PHONE.

$$y' = \bar{y} + \frac{r s_y (y - \bar{x})}{s x}$$

Note that the slope of a population regression line is frequently symbolized by β_1 or β, and the Y-intercept of is frequently symbolized by β_0 or α. Further, the slope of a sample regression line is frequently symbolized by $\hat{\beta}_1$, b_1, or b, and the y-intercept is frequently symbolized by $\hat{\beta}_0$, b_0, or a.

The assumptions required to make statistical inferences about simple linear regressions are:

1. A linear model is appropriate for the data.
2. The conditional probability distributions of Y given X are normal.
3. The conditional standard deviations of Y given X are all equal (this is called homoscedasticity).
4. The random error terms are independent.
5. The values of the independent variable are set at fixed levels.

Multiple regression analysis is an extension of simple linear regression analysis in which two or more independent variables are included in the model. Once again, formulas will not be presented due to a lack of notational uniformity. Note that the slope of the ith independent variable is frequently symbolized by β_i for a population model and $\hat{\beta}_i$ or b_i for a sample model.

Two more assumptions are required to make statistical inferences about a multiple linear regression. They are:

6. The number of observations must be larger than the number of regression coefficients to be estimated.
7. There must not be an exact linear relationship between any of the independent variables (this is called multicollinearity).

There are several common errors associated with the use of regression analysis that should be pointed out:

1. Estimating the dependent variable outside the observed range of the independent variable.
2. Assuming a cause-and-effect relationship exists if a statistically significant regression model is computed.
3. Assuming that conditions present in the past will remain constant in the future.
4. Estimating spurious relationships between variables (finding variables that are mathematically related by chance but have no practical relationship).

Estimating the dependent variable outside the observed range of the independent variable can cause errors because the linear relationship that exists between Y and X in the observed range of X may not be valid outside the observed range of X. An example of this type of error can be seen below.

Example 9–2   Forecasting the number of cars in a household from the size of a household

Problem   Ms. Sharon Lowe, division head of the Stat City Department of Traffic, would like to know if it is possible to forecast the number of cars in a household as of January 1985 (CARS) from the numbers of people in a household as of January 1985 (PEPLE). Further, she wants to know the estimated number of cars per household for households ranging in size from 1 to 10 people.

You have been retained by Ms. Lowe to provide this information. Conduct the required study and type a memorandum reporting your findings. Present a scatter diagram, a coefficient of determination, a regression line, and a chart indicating the estimated number of cars for households of varying sizes.

Please use the following randomly selected identification numbers (variable 4) to draw your sample:

58, 127, 584, 615, 767, 806, 1121, 1160, 1164, 1366.

The random sample is shown in Table 9–2.

Solution

Table 9–2

$r = .87$

ID	CARS (Y)	PEPLE (X)
58	4	6
127	1	2
584	3	4
615	2	3
767	2	4
806	3	4
1121	4	6
1160	1	3
1164	2	2
1366	2	4

Exhibit 9–4 shows the relevant statistics for the problem; see Exhibit 9–5 for details.

Exhibit 9–4

CANS
mean = 2.4
$S = 1.02$
$S^2 = 1.04$
$\hat{\sigma} = 1.07$
$\hat{\sigma}^2 = 1.16$

PEPLE(x) People (y)	CARS(y) CANS (X)
$\Sigma x = 38$	$\Sigma y = 24$
$\Sigma(x^2) = 162$	$\Sigma(y^2) = 68$
$\Sigma(x - \bar{x})^2 = 17.6$	$\Sigma(y - \bar{y})^2 = 10.4$
$\bar{x} = 3.8$	$\bar{y} = 2.4$
$s_x^2 = 1.96$	$s_y^2 = 1.16$
Minimum = 1	Minimum = 2
Maximum = 4	Maximum = 6

$$\Sigma(xy) = 103$$
$$\Sigma(x - \bar{x})(y - \bar{y}) = 11.8$$
$$n = 10$$
$$\text{slope} = .67$$
$$y - \text{intercept} = -.15$$
$$\text{CARS} = -.15 + .16(\text{PEPLE})$$
$$\text{CARS} = -.15 + .67(10) = 6.55$$

People
mean = 3.8
$S = 1.32$
$S^2 = 1.76$
$\hat{\sigma} = 1.398$
$\hat{\sigma}^2 = 1.96$

Exhibit 9–5

#	x	x^2	$(x - x)$	$(x - \bar{x})^2$	y	y^2	$(y - \bar{y})$	$(y - \bar{y})^2$	$(x)(y)$	$(x - \bar{x})(y - \bar{y})$
1	6	36	2.2	4.84	4	16	1.6	2.56	24	3.52
2	2	4	−1.8	3.04	1	1	−1.4	1.96	2	2.52
3	4	16	.2	.04	3	9	.6	.36	12	.12
4	3	9	−.8	.64	2	4	−.4	.16	6	.32
5	4	16	.2	.04	2	4	−.4	.16	8	−.08
6	4	16	.2	.04	3	9	.6	.36	12	.12
7	6	36	2.2	4.84	4	16	1.6	2.56	24	3.52
8	3	9	−.8	.64	1	1	−1.4	1.96	3	1.12
9	2	4	−1.8	3.24	2	4	−.4	.16	4	.72
10	4	16	.2	.04	2	4	−.4	.16	8	−.08

$y = $ People

The memorandum appears in Exhibit 9–6.

Exhibit 9–6

ROSA OPPENHEIM, PH.D.
STATISTICAL CONSULTANT

MEMORANDUM

TO: Ms. Sharon Lowe, Division Head
        Stat City Department of Traffic

FROM: Rosa Oppenheim

DATE: May 15, 1985

RE: Forecasting the number of cars in a household from
        the number of people in a household.

   As per your request and using commonly accepted
statistical techniques, I have investigated the
relationship between "the number of cars in a household as
of January 1985 (CARS)" and "the number of people in a
household as of January 1985 (PEPLE)." A plot of CARS
versus PEPLE is shown below.[1]

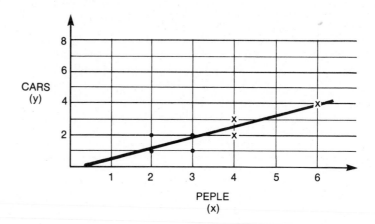

Note: (x) indicates more than one observation at that
value.

Exhibit 9–6 (*concluded*)

As you can see, CARS and PEPLE tend to increase simultaneously. Approximately 76 percent of the variation in CARS is explained by PEPLE.[2] This strong relationship between CARS and PEPLE is expressed in the following equation:[3]

$$CARS = -0.15 + 0.67 \, (PEPLE)$$

The above equation can be used to estimate the expected number of cars for different family sizes; see the chart below.

Estimated number of cars per household as of January 1985	Number of people per household as of January 1985
.52	1
1.19	2
1.86	3
2.53	4
3.20	5
3.87	6
4.54	7
5.21	8
5.88	9
6.55	10

I hope that the above information meets your needs. If you have any further questions, please call me at 305–999–9999.

Footnotes

[1] Insert standard footnote 6, using a sample size of 10.

[2] The coefficient of determination (r^2) is 76 percent. This indicates that the variability in CARS is reduced by 76 percent when PEPLE is considered.

[3] The strong proportional reduction in the variability of CARS (76 percent) when considering PEPLE suggests that the regression relation may provide a useful forecasting model for CARS based on PEPLE.

The above forecast is subject to error because the relationship between PEPLE and CARS is valid only in the observed range of CARS (in this example the observed range of cars is between one and four). For example, the forecasted number of cars for a family of zero people is -0.15 cars (rather than zero cars), which is clearly absurd.

It is important to realize that statistics cannot imply cause-and-effect relationships. Three conditions must be present to imply cause-and-effect relationships: (1) a correlation must exist between two variables, (2) a sequence of events must exist between two variables—the causal factor must occur first, and (3) no other causal factors must exist (no other factors can explain the relationship being investigated). To clarify, the existence of a correlation between two variables is only one of the three requirements necessary to imply that cause-and-effect relationship exists.

Users of regression analysis must be careful not to assume that conditions which existed in the past will remain constant in the future. The above caveat is especially important if the independent variable is time. For example, if the purchase of automobiles is being forecasted over time and a war is declared, the old relationship between automobile purchases and time will be invalid due to the war effort.

Finally, it is important that users of statistics do not give too much importance to spurious mathematical correlations. For example, prior to 1940 there was a .99 correlation between the number of mules in the United States and the number of Ph.D.s in the United States. This correlation, although humorous, was spurious; it was an accident and should be treated as such. It is obviously ridiculous to attempt to increase the number of Ph.D.s in the United States by increasing the mule population.

The following sections of this chapter are devoted to Stat City-related examples, which are worked out in detail, and additional problems. Keep the assumptions and the caveats of regression and correlation analysis in mind when you work on the cases in this chapter.

Example 9–3 Forecasting the average number of cars in a household of four people

Problem   Ms. Sharon Lowe, division head of the Stat City Department of Traffic, received your memorandum on forecasting the number of cars in a household from the number of people in a household (see Example 9–2). She liked the general format of the memo but wants you to rewrite the document to include information on the statistical significance of your findings and add any footnotes that are needed to fully explain your work. Further, she has decided that she is only interested in forecasting the average number of cars for a family with four people (see Exhibit 9–7).

The random sample that was drawn and several statistics appear in Table 9–2 and Exhibit 9–4.

Solution   To meet the new requirements of the Stat City Department of Traffic, the following items had to be added to the original memorandum:

1. A test of the statistical significance of the regression line.
2. Construction of point-and-interval estimates of the average number of cars for a family with four people.
3. Insertion of appropriate footnotes as to the dangers of statistical surveys.

Statistical significance of the regression line   The statistical significance of the regression line is tested below:

$$H_0: \beta_1 = 0$$
$$H_A: \beta_1 \neq 0$$
$$\alpha = .05$$
$$n = 10$$
$$\text{Slope} = .67$$
$$\text{Standard error of slope} = .13$$
$$y\text{-intercept} = -.15$$
$$\text{CARS} = -.15 + .67(\text{PEPLE})$$
$$r = .872$$
$$r^2 = .761$$

$$\left. \begin{array}{l} t = \dfrac{.67 - 0}{.13} = 5.15 \\ t(1 - \alpha/2 = .975, 8) = 2.306 \end{array} \right\} \text{ or } \left\{ \begin{array}{l} F = 25.432 \\ F(1 - \alpha = .95, 1, 8) = 5.32 \end{array} \right.$$

Since, $|5.15| > 2.306$ (or $25.432 > 5.32$), you can conclude that the null hypothesis ($H_0: \beta_1 = 0$) is not tenable. Consequently, β_1 is not equal to zero, and PEPLE is a statistically significant predictor of CARS.

Point and interval estimates of the average number of cars for a family of four people   A family of four people will have an average of 2.53 cars; CARS = $-0.15 + .67(4) = 2.53$. The 95 percent confidence interval for the average number of cars for a family of four people is constructed below.

$$\hat{y}_h \pm t(1 - \alpha/2, n - 2)s(\hat{y}_h)$$

where

$\hat{y}_h$ = the mean value of y at the hth level of x
$s(\hat{y}_h)$ = the standard error of the mean value of
$\quad\quad y$ at the hth level of $x(x = 4$ in this example)[1]

$$2.53 \pm 2.306(.1784)$$
$$2.53 \pm .4111$$
$$2.12 \text{ to } 2.94$$
$$2.12 \le E(\hat{y}_h) \le 2.94$$

Appropriate footnotes   If the memorandum is to meet professional standards, it must include footnotes explaining:

1. The sampling plan used.
2. Sampling and nonsampling errors.
3. Standard errors.
4. The interpretation of confidence intervals.

The memorandum appears in Exhibit 9–7.

---

[1] $s(\hat{y}_h) = \sqrt{\dfrac{\Sigma(y - \hat{y})^2}{n - 2}} \quad \sqrt{\dfrac{1}{n} + \dfrac{(x_h - \bar{x})^2}{\sum (x^2) - \dfrac{(\Sigma x)^2}{n}}}$

$\quad\quad\quad = \sqrt{\dfrac{2.489}{8}} \quad \sqrt{\dfrac{1}{10} + \dfrac{(4 - 3.8)^2}{162 - \dfrac{(38)^2}{10}}} = .55783(.3198)$

$\quad\quad\quad = .1784$

Exhibit 9–7

ROSA OPPENHEIM, PH.D.
STATISTICAL CONSULTANT

MEMORANDUM

TO: Ms. Sharon Lowe, Division Head
        Stat City Department of Traffic

FROM: Rosa Oppenheim

DATE: August 13, 1985

RE: Predicting the average number of cars in a household
        of four people as of January 1985

As per our consulting contract and using commonly
accepted statistical techniques, I have investigated the
relationship between "the number of cars in a household as of
January 1985 (CARS)" and "the number of people in a household
as of January 1985 (PEPLE)."[1] I have found that 76 percent of
the variation in CARS is explained by PEPLE.[2] This strong
relationship[3] between CARS and PEPLE is expressed in the
following equation:

$$CARS = -0.15 + .67 \; (PEPLE)$$

The above equation can be used to estimate the average number
of CARS for different-sized families. An average family of
four people (PEPLE = 4) will have 2.53 cars;

$$CARS = -.015 + .67 \; (4) = 2.53$$

Allowing for the errors in survey estimates, you can have "95
percent confidence" that the average number of cars for a
family with four people is contained in the range from 2.12
cars to 2.94 cars.[4,5,6]

I hope the above information meets your needs. If you have
any questions, please call me at 305-999-9999.

_____

Footnotes

[1] Insert standard footnote 1, using a sample size of 10 drawn
via simple random sampling.
[2] The coefficient of determination (r^2) is 76 percent. This
indicates that the variability in CARS is reduced by 76 percent when
PEPLE is considered.
[3] Insert standard footnote 2.
[4] Insert standard footnote 3A. (Insert the following numbers
into the footnote: 10, 10, 1, 1, 1, 1, 2.306, 2.306, 2.306, 2.306.)
[5] The standard error of the slope of PEPLE (.67 cars) is 0.13
CARS. The 95 percent confidence interval for the estimated slope
of PEPLE ranges from 0.37 cars to 0.97 cars. Since the confidence
interval does not contain zero, you can say that PEPLE is a
statistically significant predictor of cars at the 95 percent level
of confidence.
[6] The standard error of the predicted number of CARS for a family
with four PEPLE is 0.1784 cars. The 95 percent confidence
interval for the predicted number of cars for a family with four
people ranges from 2.12 to 2.94 cars.

Example 9–4 Development of a model to forecast gasoline consumption from electric consumption

Ms. Shelly Dimmerman, division head of the Stat City Department of Energy, believes that families which consume large quantities of electricity are also large consumers of automobile gasoline—these are called "fuelish families." You have been retained by Ms. Dimmerman to ascertain if her "fuelish family" hypothesis is tenable and to develop an equation to forecast gasoline consumption from electric consumption (see Exhibit 9–8). Ms. Dimmerman has asked you to use the following operational definitions for gasoline consumption and electric consumption:

Problem

GAS = Family's average bimonthly automobile gasoline bill as of January 1985.

ELEC = Family's average monthly electric bill as of January 1985.

Ms. Dimmerman wants to forecast GAS from ELEC because GAS is difficult to measure while ELEC is metered monthly and can be obtained from the Stat City Electric Company.

Use Table 9–3 for a random sample of Stat City families.

Table 9–3

CASE-N	ID	ELEC	GAS	CASE-N	ID	ELEC	GAS
1	19.	84.	185.	57	736.	38.	120.
2	26.	86.	206.	58	748.	61.	120.
3	28.	62.	159.	59	751.	48.	120.
4	39.	74.	180.	60	764.	19.	120.
5	56.	37.	65.	61	779.	38.	120.
6	62.	76.	184.	62	786.	42.	65.
7	65.	57.	123.	63	790.	51.	120.
8	77.	90.	204.	64	821.	50.	65.
9	81.	87.	193.	65	845.	45.	120.
10	101.	71.	164.	66	848.	50.	120.
11	104.	53.	136.	67	858.	21.	120.
12	106.	85.	200.	68	872.	19.	65.
13	119.	57.	65.	69	878.	45.	120.
14	138.	65.	139.	70	892.	68.	65.
15	142.	75.	202.	71	905.	35.	120.
16	148.	84.	156.	72	1013.	36.	167.
17	180.	48.	65.	73	1016.	46.	185.
18	181.	86.	199.	74	1027.	67.	152.
19	201.	75.	142.	75	1037.	50.	120.
20	223.	83.	150.	76	1039.	21.	65.
21	228.	63.	105.	77	1044.	47.	120.
22	258.	73.	127.	78	1047.	21.	120.
23	260.	85.	195.	79	1076.	48.	218.
24	274.	66.	144.	80	1077.	38.	120.
25	295.	42.	99.	81	1087.	65.	120.
26	298.	127.	120.	82	1106.	36.	65.
27	330.	60.	175.	83	1114.	39.	65.
28	383.	85.	207.	84	1119.	54.	168.
29	386.	72.	160.	85	1136.	43.	120.
30	390.	111.	220.	86	1140.	51.	208.
31	402.	52.	202.	87	1149.	69.	120.
32	410.	51.	182.	88	1186.	45.	120.
33	411.	46.	158.	89	1189.	47.	120.
34	415.	37.	65.	90	1218.	35.	65.
35	416.	37.	174.	91	1242.	47.	168.
36	435.	72.	155.	92	1246.	47.	195.
37	439.	62.	145.	93	1256.	29.	120.
38	452.	63.	156.	94	1272.	36.	65.
39	476.	35.	119.	95	1289.	48.	120.
40	492.	76.	130.	96	1300.	37.	110.
41	498.	44.	231.	97	1307.	48.	218.
42	500.	51.	216.	98	1348.	65.	65.
43	521.	80.	152.	99	1352.	38.	120.
44	533.	88.	204.	100	1362.	45.	65.
45	537.	75.	175.				
46	549.	97.	196.				
47	565.	60.	130.				
48	573.	66.	145.				
49	600.	65.	120.				
50	616.	84.	217.				
51	624.	69.	208.				
52	640.	65.	148.				
53	651.	46.	155.				
54	653.	45.	217.				
55	693.	22.	120.				
56	720.	39.	65.				

Solution   The following summary statistics were computed from the sample data:

$y = GAS$	$x = ELEC$
$\Sigma y = 14{,}118$	$\Sigma x = 5{,}674$
$\bar{y} = 141.18$	$\bar{x} = 56.74$
$s_y = 47.26$	$s_x = 20.62$

$$r = .49$$
$$r^2 = .24$$

y − intercept = \$77.45	Standard error of the slope = .20
Slope = \$ 1.12	Standard error of the estimate = 41.40
Mean square error = 1,714.12	

The sample regression line is:

$$GAS = \$77.45 + \$1.12(ELEC)$$

The regression line is significant at the 5 percent level of significance (or 95 percent level of confidence); see test below.

$$H_0: \beta_1 = 0$$
$$H_A: \beta_1 > 0$$
$$\alpha = .05$$

$$\left.\begin{array}{l} t = \dfrac{1.12 - 0}{.20} = 5.6 \\[4pt] t(.95, 98) = 1.66 \end{array}\right\} \text{ or } \left\{\begin{array}{l} F = 31.36 \\[4pt] F(.90, 1, 98) = 2.76 \end{array}\right.$$

Since $|5.6| > 1.66$ (or $31.36 > 2.76$), the null hypothesis ($\beta_1 = 0$) is not tenable. Consequently, ELEC is a statistically significant predictor of GAS at the 5 percent level of significance.

The "fuelish family" theory has validity because β_1 is significantly greater than zero. This means that GAS increases as ELEC increases. It is important to realize that a one-tail test was called for here to test the "fuelish family" hypothesis. Consequently, all 5 percent of α is in the upper tail so the t-statistic is 1.66, not 1.99 (the two-tail t-statistic). Be sure you note that the qualitative statement of the "fuelish family" hypothesis has a quantitative, statistically testable counterpart:

Families with large electric bills have large gasoline bills.
or
GAS is directly related to ELEC.
or
There is a positive correlation between GAS and ELEC ($\rho > 0.0$).
or
$\beta_1 > 0$ which implies GAS = $\beta_0 + \beta_1$ ELEC, that is, ELEC is useful
  in forecasting GAS.

The memorandum appears in Exhibit 9–8.

Exhibit 9–8

HOWARD S. GITLOW, PH.D.
STATISTICAL CONSULTANT
———

MEMORANDUM

TO: Ms. Shelly Dimmerman, Division Head
        Stat City Department of Energy

FROM: Howard Gitlow

DATE: August 15, 1985

RE: Development of a model to forecast gasoline
        consumption from electric consumption as of
        January 1985.

As per your request and using commonly accepted
statistical techniques, I have investigated the
relationship between "average bimonthly automobile
gasoline bills as of January 1985 (GAS)" and average
monthly electric bills as of January 1985 (ELEC)."[1] I have
found that 24 percent of the variation in GAS is explained
by ELEC.[2] The relationship between GAS and ELEC is
expressed in the following equation:[3]

$$GAS = \$77.45 + \$1.12 \ (ELEC)$$

The above forecasting equation can be used to estimate
average bimonthly gasoline bills from average monthly
electric bills. Further, the forecasting equation is
statistically significant at the "95 percent level of
confidence."[4,5,6] The practical implication of the
statistical significance of the forecasting equation is
that it lends support to the "fuelish family" hypothesis.
        If you have any questions, please call me at
305–999–9999.

———————

Footnotes

[1] Insert standard footnote 1, using a sample of 100 drawn via
simple random sampling.
[2] The coefficient of determinations (r^2) is 24 percent. This
indicates that the variability in GAS is reduced by 24 percent when
ELEC is considered.
[3] Insert standard footnote 2.
[4] Insert standard footnote 3A. (Insert the following numbers
into the footnote: 100, 100, 1, 1, 1, 1, 1.66, 1.66, 1.66, 1.66.)
[5] The standard error of the slope of ELEC ($1.12) is $0.20. The
90 percent confidence interval for the estimated slope of ELEC
ranges from $0.72 and $1.52. Since the confidence interval does not
contain zero, ELEC is a statistically significant predictor of
GAS at the 90 percent level of confidence.
[6] If the regression line had not been statistically significant,
the forecasting equation would have been GAS = $77.45. To restate,
ELEC would have no value in forecasting GAS (it is not in the
equation). However, since the forecasting equation includes an
ELEC term, and the sign of that term is positive, the "fuelish
family" hypothesis is supported; big gasoline users are also big
electricity users.

Example 9–5 Development of a model to forecast electric consumption in Stat City

Problem   Mr. Saul Reisman, chief financial officer of the Stat City Electric Company, would like to develop an equation to forecast a dwelling unit's average monthly electric bill. You have been retained by Mr. Reisman to construct the equation. In your discussions with energy experts, you have isolated two categories of variables which may influence electric consumption: "fuelishness" of a household and size of a household. The energy experts have told you that the "fuelishness" of a household is measured vis-à-vis a household's yearly heating bill and bimonthly automobile gasoline bill. The size of a dwelling unit is measured by the number of rooms in the household.

Mr. Reisman has asked you to use the following operational definitions for the relevant variables:

ELEC = Average monthly electric bill as of January 1985.

HEAT = Average yearly heating bill as of January 1985.

GAS = Average bimonthly automobile gasoline bill as of January 1985.

ROOMS = Number of rooms in a dwelling unit as of January 1985.

Mr. Reisman has provided you with sufficient funds to draw a simple random sample of 144 Stat City dwelling units. Conduct the desired survey and type a memorandum to Mr. Reisman (see Exhibit 9–9).

Use the simple random sample of Stat City families given in Table 9–4.

Table 9-4	CASE-N	ID	ROOMS	HEAT	ELEC	GAS
	1	5.	6.	674.	54.	133.
	2	9.	9.	1101.	85.	177.
	3	21.	8.	983.	69.	146.
	4	41.	4.	259.	27.	120.
	5	43.	7.	876.	64.	183.
	6	54.	7.	802.	64.	124.
	7	55.	9.	1065.	83.	187.
	8	63.	7.	857.	65.	134.
	9	67.	9.	1060.	81.	202.
	10	76.	11.	1253.	100.	219.
	11	93.	9.	1019.	82.	231.
	12	100.	9.	1061.	85.	170.
	13	123.	4.	483.	32.	65.
	14	126.	5.	628.	40.	139.
	15	128.	9.	1097.	84.	150.
	16	135.	7.	987.	58.	180.
	17	143.	9.	1265.	91.	162.
	18	161.	8.	1053.	75.	173.
	19	173.	8.	1152.	74.	179.
	20	174.	9.	1309.	80.	211.
	21	175.	9.	1272.	79.	156.
	22	193.	9.	1357.	85.	175.
	23	203.	5.	656.	40.	74.
	24	213.	7.	953.	66.	148.
	25	218.	7.	989.	59.	112.
	26	223.	9.	1249.	83.	150.
	27	224.	7.	923.	62.	130.
	28	238.	8.	1201.	80.	159.
	29	249.	7.	1018.	63.	136.
	30	261.	6.	780.	53.	125.
	31	263.	9.	1256.	80.	162.
	32	267.	7.	990.	54.	105.
	33	283.	8.	1162.	82.	155.
	34	285.	7.	916.	57.	177.
	35	295.	5.	538.	42.	99.
	36	300.	8.	786.	73.	177.
	37	324.	7.	687.	53.	168.
	38	328.	8.	871.	73.	205.
	39	359.	5.	524.	40.	65.
	40	365.	10.	949.	96.	198.
	41	376.	8.	779.	75.	164.
	42	378.	8.	848.	69.	164.
	43	379.	7.	698.	58.	159.
	44	395.	10.	1035.	97.	120.
	45	401.	9.	859.	78.	169.
	46	405.	8.	678.	63.	132.
	47	408.	6.	588.	48.	165.
	48	433.	10.	835.	95.	124.
	49	438.	6.	584.	48.	120.
	50	446.	6.	562.	47.	120.

CASE-N	ID	ROOMS	HEAT	ELEC	GAS
51	451.	6.	528.	48.	172.
52	475.	6.	524.	51.	120.
53	484.	5.	427.	41.	132.
54	485.	8.	799.	64.	179.
55	489.	8.	693.	76.	142.
56	504.	8.	729.	62.	138.
57	519.	5.	464.	41.	116.
58	526.	11.	1105.	109.	120.
59	528.	8.	779.	71.	199.
60	555.	8.	809.	74.	146.
61	574.	7.	625.	62.	166.
62	585.	8.	825.	75.	136.
63	598.	9.	952.	81.	205.
64	604.	7.	730.	65.	159.
65	623.	8.	850.	75.	203.
66	629.	5.	395.	37.	120.
67	635.	8.	637.	67.	215.
68	639.	8.	621.	64.	120.
69	652.	6.	481.	41.	197.
70	667.	6.	470.	44.	65.
71	669.	7.	598.	54.	65.
72	697.	5.	422.	32.	120.
73	714.	6.	521.	48.	120.
74	738.	6.	489.	45.	120.
75	740.	7.	523.	62.	120.
76	752.	5.	393.	38.	65.
77	760.	5.	384.	40.	65.
78	778.	5.	391.	38.	65.
79	809.	7.	538.	63.	120.
80	829.	4.	268.	30.	65.
81	835.	7.	601.	59.	120.
82	838.	5.	391.	33.	120.
83	844.	7.	556.	54.	120.
84	846.	5.	374.	41.	120.
85	866.	5.	436.	41.	65.
86	871.	5.	386.	37.	120.
87	878.	6.	533.	45.	120.
88	879.	6.	529.	47.	65.
89	889.	5.	396.	36.	65.
90	894.	6.	483.	53.	120.
91	902.	4.	334.	30.	120.
92	908.	6.	451.	48.	120.
93	938.	7.	569.	56.	65.
94	941.	7.	585.	55.	120.
95	945.	3.	280.	18.	120.
96	947.	5.	388.	37.	120.
97	957.	7.	576.	45.	184.
98	958.	6.	553.	49.	165.
99	969.	8.	616.	68.	126.
100	979.	5.	369.	37.	65.
101	981.	9.	721.	77.	142.
102	987.	6.	503.	48.	185.
103	999.	5.	397.	38.	120.
104	1003.	5.	393.	35.	120.
105	1016.	6.	512.	46.	185.
106	1019.	7.	549.	54.	120.
107	1024.	7.	568.	53.	120.
108	1031.	5.	404.	35.	65.
109	1049.	5.	411.	41.	65.
110	1051.	8.	571.	63.	65.
111	1052.	5.	423.	37.	120.
112	1070.	4.	361.	30.	120.
113	1099.	9.	719.	76.	190.
114	1100.	6.	498.	43.	206.
115	1136.	6.	458.	43.	120.
116	1148.	8.	609.	64.	120.
117	1158.	4.	260.	29.	65.
118	1168.	5.	452.	38.	120.
119	1181.	7.	608.	55.	120.
120	1182.	6.	479.	43.	120.
121	1184.	5.	433.	35.	120.
122	1187.	4.	312.	31.	120.
123	1194.	3.	220.	19.	120.
124	1204.	3.	210.	22.	65.
125	1217.	5.	384.	39.	77.
126	1222.	5.	402.	36.	85.
127	1224.	6.	488.	46.	169.
128	1231.	7.	586.	59.	65.
129	1233.	10.	832.	79.	65.
130	1250.	7.	554.	60.	65.
131	1258.	5.	395.	37.	120.
132	1272.	5.	422.	36.	65.
133	1273.	5.	411.	32.	120.
134	1288.	6.	487.	42.	65.
135	1292.	6.	503.	48.	65.
136	1294.	6.	503.	40.	120.
137	1296.	5.	382.	38.	177.
138	1297.	8.	647.	70.	65.
139	1315.	5.	385.	38.	65.
140	1317.	5.	423.	40.	65.
141	1320.	6.	470.	46.	120.
142	1335.	5.	398.	36.	65.
143	1345.	5.	342.	38.	65.
144	1373.	5.	374.	38.	65.

9 / Regression and correlation

Solution

The bivariate correlation matrix for all four variables used in this study is shown in Table 9–5.

The correlation matrix reveals an extremely high correlation between ROOMS and ELEC ($r = .97$); approximately 94 percent ($r^2 = .94$) of the variation in ELEC is explained by ROOMS. Further, ELEC is highly correlated with HEAT and GAS, and ROOMS is highly correlated with HEAT and GAS. All of the above indicates redundancy among the independent variables; each independent variable is a good predictor of ELEC, and the independent variables are also good predictors of each other.[2] Consequently, it is both efficient and expedient to use only the best independent variable to predict ELEC and to delete the others from the model. The other variables should be eliminated because they bring little or no new information which would enhance the model's forecasting ability.

Just for your information, the full regression model is:

$$ELEC = -\$11.73 + \$0.016(HEAT) - \$0.005(GAS) + \$8.65(ROOMS)$$
$$(.002) \qquad (.009) \qquad (.349)$$

The interpretation of the above regression coefficients is as follows: the y-intercept ($-\$11.73$) indicates the average monthly electric bill if the average yearly heating bill is zero, the average bimonthly automobile gasoline bill is zero, and the number of rooms in a dwelling unit is zero (this is obviously an absurd situation and occurs because a forecast is being made outside the ranges of the independent variables). The slope of the average yearly heating bill ($\$0.016$) indicates that for a dwelling unit with a given average biweekly automobile gasoline bill and number of rooms, the average monthly electric bill will increase by $\$0.016$ for every dollar spent on heat. The slope of the average bimonthly automobile gasoline bill ($-\$0.005$) indicates that for a dwelling unit with a given average yearly heating bill and number of rooms, the average monthly electric bill will decrease by $\$0.005$ for every dollar spent on gasoline. The slope of the number of rooms ($\$8.65$) indicates that for a dwelling unit with a given average yearly heating bill and an average bimonthly automobile gasoline bill, the average monthly electric bill will increase by $\$8.65$ for every additional room.

The numbers in parentheses under the regression coefficients are the standard errors of the regression coefficients. It is extremely important that t-tests are not performed to test the statistical significance of each individual regression coefficient. The warning is appropriate whenever there is any correlation between the independent variables. If correlations do exist between the independent variables (this is called multicollinearity), then it is impossible to extract the portion of the variability in the dependent variable that is uniquely attributable to a particular independent variable. Consequently, t-tests are meaningless in a multiple regression context in which multicollinearity is present. The bivariate correlation matrix in this problem clearly indicates the large degree of multicollinearity present in this example; hence, t-tests should not be conducted to test the statistical significance of the independent variables. Instead, an analysis of variance table can be computed which allows all of the regression coefficients to be tested for significance simultaneously (H_0: $\beta_1 = \beta_2 = \beta_3 = 0$). Unfortunately, if the null hypothesis is rejected, you will not know which of the regression coefficients are in fact statistically different from zero. This is the method you must use to circumvent the problem of multicollinearity

Table 9–5

	ELEC	HEAT	GAS	ROOMS
ELEC	1.00	.89	.54	.97
HEAT		1.00	.58	.85
GAS			1.00	.54
ROOMS				1.00

---

[2] High correlation between the independent variables is called multicollinearity and can cause severe problems in regression analysis. One way to deal with multicollinearity is to drop all but one of the variables which are highly correlated out of the model so that only the best predictor variable(s) remain.

and still be able to test the statistical significance of your regression model. The ANOVA table for this problem is shown in Table 9–6.

Table 9–6

Source	df	SS	MS	F
Regression	3	48,304.44	16,101.48	1,159.24
Error	140	1,944.55	13.89	
Total	143			

The coefficient of multiple determination measures the percent of variation in the dependent variable that is explained by the independent variables. In this example, the coefficient of multiple determination is 0.96 (96 percent of the variation in ELEC is explained by ROOMS, HEAT, and GAS). The problem of multicollinearity can also be seen when examining the coefficient of multiple determination. Table 9–7 depicts the change in the coefficient of multiple determination as independent variables are added into the model.

Table 9–7

Independent variable	r^2	Change in r^2 due to the addition of another independent variable
ROOMS	.9473	.9473
HEAT	.9612	.0139
GAS	.9613	.0001

Once ROOMS has been included in the model, none of the other independent variables makes a substantive contribution to explaining the variation in ELEC.

The best strategy would be to revise the model and to use only ROOMS as a predictor of ELEC. To reiterate, the energy experts were correct concerning the importance of the "fuelish" family and size of dwelling unit variables; however, they failed to warn you about the interrelationships between the independent variables. Consequently, the experts wasted the Electric Company's money by requiring you to collect unnecessary data. Sounder planning on their part would have ensured a more cost effective study.

The reduced regression model is shown below.

$$\text{ELEC} = -\$15.86 + \$10.72(\text{ROOMS})$$
$$r^2 = .94$$

The statistical significance of the regression line is tested below:

$$H_0: \beta_1 = 0$$
$$H_A: \beta_1 \neq 0$$
$$\alpha = .05$$

$$\left. \begin{aligned} t = \frac{10.72 - 0}{.212} = 50.57 \\ t(.975,\ 142) = 1.97 \end{aligned} \right\} \text{ or } \left\{ \begin{aligned} F = 2,556.92 \\ F(.95,\ 1,\ 142) = 3.88 \end{aligned} \right.$$

Since $|50.57| > 1.97$ (or $2,556.92 > 3.88$), you can conclude that H_0 is not tenable. Consequently, $\beta_1 \neq 0$ and ROOMS is a statistically significant predictor of ELEC.

One final note: the regression model computed in this problem affords a perfect example of the dangers of forecasting the dependent variable outside the range of the independent variable. A forecast of the electric bill for a one-room dwelling unit yields a negative electric bill (ELEC = −$15.86 + $10.72(1) = −$5.14). Obviously, electric bills cannot be negative (except in the case of a rebate or some other atypical practice).

The memorandum appears in Exhibit 9–9.

Exhibit 9–9

ROSA OPPENHEIM, PH.D.
STATISTICAL CONSULTANT
———

MEMORANDUM

TO: Mr. Saul Reisman, Chief Financial Officer
        Stat City Electric Company

FROM: Rosa Oppenheim

DATE: August 19, 1985

RE: Development of a model to forecast electric
        consumption in Stat City as of January 1985

    As per your request and using commonly accepted
statistical techniques, I have developed an equation to
forecast a dwelling unit's average monthly electric bill as
of January 1985 (ELEC).[1,2] I have found that approximately 94
percent of the variation in ELEC is explained by the number
of rooms in a dwelling unit as of January 1985 (ROOMS).[3,4] The
relationship between ELEC and ROOMS is expressed in the
following equation:

$$ELEC = -\$15.86 + \$10.72(ROOMS).$$

The above forecasting equation can be used to estimate the
average monthly electric bill for a dwelling unit (as of
January 1985) from the number of rooms in a dwelling unit (as
of January 1985). Further, the forecasting equation is
statistically significant as the "95 percent level of
confidence."[5,6]

    If you have any questions, please call me at
305–999–9999.

Exhibit 9–9 (*concluded*)

Footnotes

[1] The original model to forecast ELEC included the "fuelishness" of a family and the size of a dwelling unit as measured by the following independent variables:

HEAT = Average yearly heating bill as of January 1980.

GAS = Average bimonthly automobile gas bill as of January 1980.

ROOMS = Number of rooms in a dwelling unit as of January 1980.

All of the above variables were suggested by the Stat City Electric Company energy experts. HEAT and GAS were surrogate measures of the "fuelishness" of a family. ROOMS was a measure of the size of a dwelling unit.

Preliminary correlation analysis revealed that approximately 94 percent of the variation in ELEC was explained by ROOMS. Further, ELEC was highly correlated with HEAT and GAS, and ROOMS was highly correlated with HEAT and GAS. All of the above indicated a redundancy among the predictor variables; each independent variable was a good predictor of ELEC and the independent variables were also good predictors of each other. Consequently, a simple model was constructed to forecast ELEC which used only ROOMS.

[2] Insert standard footnote 1, using a sample of 144 drawn via simple random sampling.

[3] The coefficient of determination (r^2) is 94 percent. This indicates that the variability in ELEC is reduced by 94 percent when ROOMS is considered.

[4] Insert standard footnote 2.

[5] Insert standard footnote 3A. (Insert the following numbers into the footnote: 144, 144, 1, 1, 1, 1, 1.97, 1.97, 1.97, 1.97.)

[6] The standard error of the slope of ROOMS (10.72) is 0.212. The 95 percent confidence interval for the estimated slope of ROOMS ranges from $10.30 to $11.14. Since the confidence interval does not contain zero, you can say that ROOMS is a statistically significant predictor of ELEC at the 95 percent level of confidence.

SUMMARY

This chapter dealt with many of the issues necessary to properly use regression and correlation analysis and the concept of statistical inference in regression analysis problems. The assumptions required to make statistical inferences in a regression analysis were enumerated. Finally, several caveats concerning the proper use of regression analysis were discussed.

ADDITIONAL PROBLEMS

9–6 Forecasting the assessed value of a dwelling unit from the number of rooms in a dwelling unit The Stat City Tax Assessor's Office wants to determine if the number of rooms in a dwelling unit as of January 1985 can be used to forecast the assessed value of a dwelling unit as of January 1985. If the number of rooms in a dwelling unit can be used to forecast the assessed value of the dwelling unit, then the tax assessor's office only needs to determine the number of rooms in a dwelling unit.

You have been retained by the Stat City Tax Assessor's Office to furnish this information. Conduct the required study and type a memorandum reporting your findings. Present a scattergram, a coefficient of determination, a regression line, and a chart indicating the proposed tax assessments for dwelling units containing different numbers of rooms.

Please use the following randomly selected identification numbers (variable 4) to draw your sample:

170, 358, 391, 610, 735, 785, 936, 1119, 1151, 1350.

9–7 Forecasting heating bills from the number of rooms in a dwelling unit Ms. Shelly Dimmerman, division head of the Stat City Department of Energy, needs to forecast the yearly heating bill of a dwelling unit from the number of rooms in the dwelling unit. You have been asked by Ms. Dimmerman to conduct a survey and construct the needed forecasting equation. Ms. Dimmerman has asked you to use the following operational definitions for heating bill and rooms:

HEAT = Average yearly heating bill as of January 1985 (for all types of heat).

ROOMS = Number of rooms in a dwelling unit as of January 1985.

Use the following simple random sample of Stat City families:

ID	ROOMS	HEAT
111.	8.	973.
550.	8.	851.
554.	8.	812.
640.	8.	661.
776.	5.	385.
777.	8.	691.
813.	6.	505.
1192.	6.	466.
1253.	5.	419.
1257.	8.	667.

9–8 Developing an equation to forecast monthly mortgage payments from the assessed value of a house Ms. Sharon Vigil, chairperson of the Stat City Real Estate Board, is trying to develop an equation for forecasting monthly mortgages in Stat City. Ms. Vigil has decided that using the assessed value of a house may provide an accurate indicator of monthly mortgages. Assessed value is a convenient predictor to use because it is publicly available and current as of January 1985.

You have been retained by Ms. Vigil to develop an equation for forecasting monthly mortgages from assessed values. Ms. Vigil has provided you with funds to draw a random sample of 10 Stat City dwelling units. Please use the following simple random sample of Stat City dwelling units.

ID	HCOST	ASST
91.	510.	58524.
129.	606.	70174.
137.	880.	94070.
162.	785.	95599.
218.	785.	87717.
286.	621.	76856.
394.	144.	27898.
537.	222.	44747.
561.	514.	51378.
575.	493.	49758.

9-9 Development of an equation to forecast the number of cars in a household

Ms. Sharon Lowe, division head of the Stat City Department of Traffic, would like to be able to forecast the number of cars in a household. You have been hired by her to develop the desired forecasting equation.

Ms. Lowe believes that the number of people in a household (PEPLE), the average bimonthly automobile gasoline bill (GAS), and the average monthly telephone bill (PHONE) are all important predictors of the number of cars in a household (CARS).

Ms. Lowe has provided you with the necessary funds to draw a simple random sample of 100 Stat City families. Conduct the desired survey and type a memorandum to Ms. Lowe.

Use the accompanying simple random sample of Stat City families.

Random sample table for Problem 9-9

CASE-N	ID	PHONE	PEPLE	CARS	GAS
1	20.	35.	2.	2.	137.
2	54.	55.	3.	2.	124.
3	58.	63.	6.	4.	217.
4	69.	59.	5.	4.	210.
5	70.	44.	3.	2.	149.
6	72.	39.	3.	2.	131.
7	91.	18.	4.	3.	177.
8	93.	22.	5.	5.	231.
9	94.	23.	3.	2.	158.
10	129.	21.	3.	2.	122.
11	137.	18.	3.	2.	135.
12	143.	85.	4.	2.	162.
13	162.	23.	4.	2.	157.
14	175.	21.	4.	2.	156.
15	180.	12.	1.	1.	65.
16	208.	20.	3.	2.	122.
17	214.	22.	2.	1.	104.
18	218.	67.	2.	1.	112.
19	234.	93.	2.	1.	111.
20	268.	15.	3.	2.	129.
21	286.	71.	2.	1.	104.
22	291.	68.	4.	3.	176.
23	320.	36.	5.	3.	200.
24	337.	118.	7.	2.	120.
25	342.	63.	6.	2.	120.
26	353.	21.	5.	4.	209.
27	375.	78.	5.	3.	184.
28	388.	20.	6.	5.	235.
29	394.	45.	2.	1.	105.
30	411.	59.	4.	2.	158.
31	421.	54.	5.	3.	193.
32	435.	22.	4.	2.	155.
33	437.	46.	3.	2.	120.
34	449.	16.	6.	2.	120.
35	452.	13.	4.	2.	156.
36	478.	30.	6.	4.	219.
37	483.	16.	3.	2.	139.
38	498.	19.	5.	5.	231.
39	503.	17.	5.	4.	208.
40	516.	13.	4.	2.	155.
41	547.	17.	6.	4.	218.
42	554.	22.	4.	2.	152.
43	578.	64.	5.	3.	187.
44	598.	15.	5.	4.	205.
45	607.	22.	5.	4.	209.
46	616.	48.	6.	4.	217.
47	677.	17.	3.	2.	120.
48	700.	20.	5.	2.	120.
49	735.	27.	8.	1.	65.
50	741.	86.	3.	1.	65.
51	745.	77.	2.	1.	100.
52	768.	37.	1.	2.	120.
53	813.	17.	5.	4.	201.
54	826.	16.	6.	2.	120.
55	837.	39.	3.	2.	120.
56	850.	23.	5.	2.	120.
57	868.	66.	5.	2.	120.
58	879.	71.	5.	1.	65.
59	886.	18.	5.	1.	65.
60	896.	17.	1.	2.	120.
61	918.	17.	3.	2.	120.
62	919.	29.	6.	2.	120.
63	937.	12.	8.	2.	120.
64	940.	76.	4.	2.	120.
65	947.	25.	4.	2.	120.
66	961.	13.	4.	3.	178.
67	984.	19.	6.	2.	120.
68	1020.	31.	5.	2.	120.
69	1022.	18.	4.	2.	120.
70	1040.	23.	7.	1.	65.

Table for Problem 9–9
(*concluded*)

CASE-N	ID	PHONE	PEPLE	CARS	GAS
71	1046.	98.	3.	1.	65.
72	1052.	59.	6.	2.	120.
73	1067.	44.	1.	2.	120.
74	1073.	71.	6.	2.	120.
75	1078.	91.	6.	1.	65.
76	1081.	17.	3.	2.	120.
77	1123.	102.	1.	2.	120.
78	1136.	105.	8.	2.	120.
79	1152.	40.	7.	1.	65.
80	1157.	29.	3.	1.	65.
81	1169.	51.	1.	2.	120.
82	1176.	85.	4.	1.	65.
83	1177.	41.	3.	2.	120.
84	1187.	18.	2.	2.	120.
85	1188.	107.	7.	2.	120.
86	1195.	13.	3.	2.	120.
87	1225.	53.	3.	2.	128.
88	1230.	19.	2.	2.	120.
89	1239.	22.	3.	2.	120.
90	1249.	20.	4.	2.	120.
91	1264.	42.	3.	2.	120.
92	1269.	18.	7.	1.	65.
93	1270.	62.	6.	1.	65.
94	1271.	13.	3.	2.	120.
95	1282.	11.	6.	2.	120.
96	1291.	23.	3.	2.	120.
97	1294.	14.	9.	2.	120.
98	1296.	34.	4.	3.	177.
99	1304.	26.	5.	3.	199.
100	1359.	11.	6.	2.	120.

9–10 An empirical test of the "extrafamilial nonpersonal family size" hypothesis Mr. Brennan (ID = 1041), a sociologist, believes that a family's level of extrafamilial, nonpersonal communication (nonface-to-face communication) increases with family size. The sociologist would like to be able to forecast a family's level of nonpersonal communication from its family size.

You have been employed by the sociologist to conduct a study and test out his hypothesis. Sufficient funds have been allocated to draw a simple random sample of 10 Stat City families. Further, the sociologist has instructed you to use the following operational definitions for his study:

Please use the following simple random sample of Stat City families.

Actual variable	Operational variable
Level of extra-familial non-personal communi-cation	PHONE = Average monthly telephone bill as of January 1985
Family size	PEPLE = Number of people in a family as of January 1985

ID	PHONE	PEPLE
3.	62.	4.
227.	69.	2.
300.	15.	4.
367.	57.	5.
714.	21.	6.
759.	77.	4.
821.	14.	7.
1338.	23.	3.
1347.	18.	3.
1372.	45.	5.

9–11 Developing a model to forecast a family's total 1984 income

Mr. Lee Kaplowitz, the mayor of Stat City, would like to be able to forecast a family's income (INCOM) to set donation guidelines for the City Benevolent Association. The mayor is convinced he will have a difficult time obtaining information on total family income (INCOM) because of his purpose. He would like to see if the number of people in a family (PELPE) and a family's rent/mortgage (HCOST) is correlated with family income (INCOM). He believes it will be easy to obtain information on the number of people in a family (PEPLE) and the amount of a family's rent/mortgage (HCOST). Consequently, if a relationship exists between HCOST, PEPLE, and INCOM, he can use the easily obtainable variables (HCOST and PEPLE) to forecast his variable of interest (INCOM).

The mayor has engaged you to conduct a survey and develop a model to forecast INCOM from HCOST and PEPLE. Sufficient funds have been allocated to draw a simple random sample of 100 Stat City dwelling units.

Please use the accompanying simple random sample of Stat City families.

Random sample table for Problem 9–10

CASE-N	ID	HCOST	INCOM	PEPLE	CASE-N	ID	HCOST	INCOM	PEPLE
1	41.	263.	19505.	1.	51	687.	547.	14967.	5.
2	52.	775.	24718.	6.	52	688.	572.	16856.	4.
3	63.	653.	35468.	3.	53	700.	323.	14217.	5.
4	67.	667.	55787.	5.	54	714.	353.	12078.	6.
5	81.	840.	33260.	5.	55	735.	313.	12741.	8.
6	118.	549.	51411.	4.	56	741.	564.	16674.	3.
7	119.	389.	15300.	6.	57	744.	330.	4405.	4.
8	136.	884.	32646.	5.	58	762.	455.	8857.	4.
9	148.	956.	120030.	4.	59	776.	343.	5830.	4.
10	161.	672.	67110.	4.	60	782.	327.	8142.	5.
11	164.	827.	68773.	3.	61	790.	416.	14757.	5.
12	173.	767.	69621.	4.	62	792.	528.	12063.	3.
13	186.	766.	73136.	4.	63	794.	430.	20323.	5.
14	201.	714.	96448.	3.	64	799.	252.	19071.	2.
15	205.	675.	57598.	3.	65	807.	540.	24995.	4.
16	207.	631.	34468.	2.	66	816.	603.	25718.	4.
17	214.	714.	67797.	2.	67	828.	361.	25649.	6.
18	227.	674.	85373.	2.	68	840.	501.	14305.	7.
19	230.	818.	66701.	3.	69	864.	386.	15236.	4.
20	237.	950.	61602.	4.	70	868.	450.	16001.	5.
21	241.	768.	47650.	3.	71	873.	485.	11509.	6.
22	262.	418.	52098.	1.	72	894.	448.	16307.	4.
23	273.	689.	68609.	1.	73	906.	460.	17915.	5.
24	290.	250.	30236.	2.	74	911.	411.	5375.	6.
25	306.	297.	31869.	2.	75	917.	303.	11300.	4.
26	315.	515.	58703.	5.	76	933.	382.	16051.	5.
27	340.	193.	18485.	4.	77	972.	232.	21765.	1.
28	348.	468.	54212.	5.	78	978.	446.	24772.	6.
29	360.	470.	28224.	3.	79	1005.	406.	20075.	3.
30	375.	419.	31998.	5.	80	1018.	404.	14985.	4.
31	386.	276.	32667.	4.	81	1043.	531.	8987.	4.
32	396.	298.	52318.	1.	82	1070.	307.	7936.	1.
33	400.	474.	27444.	6.	83	1071.	549.	5813.	5.
34	405.	591.	38353.	3.	84	1083.	560.	31908.	5.
35	410.	391.	27779.	4.	85	1102.	640.	19518.	4.
36	477.	379.	29803.	6.	86	1103.	526.	21607.	6.
37	490.	433.	27809.	4.	87	1114.	334.	18105.	5.
38	494.	463.	33673.	6.	88	1148.	525.	18718.	4.
39	509.	346.	27390.	4.	89	1149.	438.	19670.	4.
40	518.	172.	32458.	2.	90	1152.	412.	11171.	7.
41	534.	328.	25329.	3.	91	1177.	513.	19315.	3.
42	551.	439.	44988.	2.	92	1222.	365.	20390.	1.
43	585.	304.	50514.	3.	93	1239.	278.	13326.	3.
44	599.	366.	34509.	4.	94	1259.	567.	16406.	5.
45	612.	231.	41959.	4.	95	1271.	405.	6494.	3.
46	614.	397.	46125.	4.	96	1273.	396.	6798.	4.
47	615.	457.	57483.	3.	97	1310.	325.	28180.	4.
48	616.	484.	27071.	6.	98	1319.	289.	10349.	6.
49	669.	499.	6892.	4.	99	1337.	553.	14338.	4.
50	674.	582.	14408.	3.	100	1345.	284.	12639.	6.

The following summary statistics should be helpful to you in solving this problem.

$r^2 = .42$

Source	df	SS	MS	F
Regression	2	20,089,496,467.43	10044748233.7	35.5
Error	97	27,446,761,165.32	282956300.7	

$b_0 = 13115.43$
$b_1 = 73.47$ (slope for HCOST)
$b_2 = -4544.26$ (slope for PEPLE)
$s(b_1) = 9.83$
$s(b_2) = 1131.71$

	HCOST	INCOM	PEPLE
HCOST	1.00	.572	.017
INCOM		1.00	−.300
PEPLE			1.00

9–12 Studying the relationship between supermarket bill and number of hospital trips Ms. Arlene Davis, director of the Stat City Hospital, believes there may be a relationship between the average weekly supermarket bill per person (EATPL is a measure of nutrition) and the average number of trips a family makes to the hospital per year (HOSP) as of January 1985. A random sample of 10 Stat City families was drawn and appears below.

HOSP	EATPL
1.	23.
1.	22.
3.	22.
1.	23.
4.	18.
2.	23.
1.	29.
6.	23.
4.	21.
1.	26.

a. Using the graph below, plot a scatter diagram.
b. Assuming a linear relationship exists between EATPL and HOSP, compute the slope and y-intercept of EATPL (y) on HOSP (x) and graph the regression line below.
c. Forecast EATPL for families that make five trips to the hospital per year.
d. Compute the 90 percent confidence interval for the mean value of EATPL, given HOSP = 5.
e. Compute the 90 percent confidence interval for an individual forecasted value of EATPL, given HOSP = 5.
f. Compute the coefficient of determination.

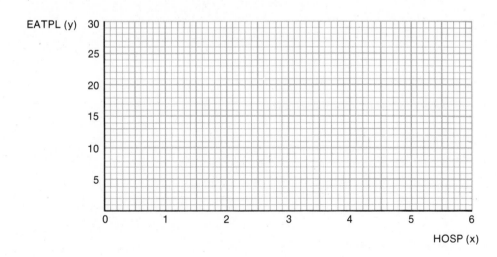

bill The mayor of Stat City believes there is a relationship between HCOST and HEAT as of January 1985. A random sample of 10 Stat City families was drawn and appears below.

HCOST	HEAT
888.	1309.
386.	752.
546.	624.
327.	851.
620.	657.
462.	475.
403.	391.
467.	413.
268.	229.
350.	481.

a. Using the graph below, plot a scatter diagram.
b. Assuming that a linear relationship exists between HCOST and HEAT, compute the slope and y-intercept of HCOST (y) on HEAT (x) and graph the regression line below.
c. Forecast HCOST for a family whose heating bill is $500 per year.
d. Compute a 90 percent confidence interval for the mean value of HCOST, given HEAT = $500.
e. Compute a 90 percent confidence interval for an individual forecasted value of HCOST, given HEAT = $500.
f. Compute the coefficient of determination.

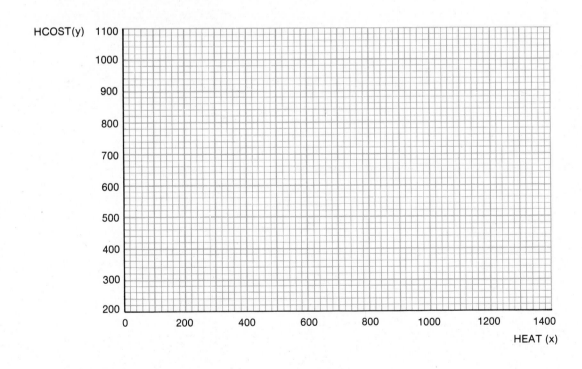

THE ANALYSIS OF VARIANCE

CHAPTER 10

INTRODUCTION

Hypotheses concerning the difference between two groups' means or proportions was introduced and explained in depth in Chapter 8. In this chapter you will be testing hypotheses concerning the differences among three or more groups' means. The name of the statistical technique used to test hypotheses concerning the differences among three or more groups' means is "analysis of variance."

The null and alternative hypotheses necessary to test if there are statistically significant differences among K groups' means are:

$$H_0: \mu_1 = \mu_2 = \ldots = \mu_K$$
$$H_A: \text{Not all the means are equal, where}$$
$$\mu_i = \text{the mean of group } i.$$

The statistics which must be computed to test the above null hypothesis are detailed in Table 10–1.[1]

Table 10–1

Source	Degrees of freedom	Sum of squares	Mean squares	F
Between group variation	$K - 1$	SS_B	$SS_B/(K - 1)$	MS_B/MS_w
Within group variation	$n_T - K$	SS_w	$SS_w/(n_T - K)$	
Total variation	$n_T - 1$	SS_T		

---

[1] $K =$ the number of groups
$n_T =$ the total sampling size
$n_j =$ the sample size for group j
$y_{ij} =$ the ith observation in group j

$$SS_T = \Sigma_{j=1}^{K} \Sigma_{i=1}^{n_j} (y_{ij}^2) - \frac{(\Sigma_{j=1}^{K} \Sigma_{i=1}^{n_j} y_{ij})^2}{n_T}$$

$$SS_B = \Sigma_{j=1}^{K} \left[\frac{(\Sigma_{i=1}^{n_j} y_{ij})^2}{n_j} \right] - \frac{(\Sigma_{j=1}^{K} \Sigma_{i=1}^{n_j} y_{ij})^2}{n_T}$$

$$SS_w = \Sigma_{j=1}^{K} \Sigma_{i=1}^{n_j} (y_{ij}^2) - \Sigma_{j=1}^{K} \left[\frac{(\Sigma_{i=1}^{n_j} y_{ij})^2}{n_j} \right]$$

If the computed F-statistic is greater than $F(1 - \alpha, K - 1, n_T - K)$, reject the null hypothesis.

The computational formulas presented in your textbook may be different from the formulas presented; either set of formulas will lead you to compute the same F-statistic.

The assumptions necessary to perform a one-way analysis of variance (fixed model) are:

1. The process being studied is in control; that is, it is repeatable.
2. The probability distribution being sampled is normal.
3. The variances in all K groups are equal (homogeneous).
4. The null hypothesis is tenable. This assumption is important because the F-statistic is valid only when the null hypothesis is tenable.

The computed F-statistic can be greater than $F(1 - \alpha, K - 1, n_T - K)$ for only two reasons: (1) the null hypothesis is tenable and the sample results happened by chance (however unlikely) *or* (2) the null hypothesis is not tenable. The above alternatives imply that the null hypothesis will be rejected 100α percent of the time when, in fact, it is tenable. This error (the type I error) represents the chance that the sample under study will lead to rejecting the null hypothesis when it is tenable.

Understanding hypothesis tests concerning three or more groups' means is more difficult than a person without formal training in statistics would suspect. Consequently, it is important that standard footnote 5 (see Exhibit 10–1), or some version thereof, appears whenever a hypothesis test concerning three or more means is reported in a memorandum, to serve as a partial disclaimer for the errors that can be encountered in hypothesis testing.

Exhibit 10–1   Standard footnote 5

Footnote

[5] The sample estimators permit us to perform a hypothesis test (for a given sample size).

To illustrate, if all possible samples were selected and each of these were surveyed under essentially the same conditions, then:

a. Approximately nine-tenths of the hypothesis tests conducted at the 90 percent level of confidence (or 10 percent level of significance) would fail to reject the null hypothesis when it was tenable (or 10 percent would reject the null hypothesis when it was tenable).

b. Approximately 19/20 of the hypothesis tests conducted at the 95 percent level of confidence (or 5 percent level of significance) would fail to reject the null hypothesis when it was tenable (or 5 percent would reject the null hypothesis when it was tenable).

Once the level of confidence (or significance) has been set, the chances of failing to reject the null hypothesis when it is false must be computed. If these probabilities are unacceptably high (type II errors), then corrective action must be taken.

There are many different types of analysis of variance (ANOVA) procedures. Discussing them all is the subject for an entire course. The only ANOVA model that will be discussed in this chapter is the model already mentioned: the one-way ANOVA model with fixed factor levels.

A final point: do not be confused by the phrase "analysis of variance." Although the phrase mentions variance and not means, ANOVA is still a technique for testing if significant differences exist among group means.

The following examples and problems deal with testing for significant differences among three or more groups' means. Pay particular attention to the assumptions stated earlier when conducting your analysis.

Example 10–1 Redefining the tax assessment policy for Stat City

Problem

Mr. Lee Kaplowitz, the mayor of Stat City, is considering changing the property tax structure of Stat City from one tax assessment rate for all Stat City families to separate tax assessment rates for families in each zone in Stat City. The mayor will enact the above change in tax assessment rates if there is a statistically significant difference in the 1984 average family incomes among the four zones in Stat City at the 5 percent level of significance.

You have been retained by the mayor to conduct a survey and determine if there is a statistically significant difference in the 1984 average family incomes among the four zones in Stat City. Sufficient monies have been budgeted for you to draw a simple random sample of 130 Stat City dwelling units (see Table 10–2 for the random sample). Conduct the survey and type a memorandum to the mayor of Stat City reporting your findings (see Exhibit 10–2).

Table 10–2

CASE-N	ID	ZONE	INCOM	CASE-N	ID	ZONE	INCOM
1	2.	1.	23280.	27	316.	3.	28784.
2	7.	1.	19469.	28	321.	3.	142251.
3	22.	1.	56263.	29	349.	3.	39124.
4	27.	1.	52135.	30	350.	3.	23889.
5	55.	1.	33910.	31	355.	3.	23584.
6	58.	1.	30830.	32	374.	3.	25648.
7	61.	1.	47583.	33	375.	3.	31998.
8	81.	1.	33260.	34	379.	3.	31622.
9	91.	1.	44914.	35	386.	3.	32667.
10	94.	1.	36165.	36	394.	3.	29338.
11	144.	2.	58321.	37	437.	3.	15515.
12	152.	2.	83484.	38	440.	3.	21729.
13	175.	2.	76434.	39	446.	3.	30436.
14	178.	2.	91058.	40	457.	3.	26552.
15	199.	2.	95761.	41	459.	3.	42806.
16	248.	2.	83195.	42	462.	3.	42231.
17	256.	2.	121005.	43	474.	3.	30138.
18	257.	2.	64797.	44	495.	3.	34054.
19	260.	2.	78070.	45	510.	3.	32901.
20	262.	2.	52098.	46	515.	3.	33009.
21	263.	2.	70732.	47	520.	3.	39430.
22	268.	2.	86760.	48	539.	3.	49245.
23	274.	2.	57519.	49	542.	3.	40026.
24	284.	2.	68773.	50	569.	3.	39785.
25	297.	3.	33745.	51	575.	3.	48159.
26	306.	3.	31869.	52	582.	3.	33408.

Table 10–2 (concluded)

CASE-N	ID	ZONE	INCOM	CASE-N	ID	ZONE	INCOM
53	590.	3.	44993.	92	992.	4.	33790.
54	592.	3.	54468.	93	998.	4.	21591.
55	593.	3.	49830.	94	1014.	4.	24059.
56	594.	3.	36794.	95	1029.	4.	12212.
57	599.	3.	34509.	96	1036.	4.	16847.
58	620.	3.	24156.	97	1045.	4.	9780.
59	626.	4.	14081.	98	1053.	4.	12661.
60	631.	4.	12037.	99	1095.	4.	27968.
61	634.	4.	17987.	100	1098.	4.	22100.
62	644.	4.	16730.	101	1107.	4.	31397.
63	647.	4.	23346.	102	1112.	4.	22603.
64	649.	4.	18425.	103	1127.	4.	18341.
65	666.	4.	27030.	104	1133.	4.	20791.
66	671.	4.	15948.	105	1148.	4.	18718.
67	679.	4.	15597.	106	1159.	4.	18141.
68	680.	4.	14055.	107	1163.	4.	19334.
69	720.	4.	16777.	108	1164.	4.	7369.
70	723.	4.	13377.	109	1168.	4.	8998.
71	725.	4.	15310.	110	1171.	4.	7406.
72	820.	4.	8473.	111	1176.	4.	15981.
73	826.	4.	23257.	112	1179.	4.	13622.
74	830.	4.	25456.	113	1181.	4.	18194.
75	835.	4.	13661.	114	1182.	4.	15918.
76	836.	4.	14572.	115	1191.	4.	7520.
77	842.	4.	15166.	116	1213.	4.	21847.
78	847.	4.	15130.	117	1214.	4.	13461.
79	851.	4.	12642.	118	1217.	4.	20175.
80	862.	4.	17522.	119	1224.	4.	25525.
81	873.	4.	11509.	120	1225.	4.	26071.
82	874.	4.	11410.	121	1234.	4.	17580.
83	895.	4.	11960.	122	1242.	4.	24743.
84	900.	4.	14657.	123	1245.	4.	22413.
85	911.	4.	5375.	124	1250.	4.	10145.
86	912.	4.	16601.	125	1265.	4.	23745.
87	922.	4.	9648.	126	1301.	4.	20073.
88	931.	4.	8987.	127	1306.	4.	29034.
89	955.	4.	8055.	128	1317.	4.	26335.
90	959.	4.	24312.	129	1365.	4.	9715.
91	980.	4.	22087.	130	1366.	4.	8967.

Solution   Before any data are collected, the hypothesis under investigation must be stated. The hypothesis of concern and its alternative are listed below:

$$H_0: \mu_1 = \mu_2 = \mu_3 = \mu_4$$
$$H_A: \text{Not all the means are equal}$$

where

μ_1 = the average 1984 family income in Zone 1
μ_2 = the average 1984 family income in Zone 2
μ_3 = the average 1984 family income in Zone 3
μ_4 = the average 1984 family income in Zone 4

The desired level of significance for the above hypothesis test is 5 percent. In layman's language, if the mayor had to revise the tax assessment 100 times, he would be willing to change the tax assessment policy erroneously 5 times.

The computations required to test the null hypothesis are shown in Table 10–3.[2]

Table 10–3

	Zone 1	Zone 2	Zone 3	Zone 4
n_j	10	14	34	72
$\sum_{i=1}^{n_j} y_{ij}$	377,089	1,088,007	1,278,693	1,236,350
$\sum_{i=1}^{n_j} (y_{ij}^2)$	15,600,504,576	88,814,668,800	61,887,816,704	24,117,963,520

The ANOVA table is shown in Table 10–4. The computed F-statistic (87.207) exceeds the critical F-value ($F_{(1-\alpha=.95,3,126)} = 2.67$). This indicates that a statistically significant proportion of the variation in income is explained by zone-to-zone variability, as opposed to variation inside the zones. The above statistics lead to rejecting the null hypothesis. Rejecting the null hypothesis indicates that there are important zone-by-zone differentials in income.

Table 10–4

Source	Degrees of freedom	Sum of squares	Mean squares	F-ratio
Between groups	3	46,246,254,737.7881	15,415,418,240.0000	87.207
Within groups	126	22,272,875,024.0000	176,768,848.0000	
Total	129	68,519,129,600.0000		

---

[2] $(\Sigma\Sigma y_{ij})^2 = (3,980,859)^2$
$\Sigma\Sigma(y_{ij}^2) = 190,420,951,040$

$$SS_T = (190,420,951,040) - \left(\frac{3,980,859^2}{130}\right) = 68,519,129,600.0000$$

$$SS_B = \left[\frac{377,809^2}{10} + \frac{1,088,007^2}{14} + \frac{1,278,693^2}{34} + \frac{1,236,350^2}{72}\right] - \left[\frac{3,980,859^2}{130}\right]$$
$$= 46,246,254,737.7861$$

$$SS_w = [190,420,951,040] - \left[\frac{377,809^2}{10} + \frac{1,088,007^2}{14} + \frac{1,278,693^2}{34} + \frac{1,236,350^2}{72}\right]$$
$$= 176,768,848.0000$$

The memorandum appears in Exhibit 10–2.

Exhibit 10–2

ROSA OPPENHEIM, PH.D.
STATISTICAL CONSULTANT

MEMORANDUM

TO: Mr. Lee Kaplowitz
        Mayor, Stat City

FROM: Rosa Oppenheim

DATE: October 21, 1985

RE: Redefining the tax assessment policy for Stat City

In accordance with our consulting contract and using commonly accepted statistical techniques, I have computed several summary measures concerning the 1984 average family income in each zone of Stat City.[1] My research indicates that the average 1984 family income for each zone in Stat City is as follows:[2]

Zone	Average 1984 family income
1	$37,780.90
2	77,714.79
3	37,608.62
4	17,171.53
Overall	$30,621.99

Allowing for the errors in survey estimates, you can have "95 percent confidence" that there is a statistically significant difference in the average 1984 family incomes among the four residential zones in Stat City.[3]

I hope the above information meets your needs. If you have any questions, please do not hesitate to call me at 305–999–9999.

Footnotes

[1] Insert standard footnote 1, using a sample of 130 drawn via simple random sampling.
[2] Insert standard footnote 2.
[3] Insert standard footnote 5.

In this chapter you learned how to extend hypothesis tests concerning means from two groups (populations) to three or more groups (populations). In this chapter, the F-distribution was introduced for the first time, and you were made aware of the severe limitations as to the choices of α when using an F-statistic. These limitations occur because F-statistics are usually tabulated for alphas of 0.01, 0.05, and 0.10. However, computer programs and hand calculators are now available that will compute F-statistics for any value of α.

This chapter will not deal with computing the power curve or sample size requirements for a one-way analysis of variance. These topics are considered beyond the scope of this book.

SUMMARY

10–2 Special promotional campaign to boost Stat City gas station retail sales Mr. Paul Sugrue, chairman of the Stat City Gas Station Association, needs to determine if the average number of trips to the gas station per month are the same for families who perform their own automotive repairs, have their repairs performed by a gas station, or have their repairs performed by a dealer (as of January 1985). If the average number of trips to the gas station per month are significantly different among the above types of families, the association will pay for a special advertising campaign to boost the average number of trips to the gas station per month for the category of family that is the lowest.

You have been retained by Mr. Sugrue to conduct the necessary survey at the 10 percent level of significance. He has advanced sufficient funds to draw a sample of 100 Stat City families. Perform the necessary survey and type a memorandum to Mr. Sugrue reporting your findings.

Use the accompanying simple random sample of 100 Stat City families.

ADDITIONAL PROBLEMS

Random sample for Problem 10–2

CASE-N	ID	GASTR	REPAR	CASE-N	ID	GASTR	REPAR
1	8.	3.	2.	51	766.	6.	0.
2	18.	5.	1.	52	768.	5.	0.
3	41.	8.	1.	53	775.	3.	1.
4	42.	4.	2.	54	777.	6.	1.
5	78.	4.	2.	55	779.	5.	1.
6	90.	3.	2.	56	784.	4.	1.
7	95.	6.	2.	57	791.	6.	0.
8	117.	5.	1.	58	848.	7.	0.
9	121.	4.	1.	59	857.	4.	0.
10	128.	4.	2.	60	873.	7.	1.
11	146.	4.	2.	61	892.	5.	0.
12	168.	4.	2.	62	916.	7.	1.
13	182.	6.	2.	63	930.	5.	1.
14	187.	4.	2.	64	957.	4.	2.
15	188.	6.	2.	65	965.	6.	1.
16	218.	6.	2.	66	985.	6.	1.
17	238.	5.	1.	67	1027.	4.	2.
18	253.	9.	2.	68	1030.	4.	0.
19	298.	5.	2.	69	1035.	12.	0.
20	312.	8.	1.	70	1039.	3.	0.
21	321.	2.	2.	71	1042.	6.	0.
22	326.	7.	2.	72	1043.	7.	1.
23	371.	6.	0.	73	1047.	6.	1.
24	373.	7.	0.	74	1055.	5.	1.
25	405.	6.	1.	75	1057.	3.	0.
26	432.	3.	1.	76	1059.	6.	0.
27	435.	4.	0.	77	1072.	2.	1.
28	448.	5.	2.	78	1086.	5.	1.
29	468.	6.	1.	79	1091.	9.	0.
30	475.	10.	2.	80	1124.	4.	0.
31	479.	6.	1.	81	1125.	9.	1.
32	499.	8.	0.	82	1147.	7.	0.
33	502.	7.	1.	83	1151.	4.	0.
34	519.	5.	1.	84	1167.	9.	1.
35	548.	4.	0.	85	1188.	8.	1.
36	573.	10.	2.	86	1217.	5.	2.
37	585.	5.	2.	87	1244.	2.	1.
38	596.	5.	1.	88	1262.	5.	1.
39	601.	5.	1.	89	1263.	6.	1.
40	637.	6.	1.	90	1272.	5.	0.
41	647.	8.	2.	91	1291.	5.	1.
42	652.	9.	2.	92	1328.	5.	1.
43	669.	7.	0.	93	1331.	3.	1.
44	671.	9.	0.	94	1334.	5.	0.
45	690.	6.	0.	95	1339.	5.	0.
46	711.	6.	0.	96	1340.	4.	1.
47	718.	6.	1.	97	1349.	4.	2.
48	721.	7.	1.	98	1358.	6.	1.
49	744.	3.	0.	99	1363.	7.	1.
50	748.	5.	0.	100	1367.	2.	0.

10–3 Survey concerning the number of people per dwelling unit Mr. Joseph Moder, division head of the Stat City Department of Public Works, needs to determine if there is a statistically significant difference in the average number of people per dwelling unit among the residential housing zones in Stat City as of January 1985. You have been employed by Mr. Moder to conduct a survey of 90 Stat City dwelling units to ascertain the required information. Conduct the survey, complete the ANOVA table below, and report your findings to Mr. Moder at the 5 percent level of significance.

Use the accompanying simple random sample of Stat City families.

ASSIGNMENT PROBLEM

Random sample for Problem 10–3

CASE-N	ID	ZONE	PEPLE	CASE-N	ID	ZONE	PEPLE
1	2.	1.	5.	71	1098.	4.	7.
2	15.	1.	4.	72	1101.	4.	5.
3	22.	1.	3.	73	1146.	4.	1.
4	34.	1.	7.	74	1149.	4.	4.
5	38.	1.	6.	75	1173.	4.	5.
6	95.	1.	1.	76	1175.	4.	5.
7	98.	1.	2.	77	1189.	4.	5.
8	125.	1.	4.	78	1196.	4.	4.
9	157.	2.	4.	79	1212.	4.	5.
10	165.	2.	6.	80	1222.	4.	1.
11	174.	2.	5.	81	1230.	4.	2.
12	185.	2.	2.	82	1234.	4.	3.
13	191.	2.	1.	83	1241.	4.	5.
14	211.	2.	3.	84	1274.	4.	4.
15	213.	2.	3.	85	1280.	4.	5.
16	222.	2.	2.	86	1294.	4.	9.
17	234.	2.	2.	87	1316.	4.	1.
18	241.	2.	3.	88	1325.	4.	4.
19	267.	2.	2.	89	1351.	4.	4.
20	292.	3.	4.	90	1372.	4.	5.
21	302.	3.	2.				
22	343.	3.	6.				
23	366.	3.	2.				
24	368.	3.	4.				
25	379.	3.	3.				
26	389.	3.	4.				
27	409.	3.	4.				
28	430.	3.	2.				
29	442.	3.	7.				
30	472.	3.	1.				
31	482.	3.	5.				
32	488.	3.	5.				
33	509.	3.	4.				
34	510.	3.	5.				
35	511.	3.	4.				
36	540.	3.	4.				
37	558.	3.	3.				
38	559.	3.	3.				
39	592.	3.	5.				
40	594.	3.	4.				
41	600.	3.	4.				
42	610.	3.	3.				
43	619.	3.	1.				
44	655.	4.	8.				
45	706.	4.	4.				
46	717.	4.	5.				
47	726.	4.	4.				
48	732.	4.	9.				
49	755.	4.	4.				
50	781.	4.	5.				
51	825.	4.	4.				
52	841.	4.	1.				
53	857.	4.	6.				
54	868.	4.	5.				
55	886.	4.	5.				
56	917.	4.	4.				
57	936.	4.	2.				
58	952.	4.	4.				
59	982.	4.	3.				
60	983.	4.	4.				
61	1012.	4.	5.				
62	1026.	4.	3.				
63	1028.	4.	1.				
64	1029.	4.	4.				
65	1037.	4.	5.				
66	1039.	4.	5.				
67	1064.	4.	5.				
68	1080.	4.	6.				
69	1091.	4.	4.				
70	1093.	4.	6.				

Source	df	SS	MS	F
Between group variation				
Within group variation				
Total variation				

PART SIX

STATISTICS AND THE COMPUTER

CHAPTER **11**

You have spent this semester becoming acquainted with and computing statistics. By this time, you must realize that statistical computations can be very time-consuming. You also know the frustration of getting the wrong answer to a problem because you made a careless math error or pushed the wrong button on your calculator. Well, take heart, because help with statistical computations is available with computers.

Computers are your friends, not your enemies. Once you become familiar with computers, you will achieve a whole new level of statistical proficiency and awareness. Statistical computations with a computer are no longer difficult or time-consuming. Computers virtually erase the difficulties encountered in statistical computations (clients always want accurate answers "yesterday"). In this chapter, you will learn about two commonly available computer packages which are very useful for statistical analyses, SPSS and Minitab.

SPSS (*S*tatistical *P*ackage for the *S*ocial *S*ciences) is a set of prewritten computer instructions which can calculate most of the statistics you will ever need. It is easy to use SPSS. All SPSS programs have exactly the same structure. As a matter of fact, the statements in an SPSS program vary very little from program to program.

SPSS programs are made up of control cards. Each control card has a specific command that the computer must perform. Control cards have two parts, or fields: a control field and a specification field. The control field occupies the first 15 columns of a control card and contains one or more words indicating the task the computer is being instructed to perform. The specification field occupies card columns 16 through 80 of the control card. The specification field enumerates the specifics of the operation to be performed that is stated in the control field.

The cards in Exhibit 11–1 are a complete listing of an SPSS program designed to analyze a sample survey of 10 Stat City dwelling units. The control cards exemplify what control and specification fields of control cards look like.

A brief discussion of the program in Exhibit 11–1 follows. The RUN NAME card simply gives the SPSS program a name so that it can be easily recognized at some future date. The VARIABLE LIST card names

INTRODUCTION

SPSS

Exhibit 11–1

IBM

FORTRAN CODING FORM

X28-7327-6 U/M 050
Printed in U.S.A.

| PROGRAM | STAT CITY SURVEY | | | GRAPHIC | | | | | PAGE 1 OF 1 |
| PROGRAMMER | HOWARD GITLOW | DATE DEC. 3, 1985 | PUNCHING INSTRUCTIONS | PUNCH | | | | | CARD ELECTRO NUMBER* |

CONTROL FIELD	SPECIFICATION FIELD — FORTRAN STATEMENT	IDENTIFICATION SEQUENCE
RUN NAME	STAT CITY SURVEY OF THE # OF ROOMS # PEOPLE PER DWELLING UNIT	
VARIABLE LIST	ROOMS, PEOPLE, DWELL	
INPUT FORMAT	FIXED(F2.0, F2.0, F1.0)	
N OF CASES	10	
INPUT MEDIUM	CARD	
VAR LABELS	ROOMS, # OF ROOMS IN A HOUSEHOLD AS OF 1-85/	
	PEOPLE, # OF PEOPLE IN A HOUSEHOLD AS OF 1-85/	
	DWELL, TYPE OF DWELLING UNIT	
VALUE LABELS	DWELL (0)APARTMENT (1)HOUSE	
CONDESCRIPTIVE	ROOMS, PEOPLE	
STATISTICS	1, 2, 5, 9, 10, 11	
READ INPUT DATA		
09041		
09051		
08031		
05040		
07031		
10061		
05050		
03020		
08030		
05070		
FREQUENCIES	GENERAL=DWELL	
STATISTICS	4	
FINISH		

the variables that are to be analyzed. The INPUT FORMAT card tells the computer where each piece of data to be analyzed is located so the computer can properly read the data. In this example, the number of rooms in a dwelling unit is located in data card columns 1 and 2, the number of people in a dwelling unit is located in data card columns 3 and 4, and the type of dwelling unit is located in data card column 5. Further, this tells us that there is only one data card per dwelling unit. The N of CASES card indicates the number of data points (questionnaires) to be analyzed. The INPUT MEDIUM card indicates that the data is located on computer cards. The VAR LABELS card gives additional details on the exact definition of each variable under analysis. The VALUE LABELS card verbally defines the numeric codes assigned to nominal or ordinal variables. The CONDESCRIPTIVE and STATISTICS cards indicate that the mean, standard error, standard deviation, range, minimum, and maximum should be computed for ROOMS and PEPLE. The READ INPUT DATA card tells the computer that the data previously defined are about to be read into the computer. The first three lines of data are interpreted as follows:

09041　The first family selected in the sample lives in a *house* (1) with *9 rooms* and is comprised of *4 people*.

09051　The second family selected in the sample lives in a *house* (1) with *9 rooms* and is comprised of *5 people*.

08031　The third family selected in the sample lives in a *house* (1) with *8 rooms* and is comprised of *3 people*.

Please note that the READ INPUT DATA card and the data must immediately follow the first statistical task to be performed. The FREQUENCIES and STATISTICS cards indicate that a frequency distribution and mode should be constructed for the type of dwelling unit. The FINISH card signifies the end of the SPSS program.

The exact meanings of the control cards are explained in the following section of this chapter. Of course, the *SPSS* manual should be consulted for any detailed information that might be needed. The *SPSS* manual can be obtained from your college bookstore or the publisher, McGraw-Hill Book Company (New York, St. Louis, or San Francisco). The authors of the *SPSS* manual are: Norman Nie, C. Hadlai Hull, Jean Jenkins, Karen Steinbrunner, and Dale Brent.

All SPSS programs are composed of three basic parts: (1) a data definition section, (2) a data modification section, and (3) a task definition section.

The data definition section is comprised of SPSS control cards that explain the data being studied. The data definition section of an SPSS program:

1. Names the analysis being performed.
2. Names the variables under investigation.
3. Defines the location of each variable on a data card.
4. Enumerates how many cases (dwelling units) are going to be analyzed.
5. Defines how the data will be read into the computer (cards, tapes, disks, and so on).
6. Affords expanded definitions for the variables under investigation.
7. Gives verbal descriptions of all the categories of nominal or ordinal variables.

The data definition section of an SPSS program used to analyze Stat City data can be seen in Exhibit 11–2.

The data modification section of an SPSS program is comprised of control cards that modify the coding scheme for one or more variables in the data file. There are several types of data modification cards in SPSS.

1. The RECODE card. The RECODE card allows a researcher to replace (recode) any value or set of values for a variable with a new value of his/her choice. For example, the data modification card required to recode average monthly electric bills into the following categories: $10 to less than $40, $40 to less than $70, $70 to less than $100, and $100 to less than $130, is:

```
Card       Card
column 1   column 16
   ↓          ↓
   RECODE    ELEC (10 THRU 39=0) (40 THRU 69=1) (70 THRU 99=2)
             (100 THRU 130=3)
```

A researcher may want to perform the above transformation to obtain the percentage of Stat City families with average monthly electric bills in the above categories. See Example 3–3.

2. The COMPUTE card. The COMPUTE card allows a researcher to assign a variable or value according to some mathematical formula for every case in his/her data file. For example, the following modification card is required if a researcher wanted to transform all Stat City incomes into Z-scores $[Z = (X - \mu)/\sigma]$:

Card   Card
column 1   column 16
  ↓   ↓

COMPUTE   ZINCOM = (INCOM − 29394.572)/22847.118

The above transformation requires knowing that $\mu = \$29{,}394.572$ and $\sigma = \$22{,}847.118$. The above type of transformation is useful in avoiding rounding errors when working with SPSS.

Exhibit 11–2

```
SPSS BATCH SYSTEM

SPSS FOR SPERRY UNIVAC 1100 EXEC 8, VERSION H, RELEASE 8.0-UW1.0, JUNE 1985

   SPACE ALLOCATION..            ALLOWS FOR..    20 TRANSFORMATIONS
   WORKSPACE     4375 WORDS                      82 RECODE VALUES + LAG VARIABLES
   TRANSPACE      625 WORDS                     164 IF/COMPUTE OPERATIONS
```

Assigns a name to the analysis being performed. → 1.

Lists the variables under study. →

Defines the location of each variable on a computer card. →

```
  1.   RUN NAME        STAT CITY SPSS DRIVER PROGRAM
  2.   VARIABLE LIST   NAME1,NAME2,NAME3,NAME4,ADDR1,ADDR2,ADDR3,ADDR4,
  3.                   BLOCK,ID,ZONE,DWELL,HCOST,ASSES,ROOMS,HEAT,ELEC,
  4.                   PHONE,INCOM,PEPLE,CARS,GAS,GASCA,GASTR,REPAR,
  5.                   FAVGA,HOSP,EAT,EATPL,FEAT,LSODA,HSODA,BEER
  6.   INPUT FORMAT    FIXED (1X,4A3,1X,4A4,F3.0,2X,F4.0,1X,F1.0,1X,
  7.                   F1.0,1X,F4.0,1X,F6.0,1X,F2.0,1X,F4.0,1X,
  8.                   F4.0,1X,F3.0,1X,F6.0,1X/F2.0,1X,F1.0,1X,F3.0,1X,
  9.                   F3.0,1X,F2.0,1X,F1.0,1X,F1.0,1X,
 10.                   F2.0,1X,F3.0,1X,F2.0,1X,F1.0,1X,F2.0,
 11.                   1X,F2.0,1X,F2.0)


       ACCORDING TO YOUR INPUT FORMAT, VARIABLES ARE TO BE READ AS FOLLOWS

       VARIABLE   FORMAT   RECORD      COLUMNS

       NAME1      A  3        1         2-   4
       NAME2      A  3        1         5-   7
       NAME3      A  3        1         8-  10
       NAME4      A  3        1        11-  13
       ADDR1      A  4        1        15-  18
       ADDR2      A  4        1        19-  22
       ADDR3      A  4        1        23-  26
       ADDR4      A  4        1        27-  30
       BLOCK      F  3. 0      1        31-  33
       ID         F  4. 0      1        36-  39
       ZONE       F  1. 0      1        41-  41
       DWELL      F  1. 0      1        43-  43
       HCOST      F  4. 0      1        45-  48
       ASSES      F  6. 0      1        50-  55
       ROOMS      F  2. 0      1        57-  58
       HEAT       F  4. 0      1        60-  63
       ELEC       F  4. 0      1        65-  68
       PHONE      F  3. 0      1        70-  72
       INCOM      F  6. 0      1        74-  79
       PEPLE      F  2. 0      2         1-   2
       CARS       F  1. 0      2         4-   4
       GAS        F  3. 0      2         6-   8
       GASCA      F  3. 0      2        10-  12
       GASTR      F  2. 0      2        14-  15
       REPAR      F  1. 0      2        17-  17
       FAVGA      F  1. 0      2        19-  19
       HOSP       F  2. 0      2        21-  22
       EAT        F  3. 0      2        24-  26
       EATPL      F  2. 0      2        28-  29
       FEAT       F  1. 0      2        31-  31
       LSODA      F  2. 0      2        33-  34
       HSODA      F  2. 0      2        36-  37
       BEER       F  2. 0      2        39-  40
```

```
       THE INPUT FORMAT PROVIDES FOR  33 VARIABLES.   33 WILL BE READ
       IT PROVIDES FOR  2 RECORDS ('CARDS') PER CASE.  A MAXIMUM OF  80 'COLUMNS' ARE USED ON A RECORD.
```

States how many cases will be analyzed. →

Defines how the computer will read the data. →

Affords expanded definitions of the variables under study. →

```
 12.   N OF CASES      1373
 13.   INPUT MEDIUM    CARD
 14.   COMMENT         THE ENTIRE STAT CITY DATA BASE WAS COLLECTED VIA
 15.   COMMENT         QUESTIONNAIRE SURVEY DURING FEBRUARY AND MARCH OF
 16.   COMMENT         1985.  THE DATA BASE REFLECTS THE CHARACTERISTICS
 17.   COMMENT         OF STAT CITY DWELLING UNITS AS OF JANUARY 1985.
 18.   COMMENT
 19.   COMMENT
 20.   COMMENT
 21.   COMMENT         THE ELEMENTARY UNIT FOR ANALYSIS IN STAT CITY IS THE
 22.   COMMENT         DWELLING UNIT, NOT THE INDIVIDUAL.  ALL PROBLEMS WILL
 23.   COMMENT         CONSIDER THE DWELLING UNIT (FAMILY OR HOME) AS THE
 24.   COMMENT         BASIC UNIT FOR ANALYSIS.
 25.   COMMENT
 26.   COMMENT
 27.   COMMENT
 28.   VAR LABELS      NAME1,FIRST 4 LETTERS OF LAST NAME/
```

Exhibit 11–2 (*concluded*)

STAT CITY SPSS DRIVER PROGRAM

```
29.                          NAME2,SECOND 4 LETTERS OF LAST NAME/
30.                          NAME3,THIRD 4 LETTERS OF LAST NAME/
31.                          NAME4,FOURTH 4 LETTERS OF LAST NAME/
32.                          ADDR1,FIRST 4 LETTERS OF ADDRESS/
33.                          ADDR2,SECOND 4 LETTERS OF ADDRESS/
34.                          ADDR3,THIRD 4 LETTERS OF ADDRESS/
35.                          ADDR4,FOURTH 4 LETTERS OF ADDRESS/
36.                          BLOCK,BLOCK DWELLING UNIT IS LOCATED ON/
37.                          ID,IDENTIFICATION NUMBER/
38.                          ZONE,RESIDENTIAL HOUSING ZONE/
39.                          DWELL,DWELLING TYPE/
40.                          HCOST,HOUSING COST AS OF 1-85/
41.                          ASSES,ASSESSED VALUE OF HOUSE AS OF 1-85/
42.                          ROOMS,# OF ROOMS IN DWELLING UNIT AS OF 1-85/
43.                          HEAT,AVERAGE YEARLY HEATING BILL AS OF 1-85/
44.                          ELEC,AVERAGE MONTHLY ELECTRIC BILL AS OF 1-85/
45.                          PHONE,AVERAGE MONTHLY PHONE BILL AS OF 1-85/
46.                          INCOM,TOTAL FAMILY INCOME IN 1984/
47.                          PEPLE,NUMBER OF PEOPLE IN HOUSEHOLD AS OF 1-85/
48.                          CARS,NUMBER OF CARS IN HOUSEHOLD AS OF 1-85/
49.                          GAS,AVERAGE BIMONTHLY GAS BILL AS OF 1-85/
50.                          GASCA,AVE. BIMONTHLY GAS PER CAR AS OF 1-85/
51.                          GASTR,AVE. MONTHLY TRIPS FOR GAS AS OF 1-85/
52.                          REPAR,FAVORITE PLACE FOR CAR REPAIRS AS OF 1-85/
53.                          FAVGA,FAVORITE GAS STATION AS OF 1-85/
54.                          HOSP,AVE YRLY HOSP. TRIPS PER HOME AS OF 1-85/
55.                          EAT,AVERAGE WEEKLY FOOD BILL AS OF 1-85/
56.                          EATPL,AVE WKLY FOOD BILL PER PERSON AS OF 1-85/
57.                          FEAT,FAVORITE SUPERMARKET AS OF 1-85/
58.                          LSODA,AVE WKLY DIET SODA PURCHASE AS OF 1-85/
59.                          HSODA,AVE WKLY REG. SODA PURCHASE AS OF 1-85/
60.                          BEER,AVE WKLY BEER PURCHASE AS OF 1-85
61.      COMMENT             LSODA,HSODA, AND BEER PURCHASES ARE IN SIX PACK UNITS.
62.      VALUE LABELS        ZONE (1) ZONE 1 (2) ZONE 2 (3) ZONE 3 (4) ZONE 4/
63.                          DWELL (0)APARTMENT (1)HOUSE/
64.                          REPAR (0)PERFORMS OWN REPAIRS (1)REPAIRS DONE BY STATION
65.                                (2)REPAIRS DONE BY DEALER/
66.                          FAVGA (1)PAUL'S TEXACO STATION (2)HOWIE'S GULF STATION/
67.                          FEAT (1)FOODFAIR (2)GRAND UNION (3)A & P
```

Gives verbal descriptions of all the categories of nominal and ordinal variables.

3. The IF card. The IF card allows a researcher to perform logical operations. The IF card is especially useful in constructing new variables. For example, if a researcher needs a variable to indicate whether a Stat City family purchased beer, the following control card sequence could be used:

```
Card        Card
column 1    column 16
    ↓           ↓
   IF       (BEER EQ 0) BEER1 = 0
   IF       (BEER GT 0) BEER1 = 1
```

The above control card sequence has created a new variable, BEER1, which indicates if a family purchased beer (BEER1 = 1) or not (BEER1 = 0). The above data modification could also have been done with the following RECODE card:

```
Card        Card
column 1    column 16
    ↓           ↓
  RECODE    BEER (0 = 0) (1 THRU HIGHEST = 1)
```

Please note, the RECODE card transforms the variable BEER, while the IF card sequence creates a new variable BEER1 and leaves the variable BEER intact.

4. The SELECT IF card. The SELECT IF card allows a researcher to include only those cases for analysis which pass a test. For example, if a researcher wanted to perform some analysis on only Zone 1 residents, the following card would be used:

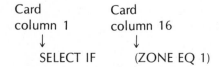

```
Card            Card
column 1        column 16
    ↓               ↓
  SELECT IF     (ZONE EQ 1)
```

The task definition section is comprised of control cards that define the statistical procedures to be performed. There are many statistical procedures that can be easily used via SPSS. However, only four statistical procedures will be discussed in this chapter.

1. FREQUENCIES. The FREQUENCIES control card initiates a statistical program that computes and presents frequency distributions for one or more variables (usually nominal or ordinal variables). The FREQUENCIES card is capable of presenting frequency distributions, means, medians, modes, standard deviations, variances, standard errors, minimums, maximums, and ranges. It is important to remember that not all of the above statistics are meaningful for the types of variables that are commonly analyzed with FREQUENCIES (nominal or ordinal data).

An example of the format of the FREQUENCIES card is:

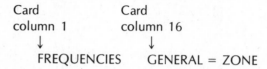

Card
column 1
↓
FREQUENCIES

Card
column 16
↓
GENERAL = ZONE

The above card will cause a frequency distribution to be printed indicating the numbers and percent of Stat City families in each residential housing zone.

An SPSS printout that computes and presents the above frequency distribution can be seen in Exhibit 11–3. Note that the printout through line 67 is as shown in Exhibit 11–2. In subsequent SPSS Exhibits in this chapter, this portion of the printout will not be shown, although you should remember that it is always part of the program.

2. CONDESCRIPTIVE. The CONDESCRIPTIVE control card initiates a statistical program that computes a set of statistics for one or more variables (usually interval or ratio variables). The statistics that are available can be selectively called via the following numbering scheme:

1	Mean	8	Skewness
2	Standard error	9	Range
5	Standard deviation	10	Minimum
6	Variance	11	Maximum
7	Kurtosis		

An example of the format of the CONDESCRIPTIVE card is:

Card
column 1
↓
CONDESCRIPTIVE
STATISTICS

Card
column 16
↓
HOSP
1, 5, 9, 10, 11

The above card sequence will cause the mean, standard deviation, range, minimum, and maximum of HOSP to be computed and printed.

An SPSS printout that computes the above parameters for HOSP, in Zone 1, can be seen in Exhibit 11–4. We have included only Zone 1 dwelling units just to show you how a SELECT IF card is utilized in conjunction with a statistical routine.

Exhibit 11-3

```
SPSS BATCH SYSTEM

SPSS FOR SPERRY UNIVAC 1100 EXEC 8, VERSION H, RELEASE 8.0-UW1.0, JUNE 1979

SPACE ALLOCATION..        ALLOWS FOR..    20 TRANSFORMATIONS
WORKSPACE    4375 WORDS                   82 RECODE VALUES + LAG VARIABLES
TRANSPACE     625 WORDS                  164 IF/COMPUTE OPERATIONS

         1.  RUN NAME        STAT CITY SPSS DRIVER PROGRAM
         2.  VARIABLE LIST   NAME1,NAME2,NAME3,NAME4,ADDR1,ADDR2,ADDR3,ADDR4,
         3.                  BLOCK,ID,ZONE,DWELL,HCOST,ASSES,ROOMS,HEAT,ELEC,
         4.                  PHONE,INCOM,PEPLE,CARS,GAS,GASCA,GASTR,REPAR,
         5.                  FAVGA,HOSP,EAT,EATPL,FEAT,LSODA,HSODA,BEER
         6.  INPUT FORMAT    FIXED (1X,4A3,1X,4A4,F3.0,2X,F4.0,1X,F1.0,1X,
         7.                  F1.0,1X,F4.0,1X,F6.0,1X,F2.0,1X,F4.0,1X,
         8.                  F4.0,1X,F3.0,1X,F6.0,1X/F2.0,1X,F1.0,1X,F3.0,1X,
         9.                  F3.0,1X,F2.0,1X,F1.0,1X,F1.0,1X,
        10.                  F2.0,1X,F3.0,1X,F2.0,1X,F1.0,1X,F2.0,
        11.                  1X,F2.0,1X,F2.0)

         ACCODING TO YOUR INPUT FORMAT, VARIABLES ARE TO BE READ AS FOLLOWS

         VARIABLE   FORMAT  RECORD    COLUMNS

         NAME1      A  3      1        2-    4
         NAME2      A  3      1        5-    7
         NAME3      A  3      1        8-   10
         NAME4      A  3      1       11-   13
         ADDR1      A  4      1       15-   18
         ADDR2      A  4      1       19-   22
         ADDR3      A  4      1       23-   26
         ADDR4      A  4      1       27-   30
         BLOCK      F  3. 0   1       31-   33
         ID         F  4. 0   1       36-   39
         ZONE       F  1. 0   1       41-   41
         DWELL      F  1. 0   1       43-   43
         HCOST      F  4. 0   1       45-   48
         ASSES      F  6. 0   1       50-   55
         ROOMS      F  2. 0   1       57-   58
         HEAT       F  4. 0   1       60-   63
         ELEC       F  4. 0   1       65-   68
         PHONE      F  3. 0   1       70-   72
         INCOM      F  6. 0   1       74-   79
         PEPLE      F  2. 0   2        1-    2
         CARS       F  1. 0   2        4-    4
         GAS        F  3. 0   2        6-    8
         GASCA      F  3. 0   2       10-   12
         GASTR      F  2. 0   2       14-   15
         REPAR      F  1. 0   2       17-   17
         FAVGA      F  1. 0   2       19-   19
         HOSP       F  2. 0   2       21-   22
         EAT        F  3. 0   2       24-   26
         EATPL      F  2. 0   2       28-   29
         FEAT       F  1. 0   2       31-   31
         LSODA      F  2. 0   2       33-   34
         HSODA      F  2. 0   2       36-   37
         BEER       F  2. 0   2       39-   40

THE INPUT FORMAT PROVIDES FOR  33 VARIABLES.   33 WILL BE READ
IT PROVIDES FOR  2 RECORDS ("CARDS") PER CASE.  A MAXIMUM OF   80 "COLUMNS" ARE USED ON A RECORD.

        12.  N OF CASES      1373
        13.  INPUT MEDIUM    CARD
        14.  COMMENT         THE ENTIRE STAT CITY DATA BASE WAS COLLECTED VIA
        15.  COMMENT         QUESTIONNAIRE SURVEY DURING FEBRUARY AND MARCH OF
        16.  COMMENT         1985.  THE DATA BASE REFLECTS THE CHARACTERISTICS
        17.  COMMENT         OF STAT CITY DWELLING UNITS AS OF JANUARY 1985.
        18.  COMMENT
        19.  COMMENT
        20.  COMMENT
        21.  COMMENT         THE ELEMENTARY UNIT FOR ANALYSIS IN STAT CITY IS THE
        22.  COMMENT         DWELLING UNIT, NOT THE INDIVIDUAL. ALL PROBLEMS WILL
        23.  COMMENT         CONSIDER THE DWELLING UNIT (FAMILY OR HOME) AS THE
        24.  COMMENT         BASIC UNIT FOR ANALYSIS.
        25.  COMMENT
        26.  COMMENT
        27.  COMMENT
        28.  VAR LABELS      NAME1,FIRST 4 LETTERS OF LAST NAME/
        29.                  NAME2,SECOND 4 LETTERS OF LAST NAME/
        30.                  NAME3,THIRD 4 LETTERS OF LAST NAME/
        31.                  NAME4,FOURTH 4 LETTERS OF LAST NAME/
        32.                  ADDR1,FIRST 4 LETTERS OF ADDRESS/
        33.                  ADDR2,SECOND 4 LETTERS OF ADDRESS/
        34.                  ADDR3,THIRD 4 LETTERS OF ADDRESS/
        35.                  ADDR4,FOURTH 4 LETTERS OF ADDRESS/
        36.                  BLOCK,BLOCK DWELLING UNIT IS LOCATED ON/
        37.                  ID,IDENTIFICATION NUMBER/
        38.                  ZONE,RESIDENTIAL HOUSING ZONE/
        39.                  DWELL,DWELLING TYPE/
        40.                  HCOST,HOUSING COST AS OF 1-85/
        41.                  ASSES,ASSESSED VALUE OF HOUSE AS OF 1-85/
        42.                  ROOMS,# OF ROOMS IN DWELLING UNIT AS OF 1-85/
        43.                  HEAT,AVERAGE YEARLY HEATING BILL AS OF 1-85/
        44.                  ELEC,AVERAGE MONTHLY ELECTRIC BILL AS OF 1-85/
        45.                  PHONE,AVERAGE MONTHLY PHONE BILL AS OF 1-85/
        46.                  INCOM,TOTAL FAMILY INCOME IN 1984/
        47.                  PEPLE,NUMBER OF PEOPLE IN HOUSEHOLD AS OF 1-85/
        48.                  CARS,NUMBER OF CARS IN HOUSEHOLD AS OF 1-85/
        49.                  GAS,AVERAGE BIMONTHLY GAS BILL AS OF 1-85/
        50.                  GASCA,AVE. BIMONTHLY GAS PER CAR AS OF 1-85/
```

Exhibit 11–3 (*concluded*)

```
STAT CITY SPSS DRIVER PROGRAM
        51.                  GASTR,AVE. MONTHLY TRIPS FOR GAS AS OF 1-85/
        52.                  REPAR,FAVORITE PLACE FOR CAR REPAIRS AS OF 1-85/
        53.                  FAVGA,FAVORITE GAS STATION AS OF 1-85/
        54.                  HOSP,AVE YRLY HOSP. TRIPS PER HOME AS OF 1-85/
        55.                  EAT,AVERAGE WEEKLY FOOD BILL AS OF 1-85/
        56.                  EATPL,AVE WKLY FOOD BILL PER PERSON AS OF 1-85/
        57.                  FEAT,FAVORITE SUPERMARKET AS OF 1-85/
        58.                  LSODA,AVE WKLY DIET SODA PURCHASE AS OF 1-85/
        59.                  HSODA,AVE WKLY REG. SODA PURCHASE AS OF 1-85/
        60.                  BEER,AVE WKLY BEER PURCHASE AS OF 1-85
        61.    COMMENT       LSODA,HSODA, AND BEER PURCHASES ARE IN SIX PACK UNITS.
        62.    VALUE LABELS  ZONE (1) ZONE 1 (2) ZONE 2 (3) ZONE 3 (4) ZONE 4/
        63.                  DWELL (0)APARTMENT (1)HOUSE/
        64.                  REPAR (0)PERFORMS OWN REPAIRS (1)REPAIRS DONE BY STATION
        65.                       (2)REPAIRS DONE BY DEALER/
        66.                  FAVGA (1)PAUL'S TEXACO STATION (2)HOWIE'S GULF STATION/
        67.                  FEAT (1)FOODFAIR (2)GRAND UNION (3)A & P
        68.    FREQUENCIES   GENERAL=ZONE
GIVEN WORKSPACE ALLOWS FOR  2187 VARIABLES WITH   1093 VALUES AND     437 LABELS PER VARIABLE FOR "FREQUENCIES"
        69.    READ INPUT DATA

FILE   NONAME   (CREATION DATE = 05/18/85)

ZONE       RESIDENTIAL HOUSING ZONE
                                        RELATIVE  ADJUSTED    CUM
                              ABSOLUTE     FREQ      FREQ      FREQ
CATEGORY LABEL       CODE       FREQ      (PCT)     (PCT)     (PCT)

ZONE 1                1.         130        9.5       9.5       9.5

ZONE 2                2.         157       11.4      11.4      20.9

ZONE 3                3.         338       24.6      24.6      45.5

ZONE 4                4.         748       54.5      54.5     100.0
                               ------    ------    ------
                     TOTAL      1373      100.0     100.0


VALID CASES    1373      MISSING CASES        0

CPU TIME REQUIRED..    1.84 SECONDS


        70.  FINISH

      NORMAL END OF JOB.
        70 CONTROL CARDS WERE PROCESSED.
         0 ERRORS WERE DETECTED.
```

Exhibit 11–4

```
68.    SELECT IF      (ZONE EQ 1)
69.    CONDESCRIPTIVE HOSP

***** GIVEN WORKSPACE ALLOWS FOR   336 VARIABLES FOR CONDESCRIPTIVE PROBLEM *****

70.    STATISTICS   1,5,9,10,11
71.    READ INPUT DATA

FILE   NONAME   (CREATION DATE = 05/18/85)

VARIABLE  HOSP      AVE YRLY HOSP. TRIPS PER HOME AS OF 1-85
MEAN           3.185            STD DEV        2.141         RANGE        10.000
MINIMUM         .000            MAXIMUM       10.000

VALID OBSERVATIONS -    130                MISSING OBSERVATIONS -      0

TRANSPACE REQUIRED..       30 WORDS
      1 TRANSFORMATIONS
      0 RECODE VALUES + LAG VARIABLES
      3 IF/COMPUTE OPERATIONS

CPU TIME REQUIRED..    1.74 SECONDS


      72.  FINISH
    NORMAL END OF JOB.
      72 CONTROL CARDS WERE PROCESSED.
       0 ERRORS WERE DETECTED.
```

3. REGRESSION. The REGRESSION control card initiates a statistical program that computes and presents repression coefficients, correlation coefficients, coefficients of determination, standard errors of regression coefficients, plus other statistics.

An example of the format of the REGRESSION card is:

Card column 1 ↓	Card column 16 ↓
SAMPLE	100 FROM 1373
LIST CASES	CASES = 1373/VARIABLES = ID, CARS, PEPLE
REGRESSION	VARIABLES = CARS, PEPLE/REGRESSION = CARS WITH PEPLE (2)
STATISTICS	ALL

The above card sequence will draw and print a simple random sample of 100 Stat City dwelling units. Further, a regression model predicting CARS as a function of PEPLE will be computed; a myriad of supporting statistics will also be computed and printed.

An SPSS printout for the above REGRESSION procedure can be seen in Exhibit 11–5.

Exhibit 11-5

```
68.   SAMPLE          100 FROM 1373
69.   LIST CASES      CASES=1373/VARIABLES=ID,CARS,PEPLE
70.   REGRESSION      VARIABLES=CARS,PEPLE/
71.                   REGRESSION=CARS WITH PEPLE(2)
72.   STATISTICS      ALL
```

***** REGRESSION PROBLEM REQUIRES   48 WORDS WORKSPACE, NOT INCLUDING RESIDUALS *****

```
73.   READ INPUT DATA
```

STAT CITY SPSS DRIVER PROGRAM

FILE   NONAME   (CREATION DATE = 05/18/85)

CASE-N	ID	PEPLE	CARS	CASE-N	ID	PEPLE	CARS
1	14.	7.	2.	51	724.	2.	2.
2	98.	2.	2.	52	727.	3.	2.
3	102.	3.	2.	53	760.	8.	1.
4	108.	4.	2.	54	772.	4.	2.
5	214.	2.	1.	55	781.	5.	1.
6	231.	4.	3.	56	785.	4.	2.
7	233.	3.	2.	57	791.	5.	1.
8	238.	4.	2.	58	829.	1.	1.
9	243.	4.	2.	59	836.	4.	1.
10	250.	5.	3.	60	849.	7.	1.
11	263.	4.	2.	61	878.	3.	2.
12	270.	2.	1.	62	928.	4.	2.
13	289.	3.	2.	63	929.	3.	2.
14	305.	4.	2.	64	934.	3.	1.
15	317.	4.	3.	65	936.	2.	2.
16	326.	5.	3.	66	958.	7.	3.
17	349.	6.	4.	67	967.	6.	2.
18	352.	7.	3.	68	978.	6.	4.
19	355.	4.	2.	69	990.	4.	2.
20	358.	6.	4.	70	1008.	4.	3.
21	364.	3.	2.	71	1035.	4.	2.
22	373.	3.	1.	72	1053.	4.	1.
23	375.	5.	3.	73	1071.	5.	1.
24	377.	4.	3.	74	1075.	6.	1.
25	407.	4.	2.	75	1084.	2.	1.
26	431.	4.	3.	76	1097.	2.	2.
27	446.	8.	2.	77	1130.	6.	2.
28	475.	7.	2.	78	1134.	8.	1.
29	494.	6.	4.	79	1161.	6.	2.
30	517.	3.	2.	80	1164.	2.	2.
31	522.	3.	2.	81	1180.	3.	2.
32	529.	5.	3.	82	1190.	3.	2.
33	536.	3.	2.	83	1219.	4.	1.
34	544.	3.	2.	84	1230.	2.	2.
35	549.	5.	3.	85	1244.	3.	1.
36	583.	4.	3.	86	1261.	5.	2.
37	615.	3.	2.	87	1274.	4.	1.
38	627.	6.	2.	88	1290.	6.	1.
39	630.	5.	2.	89	1293.	2.	2.
40	641.	5.	3.	90	1294.	9.	2.
41	651.	4.	2.	91	1299.	4.	2.
42	657.	4.	4.	92	1309.	4.	2.
43	663.	5.	4.	93	1310.	4.	3.
44	665.	3.	2.	94	1312.	2.	2.
45	667.	7.	1.	95	1314.	6.	2.
46	672.	5.	1.	96	1339.	6.	2.
47	676.	5.	1.	97	1340.	4.	1.
48	687.	5.	2.	98	1348.	4.	1.
49	688.	4.	2.	99	1349.	7.	2.
50	715.	2.	2.	100	1367.	5.	1.

Exhibit 11–5 (*concluded*)

```
                                        CORRELATION  COEFFICIENTS
                                        A VALUE OF 99.00000 IS PRINTED
STAT CITY SPSS DRIVER PROGRAM           IF A COEFFICIENT CANNOT BE COMPUTED.
FILE   NONAME   (CREATION DATE = 05/18/85)

VARIABLE           MEAN      STANDARD DEV     CASES          CARS      PEPLE

CARS             2.0100        .8102           100    CARS   1.00000    .12724
PEPLE            4.3300       1.6333           100    PEPLE   .12724   1.00000


* * * * * * * * * * * * * * * * * * *  M U L T I P L E   R E G R E S S I O N  * * * * * * * * * * * * * *   VARIABLE LIST  1
                                                                                                          REGRESSION LIST  1
DEPENDENT VARIABLE..   CARS      NUMBER OF CARS IN HOUSEHOLD AS OF 1-85

VARIABLE(S) ENTERED ON STEP NUMBER  1..    PEPLE     NUMBER OF PEOPLE IN HOUSEHOLD AS OF 1-85

MULTIPLE R        .12724            ANALYSIS OF VARIANCE      DF      SUM OF SQUARES     MEAN SQUARE        F
R SQUARE          .01619            REGRESSION                 1.        1.05217          1.05217        1.61270
ADJUSTED R SQUARE .00615            RESIDUAL                  98.       63.93783           .65243
STANDARD ERROR    .80773

---------------- VARIABLES IN THE EQUATION ------------------       ------------ VARIABLES NOT IN THE EQUATION --------------

VARIABLE        B        BETA     STD ERROR B      F          VARIABLE      BETA IN   PARTIAL   TOLERANCE       F

PEPLE       .6311764-001  .12724    .04970       1.613
(CONSTANT)  .1736701+001

MAXIMUM STEP REACHED
STATISTICS WHICH CANNOT BE COMPUTED ARE PRINTED AS ALL NINES.

                                       SUMMARY TABLE

VARIABLE                                 MULTIPLE R  R SQUARE  RSQ CHANGE  SIMPLE R            B              BETA

PEPLE     NUMBER OF PEOPLE IN HOUSEHOLD AS OF 1-85   .12724     .01619     .01619     .12724    .6311764-001   .12724
(CONSTANT)                                                                                      .1736701+001


TRANSPACE REQUIRED..      30 WORDS
        1 TRANSFORMATIONS
        0 RECODE VALUES + LAG VARIABLES
        1 IF/COMPUTE OPERATIONS

CPU TIME REQUIRED..   1.92 SECONDS


      74.  FINISH
   NORMAL END OF JOB.
   74 CONTROL CARDS WERE PROCESSED.
    0 ERRORS WERE DETECTED.
```

4. ONEWAY. The ONEWAY control card initiates a statistical program that computes and prints a oneway analysis of variance table with supporting statistics.

An example of the format of the ONEWAY card is:

Card column 1 ↓	Card column 16 ↓
SAMPLE	150 FROM 1373
LIST CASES	CASES = 1373/VARIABLES = ID, INCOM, ZONE
ONEWAY	INCOM BY ZONE (1, 4)

The above card sequence will draw and print a simple random sample of 150 Stat City dwelling units. Further, an analysis of variance table will be computed to test if there is a statistically significant difference in mean incomes between the four residential housing zones in Stat City.

An SPSS printout for the above ONEWAY procedure can be seen in Exhibit 11–6.

Exhibit 11–6

```
    68.    SAMPLE           150 FROM 1373
    69.    LIST CASES       CASES=1373/VARIABLES=ID,INCOM,ZONE
    70.    ONEWAY           INCOM BY ZONE(1,4)
```

***** ONEWAY PROBLEM REQUIRES   30 WORDS WORKSPACE *****

```
    71.    READ INPUT DATA
```

STAT CITY SPSS DRIVER PROGRAM

FILE   NONAME   (CREATION DATE = 05/18/85)

CASE-N	ID	ZONE	INCOM	CASE-N	ID	ZONE	INCOM
1	19.	1.	49234.	66	690.	4.	17302.
2	47.	1.	51671.	67	709.	4.	9416.
3	56.	1.	15467.	68	719.	4.	17045.
4	65.	1.	29381.	69	722.	4.	12872.
5	69.	1.	33369.	70	726.	4.	16743.
6	74.	1.	43249.	71	735.	4.	12741.
7	90.	1.	57205.	72	743.	4.	16473.
8	107.	1.	41403.	73	763.	4.	8571.
9	110.	1.	15895.	74	775.	4.	3168.
10	113.	1.	26802.	75	779.	4.	7844.
11	117.	1.	32502.	76	781.	4.	8618.
12	118.	1.	51411.	77	797.	4.	13705.
13	125.	1.	48674.	78	799.	4.	19071.
14	129.	1.	63849.	79	800.	4.	22691.
15	136.	2.	32648.	80	812.	4.	24504.
16	149.	2.	89656.	81	824.	4.	18092.
17	158.	2.	42314.	82	838.	4.	10291.
18	203.	2.	76402.	83	841.	4.	10857.
19	210.	2.	146165.	84	846.	4.	16995.
20	220.	2.	60585.	85	859.	4.	18060.
21	222.	2.	126365.	86	863.	4.	16916.
22	236.	2.	74242.	87	876.	4.	8308.
23	237.	2.	61602.	88	885.	4.	10873.
24	238.	2.	72824.	89	886.	4.	13897.
25	239.	2.	33411.	90	918.	4.	6305.
26	241.	2.	47650.	91	935.	4.	2574.
27	248.	2.	83195.	92	950.	4.	6162.
28	255.	2.	129329.	93	959.	4.	24312.
29	264.	2.	43708.	94	994.	4.	34964.
30	272.	2.	66364.	95	995.	4.	23331.
31	281.	2.	109434.	96	996.	4.	18157.
32	293.	3.	15318.	97	997.	4.	17201.
33	338.	3.	33850.	98	1009.	4.	19477.
34	341.	3.	21742.	99	1015.	4.	18300.
35	356.	3.	25999.	100	1017.	4.	10581.
36	375.	3.	31998.	101	1018.	4.	14985.
37	394.	3.	29338.	102	1019.	4.	15404.
38	399.	3.	37679.	103	1021.	4.	14160.
39	403.	3.	31758.	104	1035.	4.	17928.
40	409.	3.	35005.	105	1040.	4.	13837.
41	451.	3.	24255.	106	1041.	4.	10987.
42	454.	3.	37265.	107	1051.	4.	12125.
43	462.	3.	42231.	108	1055.	4.	13371.
44	472.	3.	31856.	109	1060.	4.	7288.
45	490.	3.	12976.	110	1087.	4.	19207.
46	492.	3.	28104.	111	1105.	4.	25132.
47	510.	3.	32901.	112	1107.	4.	31397.
48	518.	3.	32458.	113	1108.	4.	19621.
49	530.	3.	45583.	114	1116.	4.	34432.
50	536.	3.	35394.	115	1125.	4.	17724.
51	554.	3.	44108.	116	1144.	4.	12597.
52	559.	3.	39167.	117	1145.	4.	13043.
53	561.	3.	43705.	118	1152.	4.	11171.
54	562.	3.	43309.	119	1159.	4.	18141.
55	563.	3.	21226.	120	1183.	4.	14584.
56	570.	3.	44546.	121	1184.	4.	6549.
57	603.	3.	24343.	122	1196.	4.	5985.
58	607.	3.	54820.	123	1199.	4.	8189.
59	610.	3.	29398.	124	1207.	4.	12451.
60	635.	4.	20943.	125	1210.	4.	16986.
61	649.	4.	18425.	126	1212.	4.	22910.
62	650.	4.	25052.	127	1219.	4.	19757.
63	651.	4.	27781.	128	1231.	4.	19576.
64	659.	4.	28077.	129	1232.	4.	14331.
65	667.	4.	8495.	130	1254.	4.	13827.

Exhibit 11–6 (*concluded*)

```
STAT CITY SPSS DRIVER PROGRAM
FILE   NONAME   (CREATION DATE = 05/18/85)
     CASE-N       IC      ZONE       INCOM
       131      1262.       4.      11557.
       132      1266.       4.      20685.
       133      1268.       4.      24872.
       134      1270.       4.       9179.
       135      1273.       4.       6798.
       136      1276.       4.       3342.
       137      1291.       4.      14120.
       138      1294.       4.      14656.
       139      1296.       4.      20703.
       140      1313.       4.       6971.
       141      1318.       4.      14303.
       142      1323.       4.      19772.
       143      1325.       4.      13694.
       144      1328.       4.      18497.
       145      1333.       4.       8807.
       146      1336.       4.      10960.
       147      1354.       4.       8312.
       148      1356.       4.       8544.
       149      1365.       4.       9715.
       150      1372.       4.       7732.
```

- O N E W A Y -

VARIABLE INCOM   TOTAL FAMILY INCOME IN 1984

ANALYSIS OF VARIANCE

SOURCE	D.F.	SUM OF SQUARES	MEAN SQUARES	F RATIO	F PROB.
BETWEEN GROUPS	3	57408403349.3086	19136134400.0000	98.604	.0000
WITHIN GROUPS	146	28334333632.0000	194070778.0000		
TOTAL	149	85742736384.0000			

```
TRANSPACE REQUIRED..      30 WORDS
          1 TRANSFORMATIONS
          0 RECODE VALUES + LAG VARIABLES
          1 IF/COMPUTE OPERATIONS

CPU TIME REQUIRED..    2.02 SECONDS


       72.  FINISH
     NORMAL END OF JOB.
       72 CONTROL CARDS WERE PROCESSED.
        0 ERRORS WERE DETECTED.
```

One final note on data modification and task definition control cards is that multiple statistical procedures can be run in one SPSS program. If an asterisk (*) appears before a data modification card, the data modification is only in effect for the statistical procedure that immediately follows the card. For example, if a researcher wanted to compute the average income in Stat City and in each zone of Stat City, the following task definition card sequence would be used:

Card column 1 ↓	Card column 16 ↓
CONDESCRIPTIVE	INCOM
STATISTICS	1
READ INPUT DATA	

```
*SELECT IF          (ZONE EQ 1)
CONDESCRIPTIVE      INCOM
STATISTICS          1
*SELECT IF          (ZONE EQ 2)
CONDESCRIPTIVE      INCOM
STATISTICS          1
*SELECT IF          (ZONE EQ 3)
CONDESCRIPTIVE      INCOM
STATISTICS          1
*SELECT IF          (ZONE EQ 4)
CONDESCRIPTIVE      INCOM
STATISTICS          1
FINISH
```

An SPSS printout for the above problem can be seen in Exhibit 11–7.

Exhibit 11–7

```
        68.   CONDESCRIPTIVE INCOM

***** GIVEN WORKSPACE ALLOWS FOR  336 VARIABLES FOR CONDESCRIPTIVE PROBLEM *****

        69.   STATISTICS      1
        70.   READ INPUT DATA              4

VARIABLE   INCOM     TOTAL FAMILY INCOME IN 1989
MEAN        29394.572

VALID OBSERVATIONS -    1373               MISSING OBSERVATIONS -      0

CPU TIME REQUIRED..    2.10 SECONDS


        71.   *SELECT IF     (ZONE EQ 1)
        72.   CONDESCRIPTIVE INCOM

 ***** GIVEN WORKSPACE ALLOWS FOR  336 VARIABLES FOR CONDESCRIPTIVE PROBLEM *****

        73.   STATISTICS      1

VARIABLE   INCOM     TOTAL FAMILY INCOME IN 1984
MEAN        39492.608

VALID OBSERVATIONS -     130               MISSING OBSERVATIONS -      0

TRANSPACE REQUIRED..       30 WORDS
        1 TRANSFORMATIONS
        0 RECODE VALUES + LAG VARIABLES
        3 IF/COMPUTE OPERATIONS

CPU TIME REQUIRED..     .23 SECONDS


        74.   *SELECT IF     (ZONE EQ 2)
        75.   CONDESCRIPTIVE INCOM
```

Exhibit 11-7 (*concluded*)

```
***** GIVEN WORKSPACE ALLOWS FOR  336 VARIABLES FOR CONDESCRIPTIVE PROBLEM *****

        76.   STATISTICS      1

VARIABLE  INCOM      TOTAL FAMILY INCOME IN 1984
MEAN        74483.745
VALID OBSERVATIONS -    157                    MISSING OBSERVATIONS -        0

STAT CITY SPSS DRIVER PROGRAM

TRANSPACE REQUIRED..      30 WORDS
      1 TRANSFORMATIONS
      0 RECODE VALUES + LAG VARIABLES
      3 IF/COMPUTE OPERATIONS
CPU TIME REQUIRED..      .24 SECONDS


        77.   *SELECT IF      (ZONE EQ 3)
        78.   CONDESCRIPTIVE INCOM


***** GIVEN WORKSPACE ALLOWS FOR  336 VARIABLES FOR CONDESCRIPTIVE PROBLEM *****

        79.   STATISTICS      1

FILE   NONAME   (CREATION DATE = 05/18/85)

VARIABLE  INCOM      TOTAL FAMILY INCOME IN 1984
MEAN        35438.234
VALID OBSERVATIONS -    338                    MISSING OBSERVATIONS -        0

TRANSPACE REQUIRED..      30 WORDS
      1 TRANSFORMATIONS
      0 RECODE VALUES + LAG VARIABLES
      3 IF/COMPUTE OPERATIONS
CPU TIME REQUIRED..      .30 SECONDS


        80.   *SELECT IF      (ZONE EQ 4)
        81.   CONDESCRIPTIVE INCOM


***** GIVEN WORKSPACE ALLOWS FOR  336 VARIABLES FOR CONDESCRIPTIVE PROBLEM *****

        82.   STATISTICS      1

VARIABLE  INCOM      TOTAL FAMILY INCOME IN 1984
MEAN        15444.702
VALID OBSERVATIONS -    748                    MISSING OBSERVATIONS -        0

TRANSPACE REQUIRED..      30 WORDS
      1 TRANSFORMATIONS
      0 RECODE VALUES + LAG VARIABLES
      3 IF/COMPUTE OPERATIONS
CPU TIME REQUIRED..      .45 SECONDS


        83.   FINISH
     NORMAL END OF JOB.
        83 CONTROL CARDS WERE PROCESSED.
         0 ERRORS WERE DETECTED.
```

MINITAB   Minitab is an easy-to-use, flexible, and powerful statistical computing system. It is designed especially for students and researchers who have had no previous experience with computers. Both batch and interactive modes are available, although your institution may be able to accommodate only one of these modes. Minitab runs just as well either way.

A system is said to be running in a "batch" mode if a program or data set is put onto cards (or typed on a terminal) and then submitted as an entity for processing by a computer. A system is in an "interactive" mode if it carries out each command as it is entered.

Minitab has the capability of performing many sophisticated statistical calculations and analyses. Only a few very basic operations will be considered here. For more detailed information, you should consult the *Minitab Student Handbook,* which can be obtained from your college bookstore or the publisher, Duxbury Press (Boston). The authors of the manual are: Thomas Ryan, Brian Joiner, and Barbara Ryan.

Minitab consists of a worksheet of rows and columns in which data is stored. About 150 commands will instruct the system to perform various operations on the stored data.

1. The READ command. Generally, Minitab programs begin with a READ statement. The computer is told to read data, to be given in the next statements, and put it into tabular form. For example, the input data below would be entered in the following way in Minitab.

READ THE FOLLOWING DATA INTO COLUMNS C1 C2 AND C3

9	4	1
9	5	1
8	3	1
5	4	0
7	3	1
10	6	1
5	5	0
3	2	0
8	3	0
5	7	0

END

The READ statement is one of Minitab's 150 commands. It has told the computer to enter the above 10 lines of data into Minitab's first three columns.

Minitab has a dictionary of command names it recognizes, but each command must be on a separate line. Commands and data must always appear on separate lines, but values may be entered anywhere on a line so long as they are in the correct order. Only the commands are interpreted by the computer; all other text is for the benefit of the user only. Thus, the first command could have been written as:

READ DATA IN COLUMNS C1 C2 AND C3

or just

READ C1 C2 C3

A Minitab program which illustrates some of the fundamental principles and techniques of data handling is shown in Exhibit 11–8, in which the number of rooms per person is computed. This program was executed interactively. Once you enter your computer system and get into the Minitab program, the computer will respond with MTB>, after which you enter the command you want the computer to execute (for example, READ data, NAME variables, PRINT data, etc.). If you execute the READ command, the computer will request data via DATA>. Once you have entered all your data, you tell the machine so by entering END.

Exhibit 11–8

```
MTB > READ C1 C2 C3
DATA> 9  4  1
DATA> 9  5  1
DATA> 8  3  1
DATA> 5  4  0
DATA> 7  3  1
DATA> 10 6  1
DATA> 5  5  0
DATA> 3  2  0
DATA> 8  3  0
DATA> 5  7  0
DATA> END
     10 ROWS READ
MTB > NAME C1 = 'ROOMS'
MTB > NAME C2 = 'PEPLE'
MTB > NAME C3 = 'DWELL'

MTB > LET C4 = C1/C2

MTB > NAME C4= 'RMS/PPLE'
MTB > PRINT C1-C4
 ROW  ROOMS  PEPLE  DWELL  RMS/PPLE

  1     9      4      1    2.25000
  2     9      5      1    1.80000
  3     8      3      1    2.66667
  4     5      4      0    1.25000
  5     7      3      1    2.33333
  6    10      6      1    1.66667
  7     5      5      0    1.00000
  8     3      2      0    1.50000
  9     8      3      0    2.66667
 10     5      7      0    0.71429
```

END command signifies termination of data set.

NAME command allows user to give each column (variable) a nickname. The nickname must be inside quotation marks.

LET command allows the user to perform basic arithmetic computations (for example, creating a new variable by pre-existing variables).

PRINT command creates a listing of the information in the specified columns.

2. The SAMPLE command. Random sample selection can be accomplished using the SAMPLE statement:

SAMPLE 100 ROWS FROM C1 − C7, PUT INTO C21 − C27

Minitab then randomly selects 100 households, without replacement, from the complete data base. The data contained in columns 1 through 7 for the 100 selected households would then be stored in columns 21 through 27.

For purposes of illustration and comparison, instead of entering (using the READ command) the entire Stat City data base and generating a random sample from that (using the SAMPLE command), the sample of 100 data points which you used for the SPSS regression (see Exhibit 11–5) has been entered manually (using the READ command) and will be used in several Minitab operations. Exhibit 11–9 shows these 100 values for ID number, zone, type of dwelling unit, income, number of people, number of cars, and hospital visits.

Exhibit 11–9

ROW	ID	ZONE	DWELL	INCOM	PEPLE	CARS	HOSP
1	14	1	1	28180	7	2	7
2	98	1	1	57092	2	2	1
3	102	1	1	35399	3	2	4
4	108	1	1	40120	4	2	5
5	214	2	1	67797	2	1	1
6	231	2	1	56476	4	3	3
7	233	2	1	70074	3	2	1
8	238	2	1	72824	4	2	1
9	243	2	1	46337	4	2	4
10	250	2	1	76231	5	3	4
11	263	2	1	70732	4	2	3
12	270	2	1	54379	2	1	1
13	289	3	1	17624	3	2	1
14	305	3	1	17617	4	2	1
15	317	3	1	42701	4	3	3
16	326	3	1	25287	5	3	3
17	349	3	1	39124	6	4	8
18	352	3	1	45473	7	3	3
19	355	3	1	23584	4	2	7
20	358	3	1	21129	6	4	4
21	364	3	1	25067	3	2	4
22	373	3	1	16222	3	1	4
23	375	3	1	31998	5	3	4
24	377	3	1	21821	4	3	1
25	407	3	0	30951	4	2	2
26	431	3	0	32195	4	3	4
27	446	3	0	30436	8	2	4
28	475	3	0	42768	7	2	3
29	494	3	0	33673	6	4	3
30	517	3	0	37596	3	2	3
31	522	3	1	29644	3	2	0
32	528	3	1	41557	5	3	5
33	536	3	1	35394	3	2	4
34	544	3	1	40988	3	2	4
35	549	3	1	44364	5	3	5
36	583	3	1	41420	4	3	3

37	615	3	1	57483	3	2	2
38	627	4	0	13735	6	2	3
39	630	4	0	18640	5	2	5
40	641	4	0	30195	5	3	5
41	651	4	0	27781	4	2	8
42	657	4	0	28604	4	4	1
43	663	4	0	27962	5	4	4
44	665	4	0	21450	3	2	3
45	667	4	0	8495	7	1	0
46	672	4	0	11480	5	1	1
47	676	4	0	12755	5	1	5
48	687	4	0	14967	5	1	4
49	688	4	0	16856	4	2	4
50	715	4	0	18864	2	2	1
51	724	4	0	17943	2	2	1
52	727	4	0	15458	3	2	4
53	760	4	0	10268	8	1	8
54	772	4	0	4322	4	2	1
55	781	4	0	8618	5	1	4
56	785	4	0	9765	4	2	4
57	791	4	0	12185	5	1	1
58	829	4	0	24567	1	1	2
59	836	4	0	14572	4	1	1
60	849	4	0	14868	7	1	3
61	878	4	0	13466	3	2	1
62	928	4	0	7711	4	2	2
63	929	4	0	11873	3	2	1
64	934	4	0	3509	3	1	1
65	936	4	0	4842	2	2	1
66	958	4	0	26105	7	3	2
67	967	4	0	26820	6	2	4
68	978	4	0	24772	6	4	6
69	990	4	0	19681	4	2	3
70	1008	4	0	20216	4	3	5
71	1035	4	0	17928	4	2	2
72	1053	4	0	12661	4	1	4
73	1071	4	0	5813	5	1	4
74	1075	4	0	5175	6	1	4
75	1084	4	0	49410	2	1	2
76	1097	4	0	19691	2	2	2
77	1130	4	0	19307	6	2	6
78	1134	4	0	17395	8	1	6
79	1161	4	0	11075	6	2	0
80	1164	4	0	7369	2	2	1
81	1180	4	0	11156	3	2	5
82	1190	4	0	8126	3	2	1
83	1219	4	0	19757	4	1	5
84	1230	4	0	11648	2	2	3
85	1244	4	0	16095	3	1	2
86	1261	4	0	10925	5	2	3
87	1274	4	0	11148	4	1	2
88	1290	4	0	16555	6	1	3
89	1293	4	0	13723	2	2	2
90	1294	4	0	14656	9	2	9
91	1299	4	0	29004	4	2	2
92	1309	4	0	24029	4	2	1
93	1310	4	0	28180	4	3	5
94	1312	4	0	11936	2	2	3
95	1314	4	0	14298	6	2	1
96	1339	4	0	12668	6	2	0
97	1340	4	0	10032	4	1	2
98	1348	4	0	19695	4	1	1
99	1349	4	0	22455	7	2	9
100	1367	4	0	5503	5	1	0

Exhibit 11–9 (*concluded*)

3. The DESCRIBE command. DESCRIBE, followed by one or more column numbers, will produce 10 statistics for the data in each of those columns. These statistics are: the number of data points, n; the mean; the median; a trimmed mean; the standard deviation; the standard error of the mean (the standard deviation divided by the square root of n); the maximum; the minimum; the third quartile; and the first quartile. Exhibit 11–10 shows this output for your 100 data points.

Exhibit 11–10

```
MTB > DESCRIBE C1 - C7

              ID      ZONE     DWELL     INCOM     PEPLE      CARS      HOSP
N            100       100       100       100       100       100       100
MEAN         758     3.470     0.310     25225      4.33     2.010      3.09
MEDIAN       720     4.000     0.000     19726      4.00     2.000      3.00
TMEAN        762     3.567     0.289     23795      4.27     1.956      2.97
STDEV        379     0.810     0.465     16741      1.63     0.810      2.06
SEMEAN        38     0.081     0.046      1674      0.16     0.081      0.21
MAX         1367     4.000     1.000     76231      9.00     4.000      9.00
MIN           14     1.000     0.000      3509      1.00     1.000      0.00
Q3          1094     4.000     1.000     32146      5.00     2.000      4.00
Q1           413     3.000     0.000     12690      3.00     1.000      1.00
```

4. The HISTOGRAM command. Minitab has many graphical capabilities. Exhibit 11–11 shows a histogram for the variable INCOM using the command HISTOGRAM followed by the appropriate column number. You should note that Minitab selects convenient interval midpoints and outputs the numerical frequencies as well as their graphical depiction. Other options, such as interval selection, are available and are described in the Minitab manual.

Exhibit 11–11

```
MTB > HISTOGRAM 'INCOM'

INCOM

MIDDLE OF      NUMBER OF
INTERVAL       OBSERVATIONS
      0         3    ***
  10000        31    *******************************
  20000        26    **************************
  30000        17    *****************
  40000        11    ***********
  50000         4    ****
  60000         3    ***
  70000         4    ****
  80000         1    *
```

5. The REGRESS command. Regression of one variable on another using the REGRESS statement will compute and output regression coefficients, correlation coefficients, standard deviations of the regression coefficients, and many other important statistics, just as SPSS does. Exhibit 11–12 shows the output from

REGRESS 'CARS' ON 1 PREDICTOR 'PEPLE'

This corresponds to the SPSS illustration in Exhibit 11–5 and, of course, yields the same regression equation.

```
MTB > REGRESS 'CARS' ON 1 PREDICTOR VARIABLE 'PEPLE'          Exhibit 11-12

THE REGRESSION EQUATION IS
CARS = 1.74 + 0.0631 PEPLE

                                 ST. DEV.     T-RATIO =
COLUMN         COEFFICIENT       OF COEF.     COEF/S.D.
                  1.7367          0.2299        7.56
PEPLE             0.06312         0.04970       1.27

S = 0.8077       ◄──────────────────────────  Standard error
                                               of the estimate

R-SQUARED =   1.6 PERCENT
R-SQUARED =   0.6 PERCENT, ADJUSTED FOR D.F.

ANALYSIS OF VARIANCE

 DUE TO      DF          SS         MS=SS/DF
 REGRESSION   1        1.0522        1.0522
 RESIDUAL    98       63.9378        0.6524
 TOTAL       99       64.9900
```

6. The CHOOSE command. Sometimes you want to group data by a particular attribute for analysis. For example, if you wanted to compare the number of cars per household by zone, you might want to generate a separate list of all the values for CARS in Zone 1, in Zone 2, in Zone 3, and in Zone 4. The CHOOSE command will select those households from each zone of interest, place the zone number in a new column, and place the corresponding number of cars in another new column. In Exhibit 11–13, the sample of 100 households is regrouped by zone, with the corresponding number of cars recorded in a new column. C2 is the column which contains the Stat City variable ZONE; C6 is the column which contains the Stat City variable CARS. The first CHOOSE command in Exhibit 11–13 selects all families in Zone 1 and places the Zone number (1) in column 10 (C10) and the corresponding number of cars in column 11 (C11). Thus, column C10 contains all 1s, indicating that each entry is in Zone 1; column C11 shows the number of cars in each of those households; column C12 contains all 2s, indicating that each entry is in Zone 2; column C13 shows the number of cars in each of those households, and so on.

Exhibit 11–13

```
MTB > CHOOSE 1 IN C2, CORRESPONDING ROWS OF C6 INTO C10, C11
MTB > CHOOSE 2 IN C2, CORRESPONDING ROWS OF C6 INTO C12, C13
MTB > CHOOSE 3 IN C2, CORRESPONDING ROWS OF C6 INTO C14, C15
MTB > CHOOSE 4 IN C2, CORRESPONDING ROWS OF C6 INTO C16, C17
MTB > PRINT C11, C13, C15, C17
```

ROW	C11	C13	C15	C17
1	2	1	2	2
2	2	3	2	2
3	2	2	3	3
4	2	2	3	2
5		2	4	4
6		3	3	4
7		2	2	2
8		1	4	1
9			2	1
10			1	1
11			3	1
12			3	2
13			2	2
14			3	2
15			2	2
16			2	1
17			4	2
18			2	1
19			2	2
20			3	1
21			2	1
22			2	1
23			3	1
24			3	2
25			2	2
26				2
27				1
28				2
29				3
30				2
31				4
32				2
33				3
34				2
35				1
36				1
37				1
38				1
39				2
40				2
41				1
42				2
43				2
44				2
45				2
46				1
47				2
48				1
49				2
50				1
51				1
52				2
53				2
54				2
55				2
56				3
57				2
58				2
59				2
60				1
61				1
62				2
63				1

7. The AOVONEWAY command. As you know, one-way analysis of variance is a technique for comparing the means of different groups. You saw in Exhibit 11–13 how the CHOOSE command was used to separate the sample of 100 households into four groups by zone, carrying along the number of cars with each entry. The AOVONEWAY command, followed by a column number, can be used to perform a one-way analysis of variance on groups such as these. Exhibit 11–14 illustrates this techniques for the number of cars per household for the zone groups created in Exhibit 11–13, where these values were placed in columns C11, C13, C15, and C17 using the CHOOSE command.

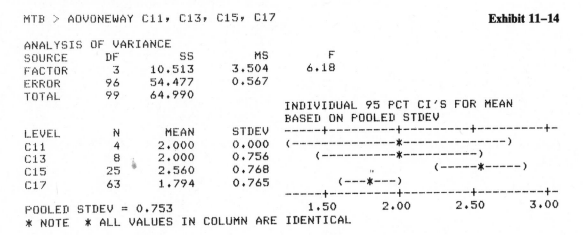

Exhibit 11–14

```
MTB > AOVONEWAY C11, C13, C15, C17

ANALYSIS OF VARIANCE
SOURCE      DF        SS        MS        F
FACTOR       3    10.513     3.504     6.18
ERROR       96    54.477     0.567
TOTAL       99    64.990
                                    INDIVIDUAL 95 PCT CI'S FOR MEAN
                                    BASED ON POOLED STDEV
                                    -----+---------+---------+---------+-
LEVEL        N      MEAN     STDEV
C11          4     2.000     0.000   (------------------*----------------)
C13          8     2.000     0.756        (-----------*-----------)
C15         25     2.560     0.768                     (-----*-----)
C17         63     1.794     0.765        (---*---)
                                    -----+---------+---------+---------+-
POOLED STDEV = 0.753                1.50      2.00      2.50      3.00
* NOTE  * ALL VALUES IN COLUMN ARE IDENTICAL
```

Please note that the assumption of equal group standard deviations was violated in the above example. Consequently, it would be dangerous to interpret the *F*-statistic.

Both SPSS and Minitab are "user-friendly" systems which allow you to practically speak English to a computer once you understand some of the basic rules. Packages like this are extremely valuable for performing the statistical analyses you have learned about in this book.

APPENDIX A   REFERENCE EXHIBITS

Number	Variable name	Variable description
1	NAME	Last name of the head of the household
2	ADDR	Street address
3	BLOCK	Block location of dwelling unit
4	ID	Dwelling unit's identification code
5	ZONE	Residential housing zone
6	DWELL	Type of dwelling unit 0 = Apartment 1 = House
7	HCOST	Housing cost as of January 1985 Rent if apartment Mortgage if house
8	ASST	Assessed value of home ($0.00 if apartment) as of January 1985
9	ROOMS	Numbers of rooms in dwelling unit as of January 1985
10	HEAT	Average total yearly heating bill as of January 1985 (includes all types of heat—electric, gas, etc.)
11	ELEC	Average monthly electric bill as of January 1985
12	PHONE	Average monthly telephone bill as of January 1985
13	INCOM	Total family income for 1984
14	PEPLE	Number of people in household as of January 1985
15	CARS	Number of cars in household as of January 1985
16	GAS	Average bimonthly automobile gas bill as of January 1985
17	GASCA	Average bimonthly automobile gas bill per car as of January 1985
18	GASTR	Average number of trips to the gas station per month as of January 1985
19	REPAR	Favorite place to have automotive repairs performed as of January 1985 0 = Performs own repairs 1 = Repairs done by service station 2 = Repairs done by dealer
20	FAVGA	Favorite gas station as of January 1985 1 = Paul's Texaco (11th St. & 7th Ave.) 2 = Howie's Gulf (11th St. & Division St.)
21	HOSP	Average yearly trips to the hospital by all members of a dwelling unit as of January 1985
22	EAT	Average weekly supermarket bill as of January 1985
23	EATPL	Average weekly supermarket bill per person as of January 1985
24	FEAT	Favorite supermarket as of January 1985 1 = Food Fair 2 = Grand Union 3 = A&P
25	LSODA	Average weekly purchase of six-packs of diet soda as of January 1985
26	HSODA	Average weekly purchase of six-packs of regular soda as of January 1985
27	BEER	Average weekly purchase of six-packs of beer as of January 1985

STAT CITY QUESTIONNAIRE
(mailed JANUARY 1985)
(To be filled out by the head of the household!)

1. Name of head of household: _____

2. Local address: _____
   _____

3. Block: _____

4. Identification code: _____
   (Assigned by Census Bureau)

5. Zone: _____

6. What type of building do you live in?
   Apartment building [0]   House [1]

7. If you live in an apartment, what is your monthly rent?
   $ _____

8. If you live in a house:
   a. What is your monthly mortgage? $ _____
   b. What is the assessed value of your house? $ _____

9. How many rooms are in your home (count all rooms including bathrooms)? _____
   _____

10. What is your average total yearly heating bill (lump all types of heat used—gas, oil, or electric)? $ _____

11. What is your year-round average monthly electric bill?
    $ _____

12. What is your year-round average monthly telephone bill?
    $ _____

13. What was your gross yearly family income in 1984? $ _____

14. How many people live in your household (include yourself)? _____

15. How many automobiles are registered at your household? _____

16. What is the year-round bimonthly gasoline bill for the automobiles registered at your household? _____

17. What is the year-round average monthly number of trips to a gas station the automobiles registered at your household make? _____

18. In general, where are simple repairs and maintenance performed on *your* automobiles?
    Perform own repairs [0]   Repairs done by service station [1]   Repairs done by dealer [2]

19. Which gas station gets more of *your* business?
    a. Pauls' Texaco
         11th St. and 7th Avenue   [station 1]
    b. Howie's Gulf
         11th St. and Division St.   [station 2]

20. What is the yearly average number of trips to the hospital made by all members of your household? _____

21. What is your year-round average weekly supermarket bill? $ _____

22. Which supermarket gets more of *your* business?
    a. Food Fair
         Park St. between 8th and 9th Ave.   [supermarket 1]
    b. Grand Union
         9th St. and 9th Ave.   [supermarket 2]
    c. A&P
         E. Division St. and 5th Ave.   [supermarket 3]

23. What is the year-round average weekly purchase of six-packs of diet soda for your household? _____ six-packs

24. What is the year-round average weekly purchase of six-packs of regular soda for your household? _____ six-packs

25. What is the year-round average weekly purchase of six-packs of beer for your household? _____ six-packs

PLEASE ENCLOSE YOUR COMPLETED QUESTIONNAIRE IN THE STAMPED, ADDRESSED ENVELOPE AND MAIL TO:
STAT CITY CHAMBER OF COMMERCE
CITY HALL
E. DIVISION ST. AND 7TH AVE.
STAT CITY
NO LATER THAN February 25, 1985.
THANK YOU FOR YOUR COOPERATION.

5347	8111	9803	1221	5952	4023	4057	3935	4321	6925
9734	7032	5811	9196	2624	4464	8328	9739	9282	7757
6602	3827	7452	7111	8489	1395	9889	9231	6578	5964
9977	7572	0317	4311	8308	8198	1453	2616	2489	2055
3017	4897	9215	3841	4243	2663	8390	4472	6921	6911
8187	8333	1498	9993	1321	3017	4796	9379	8669	9885
1983	9063	7186	9505	5553	6090	8410	5534	4847	6379
0933	3343	5386	5276	1880	2582	9619	6651	7831	9701
3115	5829	4082	4133	2109	9388	4919	4487	4718	8142
6761	5251	0303	8169	1710	6498	6083	8531	4781	0807
6194	4879	1160	8304	2225	1183	0434	9554	2036	5593
0481	6489	9634	7906	2699	4396	6348	9357	8075	9658
0576	3960	5614	2551	8615	7865	0218	2971	0433	1567
7326	5687	4079	1394	9628	9018	4711	6680	6184	4468
5490	0997	7658	0264	3579	4453	6442	3544	2831	9900
4258	3633	6006	0404	2967	1634	4859	2554	6317	7522
2726	2740	9752	2333	3645	3369	2367	4588	4151	0475
4984	1144	6668	3605	3200	7860	3692	5996	6819	6258
2931	4046	2707	6923	5142	5851	4992	0390	2659	3306
3046	2785	6779	1683	7427	0579	0290	6349	0078	3509
2870	8408	6553	4425	3386	8253	9839	2638	0283	3683
1318	5065	9487	2825	7854	5528	3359	6196	5172	1421
6079	7663	3015	4029	9947	2833	1536	4248	6031	4277
1348	4691	6468	0741	7784	0190	4779	6579	4423	7723
3491	9450	3937	3418	5750	2251	0406	9451	4461	1048
2810	0481	8517	8649	3569	0348	5731	6317	7190	7118
5923	4502	0117	0884	8192	7149	9540	3404	0485	6591
8743	8275	7109	3683	5358	2598	4600	4284	8168	2145
2904	0130	5534	6573	7871	4364	4624	5320	9486	4871
6203	7188	9450	1526	6143	1036	4205	6825	1438	7943
3885	8004	5997	7336	5287	4767	4102	8229	2643	8737
4066	4332	8737	8641	9584	2559	5413	9418	4230	0736
4058	9008	3772	0866	3725	2031	5331	5098	3290	3209
7823	8655	5027	2043	0024	0230	7102	4993	2324	0086
9824	6747	7145	6954	0116	0332	6701	9254	9797	5272
6997	7855	6543	3262	2831	6181	1459	7972	5569	9134
3984	2307	4081	0371	2189	9635	9680	2459	2620	2600
6288	8727	9989	9996	3437	4255	1167	9960	9801	4886
5613	6492	2945	5296	8662	6242	3016	7618	9531	3926
9080	5602	4899	6456	6746	6018	1297	0384	6258	9385
0966	4467	7476	3335	6730	8054	9765	1134	7877	4501
3475	5040	7663	1276	3222	3454	1810	5351	1452	7212
1215	7332	7419	2666	7808	5363	5230	0000	0570	6353
6938	0773	9445	7642	1612	0930	6741	6858	8793	3884
9335	6456	4376	4504	4493	6997	1696	0827	6775	6029
3887	3554	9956	8540	0491	6254	7840	0101	8618	2207
5831	6029	7239	6966	1247	9305	0205	2980	6364	1279
8356	1022	9947	7472	2207	1023	2157	2032	2131	5712
2806	9115	4056	3370	6451	0706	6437	2633	7965	3114
0573	7555	9316	8092	5587	5410	3480	8315	0453	8136

Exhibit A–3   2,500 four-digit random numbers

Exhibit A–3 (*continued*)

2668	7422	4354	4569	9446	8212	3737	2396	6892	3766
6067	7516	2451	1510	0201	1437	6518	1063	6442	6674
4541	9863	8312	9855	0995	6025	4207	4093	9799	9308
6987	4802	8975	2847	4413	5997	9106	2876	8596	7717
0376	8636	9953	4418	2388	8997	1196	5158	1803	5623
8468	5763	3232	1986	7134	4200	9699	8437	2799	2145
9151	4967	3255	8518	2802	8815	6289	9549	2942	3813
1073	4930	1830	2224	2246	1000	9315	6698	4491	3046
5487	1967	5836	2090	3832	0002	9844	3742	2289	3763
4896	4957	6536	7430	6208	3929	1030	2317	7421	3227
9143	7911	0368	0541	2302	5473	9155	0625	1870	1890
9256	2956	4747	6280	7342	0453	8639	1216	5964	9772
4173	1219	7744	9241	6354	4211	8497	1245	3313	4846
2525	7811	5417	7824	0922	8752	3537	9069	5417	0856
9165	1156	6603	2852	8370	0995	7661	8811	7835	5087
0014	8474	6322	5053	5015	6043	0482	4957	8904	1616
5325	7320	8406	5962	6100	3854	0575	0617	8019	2646
2558	1748	5671	4974	7073	3273	6036	1410	5257	3939
0117	1218	0688	2756	7545	5426	3856	8905	9691	8890
8353	1554	4083	2029	8857	4781	9654	7946	7866	2535
1990	9886	3280	6109	9158	3034	8490	6404	6775	8763
9651	7870	2555	3518	2906	4900	2984	6894	5050	4586
9941	5617	1984	2435	5184	0379	7212	5795	0836	4319
7769	5785	9321	2734	2890	3105	6581	2163	4938	7540
3224	8379	9952	0515	2724	4826	6215	6246	9704	1651
1287	7275	6646	1378	6433	0005	7332	0392	1319	1946
6389	4191	4548	5546	6651	8248	7469	0786	0972	7649
1625	4327	2654	4129	3509	3217	7062	6640	0105	4422
7555	3020	4181	7498	4022	9122	6423	7301	8310	9204
4177	1844	3468	1389	3884	6900	1036	8412	0881	6678
0927	0124	8176	0680	1056	1008	1748	0547	8227	0690
8505	1781	7155	3635	9751	5414	5113	8316	2737	6860
8022	8757	6275	1485	3635	2330	7045	2106	6381	2986
8390	8802	5674	2559	7934	4788	7791	5202	8430	0289
3630	5783	7762	0223	5328	7731	4010	3845	9221	5427
9154	6388	6053	9633	2080	7269	0894	0287	7489	2259
1441	3381	7823	8767	9647	4445	2509	2929	5067	0779
8246	0778	0993	6687	7212	9968	8432	1453	0841	4595
2730	3984	0563	9636	7202	0127	9283	4009	3177	4182
9196	8276	0233	0879	3385	2184	1739	5375	5807	4849
5928	9610	9161	0748	3794	9683	1544	1209	3669	5831
1042	9600	7122	2135	7868	5596	3551	9480	2342	0449
6552	4103	7957	0510	5958	0211	3344	5678	1840	3627
5968	4307	9327	3197	0876	8480	5066	1852	8323	5060
4445	1018	4356	4653	9302	0761	1291	6093	5340	1840
8727	8201	5980	7859	6055	1403	1209	9547	4273	0857
9415	9311	4996	2775	8509	7767	6930	6632	7781	2279
2648	7639	9128	0341	6875	8957	6646	9783	6668	0317
3707	3454	8829	6863	1297	5089	1002	2722	0578	7753
8383	8957	5595	9395	3036	4767	8300	3505	0710	6307

5503	8121	9056	8194	1124	8451	1228	8986	0076	7615
2552	9953	4323	4878	4922	0696	3156	2145	8819	0631
8542	7274	9724	6638	0013	0566	9644	3738	5767	2791
6121	4839	4734	3041	3939	9136	5620	7920	0533	3119
2023	0314	5885	1165	2841	1282	5893	3050	6598	2667
9577	8320	5614	5595	8978	6442	0844	4570	8036	6026
0760	1734	0114	8330	9695	6502	3171	8901	7955	4975
0064	1745	7874	3900	3602	9880	7266	5448	6826	3882
6295	8316	6150	3155	8059	4789	7236	7272	0839	3367
7935	1027	8193	2634	0806	6781	0665	8791	7416	8551
4833	6983	5904	8217	9201	5844	6959	5620	9570	8621
0584	0843	7983	5095	3205	3291	1584	1391	4136	8011
2585	0220	0730	5994	7138	7615	1126	3878	6154	2260
2527	1615	8232	7071	9808	3863	9195	4990	7625	3397
7300	2905	1760	4929	4767	9044	6891	0567	2382	8489
8131	9443	2266	0658	3814	0014	1749	5111	6145	6579
1002	4471	5983	8072	6371	6788	2510	4534	5574	6761
8467	5280	8912	3769	2089	8233	2262	0614	0577	0354
2929	5816	2185	3373	9405	8880	5460	0038	6634	6923
5177	9407	7063	4128	9058	8768	1396	5562	2367	3510
4216	5625	6077	5167	3603	7727	8521	1481	9075	2367
7835	6704	2249	5152	3116	3045	2760	4442	9638	2677
0955	5134	3386	8901	7341	8153	7739	3044	9774	1815
1577	6312	3484	0566	0615	4897	5569	6181	9176	2082
1323	9905	9375	3673	4428	4432	1572	3750	4726	1333
5058	0357	3847	7323	6761	7278	7817	1871	9909	6411
9948	5733	1063	7490	9067	1964	6990	6095	1796	3721
5467	3952	7378	4886	6983	6279	6520	6918	0557	7474
9934	7154	1024	7603	3170	7686	8890	6957	2764	0033
3549	4023	3486	5535	1284	6809	5264	3273	6701	4678
9817	2538	0384	2392	4795	1035	7011	1117	6329	9990
0267	8615	5686	0259	0164	4220	7995	3776	8234	7195
3693	4287	8163	7995	0706	4162	9680	9238	8886	6858
5685	1277	2430	7366	8426	2466	1668	0223	6602	6413
0546	2889	1427	2377	8859	1708	3388	8878	3901	5711
1502	2023	6338	7112	0662	0741	9498	3232	7942	7038
9561	0803	8146	9106	8885	5658	0122	2809	1972	7146
0902	4037	0573	5512	7429	4919	3166	4260	3036	9642
8143	9995	5246	6766	9732	6980	2124	6592	1262	9289
2143	5933	5862	9482	6548	0964	4101	8510	1611	3207
9583	7614	1163	8028	1778	9793	1282	7389	6600	2752
9981	4463	4374	9979	8682	1211	3170	0502	2815	0420
7721	3114	5054	1160	5093	0249	0918	9587	8584	7195
1326	0260	7983	6605	8027	0853	2867	3753	7053	8235
4428	7173	2662	5469	1490	5213	8111	7454	7885	3199
7052	4595	7963	5737	0505	3196	3337	1323	8566	8661
8838	1122	2508	7146	0981	4600	1906	6898	1831	7417
8316	7399	1720	7944	6409	4979	1193	4486	8697	3453
5021	7172	3385	4514	0569	2993	1282	0159	0845	5282
9768	2934	6774	8064	1362	2394	4939	8368	3730	9535

1236	2389	3150	9072	1871	8914	5859	9942	2284	0826
3889	3023	3423	2257	7442	2273	2693	4060	1078	8012
8078	5541	3977	9331	1827	2114	5208	7809	8563	8114
0239	7758	0885	2356	3354	4579	1097	4472	2478	0969
7372	7018	6911	7188	8014	7287	3898	2340	6395	4475
6138	1722	5523	1896	3900	9350	1827	4981	5280	6967
3916	4428	1497	9749	2597	3360	6014	3003	7767	4929
8090	7448	3988	1988	3731	0420	4967	3959	0105	4399
0905	6567	6366	3403	0657	8783	2812	4888	5048	5573
3342	2422	3204	6008	2041	8504	5357	3255	6409	5232
7265	6947	7364	7153	5545	1957	1555	2057	1212	5003
0414	3209	8358	6182	3548	3273	6340	9149	3719	0276
8522	1419	5221	6074	2441	5785	3188	5126	8229	7355
5488	0357	9167	5950	0861	3379	2901	8519	6226	2868
3325	5151	8203	4523	3935	3322	5946	6554	7680	1698
7597	1595	3240	8208	0221	5714	3352	4719	9452	7325
9063	7531	3538	3445	4924	1146	2510	7148	8988	9970
6506	1549	9334	3356	1942	6682	0304	9736	0815	4748
6442	0742	8223	9781	3957	0776	6584	2998	1553	9011
2717	1738	7696	7511	4558	9990	4716	5536	2566	2540
3221	3009	8727	5689	1562	3259	8066	0808	1942	8071
5420	5804	7235	8982	0270	1681	8998	3738	4403	5936
5928	6696	8484	7154	6755	3386	8301	6621	6937	2390
8387	5816	0122	9555	2219	6590	3878	0135	4748	2817
8331	5708	0336	8001	3960	4069	5643	6405	0249	5088
6454	2950	1335	7864	9262	1935	6047	5733	5213	0711
3926	0007	5548	0152	7656	2257	2032	8462	3018	4390
2976	0567	2819	6551	1195	7859	6390	2134	1921	9028
0631	0299	0146	2773	9028	1769	6451	3955	3469	0321
9754	4760	5765	5910	2185	4444	0797	5429	8467	7875
8296	8571	1161	9772	5351	5378	9894	3840	7093	1131
7687	3472	1252	9064	1692	1366	1742	8448	6830	8524
8739	7888	8723	9208	9563	6684	2290	6498	8695	5470
7404	1273	5961	3369	1259	4489	6798	7297	8979	1058
4789	4141	6643	7004	5079	4592	9656	6795	5636	4472
8777	7169	6414	5436	9211	3403	5906	6205	6204	3352
9697	6314	7221	8004	1199	4769	9562	7299	2904	8589
4382	1328	7781	8169	2993	7075	0202	3237	0055	8668
5720	8396	4009	3923	6595	5991	9141	5557	8842	4557
4906	7217	8093	0601	9032	6368	0793	9958	4901	2645
9425	8427	9579	1347	8013	2633	5516	7341	4076	4517
6814	8138	8238	1867	4045	9282	3004	3741	4342	4513
1220	9780	3361	2886	4164	1673	8886	3263	4198	8461
8831	8970	2611	1241	1943	6566	6098	5976	1141	1825
5672	8035	2961	6305	1525	4468	6468	4235	5102	7768
0713	1232	0107	1930	8704	5892	2845	8106	9397	6665
2118	6455	5561	3608	2433	8439	1602	1220	7755	7566
0215	1225	8873	4391	0365	2109	6080	6324	2684	3581
9095	8523	3277	0730	3618	4742	1968	3318	4138	0324
8010	9130	1285	4129	0032	1501	1957	9113	1272	9260

9263	7824	1926	9545	5349	2389	3770	7986	7647	6641
7944	7873	7154	4484	2610	6731	0070	3498	6675	9972
5965	7196	2738	5000	0535	9403	2928	1854	5242	0608
3152	4958	7661	3978	1353	4808	5948	6068	8467	5301
0634	7693	9037	5139	5588	7101	0920	7915	2444	3024
2870	5170	9445	4839	7378	0643	8664	6923	5766	8018
6810	8926	9473	9576	7502	4846	6554	9658	1891	1639
9993	9070	9362	6633	3339	9526	9534	5176	9161	3323
9154	7319	3444	6351	8383	9941	5882	4045	6926	4856
4210	0278	7392	5629	7267	1224	2527	3667	2131	7576
1713	2758	2529	2838	5135	6166	3789	0536	4414	4267
2829	1428	5452	2161	9532	3817	6057	0808	9499	7846
0933	5671	5133	0628	7534	0881	8271	5739	2525	3033
3129	0420	9371	5128	0575	7939	8739	5177	3307	9706
3614	1556	2759	4208	9928	5964	1522	9607	0996	0537
2955	1843	1363	0552	0279	8101	4902	7903	5091	0939
2350	2264	6308	0819	8942	6780	5513	5470	3294	6452
5788	8584	6796	0783	1131	0154	4853	1714	0855	6745
5533	7126	8847	0433	6391	3639	1119	9247	7054	2977
1008	1007	5598	6468	6823	2046	8938	9380	0079	9594
3410	8127	6609	8887	3781	7214	6714	5078	2138	1670
5336	4494	6043	2283	1413	9659	2329	5620	9267	1592
8297	6615	8473	1943	5579	6922	2866	1367	9931	7687
5482	8467	2289	0809	1432	8703	4289	2112	3071	4848
2546	5909	2743	8942	8075	8992	1909	6773	8036	0879
6760	6021	4147	8495	4013	0254	0957	4568	5016	1560
4492	7092	6129	5113	4759	8673	3556	7664	1821	6344
3317	3097	9813	9582	4978	1330	3608	8076	3398	6862
8468	8544	0620	1765	5133	0287	3501	6757	6157	2074
7188	5645	3656	0939	9695	3550	1755	3521	6910	0167
0047	0222	7472	1472	4021	2135	0859	4562	8398	6374
2599	3888	6836	5956	4127	6974	4070	3799	0343	1887
9288	5317	9919	9380	5698	5308	1530	5052	5590	4302
2513	2681	0709	1567	6068	0441	2450	3789	6718	6282
8463	7188	1299	8302	8248	9033	9195	7457	0353	9012
3400	9232	1279	6145	4812	7427	2836	6656	7522	3590
5377	4574	0573	8616	4276	7017	9731	7389	8860	1999
5931	9788	7280	5496	6085	1193	3526	7160	5557	6771
2047	6655	5070	2699	0985	5259	1406	3021	1989	1929
8618	8493	2545	2604	0222	5201	2182	5059	5167	6541
2145	6800	7271	4026	6128	1317	6381	4897	5173	5411
9806	6837	8008	2413	7235	9542	1180	2974	8164	8661
0178	6442	1443	9457	7515	9457	6139	9619	0322	3225
6246	0484	4327	6870	0127	0543	2295	1894	9905	4169
9432	3108	8415	9293	9998	8950	9158	0280	6947	6827
0579	4398	2157	0990	7022	1979	5157	3643	3349	7988
1039	1428	5218	0972	2578	3856	5479	0489	5901	8925
3517	5698	2554	5973	6471	5263	3110	6238	4948	1140
2563	8961	7588	9825	0212	7209	5718	5588	0932	7346
1646	4828	9425	4577	4515	6886	1138	1178	2269	4198

Source: Compiled from Rand Corporation, *A Million Random Digits with 100,000 Normal Deviates* (New York: The Free Press, 1955). Used with permission.

Exhibit A–4 Standard footnotes for memoranda

Standard footnote 1

The summary statistics reported are estimates derived from a sample survey of _____ units drawn via a _____ random sampling plan.

"There are two types of errors possible in an estimate based on a sample survey—sampling and nonsampling. Sampling errors occur because observations are made only on a sample, not on the entire population. Nonsampling errors can be attributed to many sources; for example, inability to obtain information about all cases in the sample, definitional difficulties, differences in the interpretation of questions, inability or unwillingness to provide correct information on the part of respondents, mistakes in recording or coding the data obtained and other errors of collection, response, processing, coverage, and estimation for missing data. Nonsampling errors also occur in complete censuses. The 'accuracy' of a survey result is determined by the joint effects of sampling and nonsampling errors." [Journal of the American Statistical Association 70, no. 351, part II (September 1975), p. 6.]

Standard footnote 2

"The particular sample used in this survey is one of a large number of all possible samples of the same size that could have been selected using the same sample design. Estimates derived from the different samples would differ from each other. The difference between a sample estimate and the average of all possible samples is called the sampling deviation. The standard or sampling error of a survey estimate is a measure of the variation among the estimates from all possible samples and thus is a measure of the precision with which an estimate from a particular sample approximates the average result of all possible samples. The relative standard error is defined as the standard error of the estimate divided by the value being estimated.

"As calculated for this report, the standard error also partially measures the effect of certain nonsampling errors but does not measure any systematic biases in the data. Bias is the difference, averaged over all possible samples, between the estimate and the desired value. Obviously, the accuracy of a survey result depends on both the sampling and nonsampling errors measured by the standard error and the bias and other types of nonsampling error not measured by the standard error." [Journal of the American Statistical Association 70, no. 351, part II (September 1975), p. 6.]

"The sample estimate and an estimate of its standard error permit us to construct interval estimates with prescribed confidence that the interval includes the average result of all possible samples (for a given sampling rate).

"To illustrate, if all possible samples were selected, if each of these were surveyed under essentially the same conditions, and if an estimate and its estimated standard error were calculated from each sample, then:

"a. Approximately two–thirds of the intervals from one standard error below the estimate to one standard error above the estimate would include the average value of all possible samples. We call an interval from one standard error below the estimate to one standard error above the estimate a two–thirds confidence interval.

"b. Approximately nine–tenths of the intervals from 1.6 standard errors below the estimate to 1.6 standard errors above the estimate would include the average value of all possible samples. We call an interval from 1.6 standard errors below the estimate to 1.6 standard errors above the estimate a 90 percent confidence interval.

"c. Approximately 19/20 of the intervals from two standard errors below the estimate to two standard errors above the estimate would include the average value of all possible samples. We call an interval from two standard errors below the estimate to two standard errors above the estimate a 95 percent confidence interval.

"d. Almost all intervals from three standard errors below the sample estimate to three standard errors above the sample estimate would include the average value of all possible samples.

"The average value of all possible samples may or may not be contained in any particular computed interval. But for a particular sample, one can say with specified confidence that the average of all possible samples is included in the constructed interval." [Journal of the American Statistical Association 70, no. 351, part II (September 1975), p. 6.]

The sample estimate and an estimate of its standard error permit us to construct interval estimates with prescribed confidence that the interval includes the average result of all possible samples for a sample of size _____ .

To illustrate, if all possible samples of size _____ were selected, if each of these were surveyed under essentially the same conditions, and if an estimate and its estimated standard error were calculated from each sample, then:

a. Approximately two–thirds of the intervals from _____ standard errors below the estimate to _____ standard errors above the estimate would include the average value of all possible samples. We call an interval from _____ standard errors below the estimate to _____ standard error(s) above the estimate a two–thirds confidence interval.

b. Approximately 19/20 of the intervals from _____ standard errors below the estimate to _____ standard errors above the estimate would include the average value of all possible samples. We call an interval from _____ standard errors below the estimate to _____ standard errors above the estimate a 95 percent confidence interval.

The average value of all possible samples may or may not be contained in any particular computed interval. But for a particular sample, one can say with specified confidence that the average of all possible samples is included in the constructed interval.

Exhibit A–4 (*concluded*)

Standard footnote 4

The hypothesized population parameter, the sample estimate, and an estimate of the standard error permit a hypothesis test to be performed for a given sample size.

To illustrate, if all possible samples were selected, if each of these were surveyed under essentially the same conditions, and if a sample estimate and its standard error were calculated, then:

a. Approximately nine-tenths of the hypothesis tests conducted at the 90 percent level of confidence (or 10 percent level of significance) would fail to reject the null hypothesis when it was tenable (or 10 percent would reject the null hypothesis when it was tenable).

b. Approximately 19/20 of the hypothesis tests conducted at the 95 percent level of confidence (of 5 percent level of significance) would fail to reject the null hypothesis when it was tenable (or 5 percent would reject the null hypothesis when it was tenable).

Once the level of confidence (or significance) has been set, the chances of failing to reject the null hypothesis when it is false (type II error) must be computed. If these probabilities are unacceptably high, then corrective action must be taken.

Standard footnote 5

The sample estimators permit us to perform a hypothesis test (for a given sample size).

To illustrate, if all possible samples were selected and each of these were surveyed under essentially the same conditions, then:

a. Approximately nine-tenths of the hypothesis tests conducted at the 90 percent level of confidence (or 10 percent level of significance) would fail to reject the null hypothesis when it was tenable (or 10 percent would reject the null hypothesis when it was tenable).

b. Approximately 19/20 of the hypothesis tests conducted at the 95 percent level of confidence (of 5 percent level of significance) would fail to reject the null hypothesis when it was tenable (or 5 percent would reject the null hypothesis when it was tenable).

Once the level of confidence (or significance) has been set, the chances of failing to reject the null hypothesis when it is false must be computed. If these probabilities are unacceptably high (type II errors), then corrective action must be taken.

Standard footnote 6

The statistics reported are based on a random sample of ___ Stat City dwelling units and are representative of parameters only on average. In other words, if many samples of size ___ were drawn, the average of the sample statistics would afford an accurate reflection of the true, but unknown, population parameters. Any given sample's statistics may not accurately reflect the population parameters.

APPENDIX B   REFERENCE TABLES

Table B–1   Binomial distribution

$$P(x) = \frac{n!}{x!\,(n-x)!}\,\pi^x(1-\pi)^{n-x}$$

n,	x	.05	.10	.15	.20	.25	π .30	.35	.40	.45	.50
1	0	.9500	.9000	.8500	.8000	.7500	.7000	.6500	.6000	.5500	.5000
	1	.0500	.1000	.1500	.2000	.2500	.3000	.3500	.4000	.4500	.5000
2	0	.9025	.8100	.7225	.6400	.5625	.4900	.4225	.3600	.3025	.2500
	1	.0950	.1800	.2550	.3200	.3750	.4200	.4550	.4800	.4950	.5000
	2	.0025	.0100	.0225	.0400	.0625	.0900	.1225	.1600	.2025	.2500
3	0	.8574	.7290	.6141	.5120	.4219	.3430	.2746	.2160	.1664	.1250
	1	.1354	.2430	.3251	.3840	.4219	.4410	.4436	.4320	.4084	.3750
	2	.0071	.0270	.0574	.0960	.1406	.1890	.2389	.2880	.3341	.3750
	3	.0001	.0010	.0034	.0080	.0156	.0270	.0429	.0640	.0911	.1250
4	0	.8145	.6561	.5220	.4096	.3164	.2401	.1785	.1296	.0915	.0625
	1	.1715	.2916	.3685	.4096	.4219	.4116	.3845	.3456	.2995	.2500
	2	.0135	.0486	.0975	.1536	.2109	.2646	.3105	.3456	.3675	.3750
	3	.0005	.0036	.0115	.0256	.0469	.0756	.1115	.1536	.2005	.2500
	4	.0000	.0001	.0005	.0016	.0039	.0081	.0150	.0256	.0410	.0625
5	0	.7738	.5905	.4437	.3277	.2373	.1681	.1160	.0778	.0503	.0312
	1	.2036	.3280	.3915	.4096	.3955	.3602	.3124	.2592	.2059	.1562
	2	.0214	.0729	.1382	.2048	.2637	.3087	.3364	.3456	.3369	.3125
	3	.0011	.0081	.0244	.0512	.0879	.1323	.1811	.2304	.2757	.3125
	4	.0000	.0004	.0022	.0064	.0146	.0284	.0488	.0768	.1128	.1562
	5	.0000	.0000	.0001	.0003	.0010	.0024	.0053	.0102	.0185	.0312
6	0	.7351	.5314	.3771	.2621	.1780	.1176	.0754	.0467	.0277	.0156
	1	.2321	.3543	.3993	.3932	.3560	.3025	.2437	.1866	.1359	.0938
	2	.0305	.0984	.1762	.2458	.2966	.3241	.3280	.3110	.2780	.2344
	3	.0021	.0146	.0415	.0819	.1318	.1852	.2355	.2765	.3032	.3125
	4	.0001	.0012	.0055	.0154	.0330	.0595	.0951	.1382	.1861	.2344
	5	.0000	.0001	.0004	.0015	.0044	.0102	.0205	.0369	.0609	.0938
	6	.0000	.0000	.0000	.0001	.0002	.0007	.0018	.0041	.0083	.0156
7	0	.6983	.4783	.3206	.2097	.1335	.0824	.0490	.0280	.0152	.0078
	1	.2573	.3720	.3960	.3670	.3115	.2471	.1848	.1306	.0872	.0547
	2	.0406	.1240	.2097	.2753	.3115	.3177	.2985	.2613	.2140	.1641
	3	.0036	.0230	.0617	.1147	.1730	.2269	.2679	.2903	.2918	.2734
	4	.0002	.0026	.0109	.0287	.0577	.0972	.1442	.1935	.2388	.2734
	5	.0000	.0002	.0012	.0043	.0115	.0250	.0466	.0774	.1172	.1641
	6	.0000	.0000	.0001	.0004	.0013	.0036	.0084	.0172	.0320	.0547
	7	.0000	.0000	.0000	.0000	.0001	.0001	.0006	.0016	.0037	.0078
8	0	.6634	.4305	.2725	.1678	.1002	.0576	.0319	.0168	.0084	.0039
	1	.2793	.3826	.3847	.3355	.2670	.1977	.1373	.0896	.0548	.0312
	2	.0515	.1488	.2376	.2936	.3115	.2065	.2587	.2090	.1569	.1094

n,	x	.05	.10	.15	.20	.25	π .30	.35	.40	.45	.50
8	3	.0054	.0331	.0839	.1468	.2076	.2541	.2786	.2787	.2568	.2188
	4	.0004	.0046	.0185	.0459	.0865	.1361	.1875	.2322	.2627	.2734
	5	.0000	.0004	.0026	.0092	.0231	.0467	.0808	.1239	.1719	.2188
	6	.0000	.0000	.0002	.0011	.0038	.0100	.0217	.0413	.0403	.1094
	7	.0000	.0000	.0000	.0001	.0004	.0012	.0033	.0079	.0164	.0312
	8	.0000	.0000	.0000	.0000	.0000	.0001	.0002	.0007	.0017	.0039
9	0	.6302	.3874	.2316	.1342	.0751	.0404	.0207	.0101	.0046	.0020
	1	.2985	.3874	.3679	.3020	.2253	.1556	.1004	.0605	.0339	.0176
	2	.0629	.1722	.2597	.3020	.3003	.2668	.2162	.1612	.1110	.0703
	3	.0077	.0446	.1069	.1762	.2336	.2668	.2716	.2508	.2119	.1641
	4	.0006	.0074	.0283	.0661	.1168	.1715	.2194	.2508	.2600	.2461
	5	.0000	.0008	.0050	.0165	.0389	.0735	.1181	.1672	.2128	.2461
	6	.0000	.0001	.0006	.0028	.0087	.0210	.0424	.0743	.1160	.1641
	7	.0000	.0000	.0000	.0003	.0012	.0039	.0098	.0212	.0407	.0703
	8	.0000	.0000	.0000	.0000	.0001	.0004	.0013	.0035	.0083	.0176
	9	.0000	0000	0000	0000	0000	0000	0001	.0003	.0008	.0020
10	0	.5987	.3487	.1969	.1074	.0563	.0282	.0135	.0060	.0025	.0010
	1	.3151	.3874	.3474	.2684	.1877	.1211	.0725	.0403	.0207	.0098
	2	.0746	.1937	.2759	.3020	.2816	.2335	.1757	.1209	.0763	.0439
	3	.0105	.0574	.1298	.2013	.2503	.2668	.2522	.2150	.1665	.1172
	4	.0010	.0112	.0401	.0881	.1460	.2001	.2377	.2508	.2384	.2051
	5	.0001	.0015	.0085	.0264	.0584	.1029	.1536	.2007	.2340	.2461
	6	.0000	.0001	.0012	.0055	.0162	.0368	.0689	.1115	.1596	.2051
	7	.0000	.0000	.0001	.0008	.0031	.0090	.0212	.0425	.0746	.1172
	8	.0000	.0000	.0000	.0001	.0004	.0014	.0043	.0106	.0229	.0439
	9	.0000	.0000	.0000	.0000	.0000	.0001	.0005	.0016	.0042	.0098
	10	.0000	.0000	.0000	.0000	.0000	.0000	.0000	.0001	.0003	.0010
11	0	.5688	.3138	.1673	.0859	.0422	.0198	.0088	.0036	.0014	.0005
	1	.3293	.3835	.3248	.2362	.1549	.0932	.0518	.0266	.0125	.0054
	2	.0867	.2131	.2866	.2953	.2581	.1998	.1395	.0887	.0513	.0269
	3	.0137	.0710	.1517	.2215	.2581	.2568	.2254	.1774	.1259	.0806
	4	.0014	.0158	.0536	.1107	.1721	.2201	.2428	.2365	.2060	.1611
	5	.0001	.0025	.0132	.0388	.0803	.1321	.1830	.2207	.2360	.2256
	6	.0000	.0003	.0023	.0097	.0268	.0566	.0985	.1471	.1931	.2256
	7	.0000	.0000	.0003	.0017	.0064	.0173	.0379	.0701	.1128	.1611
	8	.0000	.0000	.0000	.0002	.0011	.0037	.0102	.0234	.0462	.0806
	9	.0000	.0000	.0000	.0000	.0001	.0005	.0018	.0052	.0126	.0269
	10	.0000	.0000	.0000	.0000	.0000	.0000	.0002	.0007	.0021	.0054
	11	.0000	.0000	.0000	.0000	.0000	.0000	.0000	.0000	.0002	.0005
12	0	.5404	.2824	.1422	.0687	.0317	.0138	.0057	.0022	.0008	.0002
	1	.3413	.3766	.3012	.2062	.1267	.0712	.0368	.0174	.0075	.0029

n,	x	.05	.10	.15	.20	.25	π .30	.35	.40	.45	.50
12	2	.0988	.2301	.2924	.2835	.2323	.1678	.1088	.0639	.0339	.0161
	3	.0173	.0852	.1720	.2362	.2581	.2397	.1954	.1419	.0923	.0537
	4	.0021	.0213	.0683	.1329	.1936	.2311	.2367	.2128	.1700	.1208
	5	.0002	.0038	.0193	.0532	.1032	.1585	.2039	.2270	.2225	.1934
	6	.0000	.0005	.0040	.0155	.0401	.0792	.1281	.1766	.2124	.2256
	7	.0000	.0000	.0006	.0033	.0115	.0291	.0591	.1009	.1489	.1934
	8	.0000	.0000	.0001	.0005	.0024	.0078	.0199	.0420	.0762	.1208
	9	.0000	.0000	.0000	.0001	.0004	.0015	.0048	.0125	.0277	.0537
	10	.0000	.0000	.0000	.0000	.0000	.0002	.0008	.0025	.0068	.0161
	11	.0000	.0000	.0000	.0000	.0000	.0000	.0001	.0003	.0010	.0029
	12	.0000	.0000	.0000	.0000	.0000	.0000	.0000	.0000	.0001	.0002
13	0	.5133	.2542	.1209	.0550	.0238	.0097	.0037	.0013	.0004	.0001
	1	.3512	.3672	.2774	.1787	.1029	.0540	.0259	.0113	.0045	.0016
	2	.1109	.2448	.2937	.2680	.2059	.1388	.0836	.0453	.0220	.0095
	3	.0214	.0997	.1900	.2457	.2517	.2181	.1651	.1107	.0660	.0349
	4	.0028	.0277	.0838	.1535	.2097	.2337	.2222	.1845	.1350	.0873
	5	.0003	.0055	.0266	.0691	.1258	.1803	.2154	.2214	.1989	.1571
	6	.0000	.0008	.0063	.0230	.0559	.1030	.1546	.1968	.2169	.2095
	7	.0000	.0001	.0011	.0058	.0186	.0442	.0833	.1312	.1775	.2095
	8	.0000	.0001	.0001	.0011	.0047	.0142	.0336	.0656	.1089	.1571
	9	.0000	.0000	.0000	.0001	.0009	.0034	.0101	.0243	.0495	.0873
	10	.0000	.0000	.0000	.0000	.0001	.0006	.0022	.0065	.0162	.0349
	11	.0000	.0000	.0000	.0000	.0000	.0001	.0003	.0012	.0036	.0095
	12	.0000	.0000	.0000	.0000	.0000	.0000	.0000	.0001	.0005	.0016
	13	.0000	.0000	.0000	.0000	.0000	.0000	.0000	.0000	.0000	.0001
14	0	.4877	.2288	.1028	.0440	.0178	.0068	.0024	.0008	.0002	.0001
	1	.3593	.3559	.2539	.1539	.0832	.0407	.0181	.0073	.0027	.0009
	2	.1229	.2570	.2912	.2501	.1802	.1134	.0634	.0317	.0141	.0056
	3	.0259	.1142	.2056	.2501	.2402	.1943	.1366	.0845	.0462	.0222
	4	.0037	.0349	.0998	.1720	.2202	.2290	.2022	.1549	.1040	.0611
	5	.0004	.0078	.0352	.0860	.1468	.1963	.2178	.2066	.1701	.1222
	6	.0000	.0013	.0093	.0322	.0734	.1262	.1759	.2066	.2088	.1833
	7	.0000	.0002	.0019	.0092	.0280	.0618	.1082	.1574	.1952	.2095
	8	.0000	.0000	.0003	.0020	.0082	.0232	.0510	.0918	.1398	.1833
	9	.0000	.0000	.0000	.0003	.0018	.0066	.0183	.0408	.0762	.1222
	10	.0000	.0000	.0000	.0000	.0003	.0014	.0049	.0136	.0312	.0611
	11	.0000	.0000	.0000	.0000	.0000	.0002	.0010	.0033	.0093	.0222
	12	.0000	.0000	.0000	.0000	.0000	.0000	.0001	.0005	.0019	.0056
	13	.0000	.0000	.0000	.0000	.0000	.0000	.0000	.0001	.0002	.0009
	14	.0000	.0000	.0000	.0000	.0000	.0000	.0000	.0000	.0000	.0001
15	0	.4633	.2059	.0874	.0352	.0134	.0047	.0016	.0005	.0001	.0000
	1	.3658	.3432	.2312	.1319	.0668	.0305	.0126	.0047	.0016	.0005
	2	.1348	.2669	.2856	.2309	.1559	.0916	.0476	.0219	.0090	.0032

n, x		.05	.10	.15	.20	.25	π .30	.35	.40	.45	.50
15	3	.0307	.1285	.2184	.2501	.2252	.1700	.1110	.0634	.0318	.0139
	4	.0049	.0428	.1156	.1876	.2252	.2186	.1792	.1268	.0780	.0417
	5	.0006	.0105	.0449	.1032	.1651	.2061	.2123	.1859	.1404	.0916
	6	.0000	.0019	.0132	.0430	.0917	.1472	.1906	.2066	.1914	.1527
	7	.0000	.0003	.0030	.0138	.0393	.0811	.1319	.1771	.2013	.1964
	8	.0000	.0000	.0005	.0035	.0131	.0348	.0710	.1181	.1647	.1964
	9	.0000	.0000	.0001	.0007	.0034	.0116	.0298	.0612	.1048	.1527
	10	.0000	.0000	.0000	.0001	.0007	.0030	.0096	.0245	.0515	.0916
	11	.0000	.0000	.0000	.0000	.0001	.0006	.0024	.0074	.0191	.0417
	12	.0000	.0000	.0000	.0000	.0000	.0001	.0004	.0016	.0052	.0139
	13	.0000	.0000	.0000	.0000	.0000	.0000	.0001	.0003	.0010	.0032
	14	.0000	.0000	.0000	.0000	.0000	.0000	.0000	.0000	.0001	.0005
	15	.0000	.0000	.0000	.0000	.0000	.0000	.0000	.0000	.0000	.0000
16	0	.4401	.1853	.0743	.0281	.0100	.0033	.0010	.0003	.0001	.0000
	1	.3706	.3294	.2097	.1126	.0535	.0228	.0087	.0030	.0009	.0002
	2	.1463	.2745	.2775	.2111	.1336	.0732	.0353	.0150	.0056	.0018
	3	.0359	.1423	.2285	.2463	.2079	.1465	.0888	.0468	.0215	.0085
	4	.0061	.0514	.1311	.2001	.2252	.2040	.1553	.1014	.0572	.0278
	5	.0008	.0137	.0555	.1201	.1802	.2099	.2008	.1623	.1123	.0667
	6	.0001	.0028	.0180	.0550	.1101	.1649	.1982	.1983	.1684	.1222
	7	.0000	.0004	.0045	.0197	.0524	.1010	.1524	.1889	.1969	.1746
	8	.0000	.0001	.0009	.0055	.0197	.0487	.0923	.1417	.1812	.1964
	9	.0000	.0000	.0001	.0012	.0058	.0185	.0442	.0840	.1318	.1746
	10	.0000	.0000	.0000	.0002	.0014	.0056	.0167	.0392	.0755	.1222
	11	.0000	.0000	.0000	.0000	.0002	.0013	.0049	.0142	.0337	.0667
	12	.0000	.0000	.0000	.0000	.0000	.0002	.0011	.0040	.0115	.0278
	13	.0000	.0000	.0000	.0000	.0000	.0000	.0002	.0008	.0029	.0085
	14	.0000	.0000	.0000	.0000	.0000	.0000	.0000	.0001	.0005	.0018
	15	.0000	.0000	.0000	.0000	.0000	.0000	.0000	.0000	.0001	.0002
	16	.0000	.0000	.0000	.0000	.0000	.0000	.0000	.0000	.0000	.0000
17	0	.4181	.1668	.0631	.0225	.0075	.0023	.0007	.0002	.0000	.0000
	1	.3741	.3150	.1893	.0957	.0426	.0169	.0060	.0019	.0005	.0001
	2	.1575	.2800	.2673	.1914	.1136	.0581	.0260	.0102	.0035	.0010
	3	.0415	.1556	.2359	.2393	.1893	.1245	.0701	.0341	.0144	.0052
	4	.0076	.0605	.1457	.2093	.2209	.1868	.1320	.0796	.0411	.0182
	5	.0010	.0175	.0668	.1361	.1914	.2081	.1849	.1379	.0875	.0472
	6	.0001	.0039	.0236	.0680	.1276	.1784	.1991	.1839	.1432	.0944
	7	.0000	.0007	.0065	.0267	.0668	.1201	.1685	.1927	.1841	.1484
	8	.0000	.0001	.0014	.0084	.0279	.0644	.1134	.1606	.1883	.1855
	9	.0000	.0000	.0003	.0021	.0093	.0276	.0611	.1070	.1540	.1855

n,	x	.05	.10	.15	.20	.25	π .30	.35	.40	.45	.50
17	10	.0000	.0000	.0000	.0004	.0025	.0095	.0263	.0571	.1008	.1484
	11	.0000	.0000	.0000	.0001	.0005	.0026	.0090	.0242	.0525	.0944
	12	.0000	.0000	.0000	.0000	.0001	.0006	.0024	.0081	.0215	.0472
	13	.0000	.0000	.0000	.0000	.0000	.0001	.0005	.0021	.0068	.0182
	14	.0000	.0000	.0000	.0000	.0000	.0000	.0001	.0004	.0016	.0052
	15	.0000	.0000	.0000	.0000	.0000	.0000	.0000	.0001	.0003	.0010
	16	.0000	.0000	.0000	.0000	.0000	.0000	.0000	.0000	.0000	.0001
	17	.0000	.0000	.0000	.0000	.0000	.0000	.0000	.0000	.0000	.0000
18	0	.3972	.1501	.0536	.0180	.0056	.0016	.0004	.0001	.0000	.0000
	1	.3763	.3002	.1704	.0811	.0338	.0126	.0042	.0012	.0003	.0001
	2	.1683	.2835	.2556	.1723	.0958	.0458	.0190	.0069	.0022	.0006
	3	.0473	.1680	.2406	.2297	.1704	.1046	.0547	.0246	.0095	.0031
	4	.0093	.0700	.1592	.2153	.2130	.1681	.1104	.0614	.0291	.0117
	5	.0014	.0218	.0787	.1507	.1988	.2017	.1664	.1146	.0666	.0327
	6	.0002	.0052	.0301	.0816	.1436	.1873	.1941	.1655	.1181	.0708
	7	.0000	.0010	.0091	.0350	.0820	.1376	.1792	.1892	.1657	.1214
	8	.0000	.0002	.0022	.0120	.0376	.0811	.1327	.1734	.1864	.1669
	9	.0000	.0000	.0004	.0033	.0139	.0386	.0794	.1284	.1694	.1855
	10	.0000	.0000	.0001	.0008	.0042	.0149	.0385	.0771	.1248	.1669
	11	.0000	.0000	.0000	.0001	.0010	.0046	.0151	.0374	.0742	.1214
	12	.0000	.0000	.0000	.0000	.0002	.0012	.0047	.0145	.0354	.0708
	13	.0000	.0000	.0000	.0000	.0000	.0002	.0012	.0045	.0134	.0327
	14	.0000	.0000	.0000	.0000	.0000	.0000	.0002	.0011	.0039	.0117
	15	.0000	.0000	.0000	.0000	.0000	.0000	.0000	.0002	.0009	.0031
	16	.0000	.0000	.0000	.0000	.0000	.0000	.0000	.0000	.0001	.0006
	17	.0000	.0000	.0000	.0000	.0000	.0000	.0000	.0000	.0000	.0001
	18	.0000	.0000	.0000	.0000	.0000	.0000	.0000	.0000	.0000	.0000
19	0	.3774	.1351	.0456	.0144	.0042	.0011	.0003	.0001	.0000	.0000
	1	.3774	.2852	.1529	.0685	.0268	.0093	.0029	.0008	.0002	.0000
	2	.1787	.2852	.2428	.1540	.0803	.0358	.0138	.0046	.0013	.0003
	3	.0533	.1796	.2428	.2182	.1517	.0869	.0422	.0175	.0062	.0018
	4	.0112	.0798	.1714	.2182	.2023	.1491	.0909	.0467	.0203	.0074
	5	.0018	.0266	.0907	.1636	.2023	.1916	.1468	.0933	.0497	.0222
	6	.0002	.0069	.0374	.0955	.1574	.1916	.1844	.1451	.0949	.0518
	7	.0000	.0014	.0122	.0443	.0974	.1525	.1844	.1797	.1443	.0961
	8	.0000	.0002	.0032	.0166	.0487	.0981	.1489	.1797	.1771	.1442
	9	.0000	.0000	.0007	.0051	.0198	.0514	.0980	.1464	.1771	.1762
	10	.0000	.0000	.0001	.0013	.0066	.0220	.0528	.0976	.1449	.1762
	11	.0000	.0000	.0000	.0003	.0018	.0077	.0233	.0532	.0970	.1442
	12	.0000	.0000	.0000	.0000	.0004	.0022	.0083	.0237	.0529	.0961
	13	.0000	.0000	.0000	.0000	.0001	.0005	.0024	.0085	.0233	.0518
	14	.0000	.0000	.0000	.0000	.0000	.0001	.0006	.0024	.0082	.0222

Table B-1 (*concluded*)

n,	x	.05	.10	.15	.20	.25	π .30	.35	.40	.45	.50
19	15	.0000	.0000	.0000	.0000	.0000	.0000	.0001	.0005	.0022	.0074
	16	.0000	.0000	.0000	.0000	.0000	.0000	.0000	.0001	.0005	.0018
	17	.0000	.0000	.0000	.0000	.0000	.0000	.0000	.0000	.0001	.0003
	18	.0000	.0000	.0000	.0000	.0000	.0000	.0000	.0000	.0000	.0000
	19	.0000	.0000	.0000	.0000	.0000	.0000	.0000	.0000	.0000	.0000
20	0	.3585	.1216	.0388	.0115	.0032	.0008	.0002	.0000	.0000	.0000
	1	.3774	.2702	.1368	.0576	.0211	.0068	.0020	.0005	.0001	.0000
	2	.1887	.2852	.2293	.1369	.0669	.0278	.0100	.0031	.0008	.0002
	3	.0596	.1901	.2428	.2054	.1339	.0718	.0323	.0123	.0040	.0011
	4	.0133	.0898	.1821	.2182	.1897	.1304	.0738	.0350	.0139	.0046
	5	.0022	.0319	.1028	.1746	.2023	.1789	.1272	.0746	.0365	.0148
	6	.0003	.0089	.0454	.1091	.1686	.1916	.1712	.1244	.0746	.0370
	7	.0000	.0020	.0160	.0545	.1124	.1643	.1844	.1659	.1221	.0739
	8	.0000	.0004	.0046	.0222	.0609	.1144	.1614	.1797	.1623	.1201
	9	.0000	.0001	.0011	.0074	.0271	.0654	.1158	.1597	.1771	.1602
	10	.0000	.0000	.0002	.0020	.0099	.0308	.0686	.1171	.1593	.1762
	11	.0000	.0000	.0000	.0005	.0030	.0120	.0336	.0710	.1185	.1602
	12	.0000	.0000	.0000	.0001	.0008	.0039	.0136	.0355	.0727	.1201
	13	.0000	.0000	.0000	.0000	.0002	.0010	.0045	.0146	.0366	.0739
	14	.0000	.0000	.0000	.0000	.0000	.0002	.0012	.0049	.0150	.0370
	15	.0000	.0000	.0000	.0000	.0000	.0000	.0003	.0013	.0049	.0148
	16	.0000	.0000	.0000	.0000	.0000	.0000	.0000	.0003	.0013	.0046
	17	.0000	.0000	.0000	.0000	.0000	.0000	.0000	.0000	.0002	.0011
	18	.0000	.0000	.0000	.0000	.0000	.0000	.0000	.0000	.0000	.0002
	19	.0000	.0000	.0000	.0000	.0000	.0000	.0000	.0000	.0000	.0000
	20	.0000	.0000	.0000	.0000	.0000	.0000	.0000	.0000	.0000	.0000

Source: Extracted from "Tables of the Binomial Probability Distributions," Applied Mathematics Series 6 (Washington, D.C.: U.S. Department of Commerce, National Bureau of Standards, 1952).

Table B–2   Poisson distribution for selected values of μt (values in the table are for the function:

$$P(x) = \frac{(\mu t)^x e^{-(\mu t)}}{x!}$$

)

$P(x)$ for specified values of μt

x	$\mu t = 0.1$	$\mu t = 0.2$	$\mu t = 0.3$	$\mu t = 0.4$	$\mu t = 0.5$	$\mu t = 0.6$	$\mu t = 0.7$	$\mu t = 0.8$	$\mu t = 0.9$	$\mu t = 1.0$
0	.9048374	.8187308	.7408182	.6703200	.606531	.548812	.496585	.449329	.406570	.367879
1	.0904837	.1637462	.2222455	.2681280	.303265	.329287	347610	.359463	.365913	.367879
2	.0045242	.0163746	.0333368	.0536256	.075816	.098786	.121663	.143785	.164661	.183940
3	.0001508	.0010916	.0033337	.0071501	.012636	.019757	.028388	.038343	.049398	.061313
4	.0000038	.0000546	.0002500	.0007150	.001580	.002964	004968	.007669	.011115	.015328
5	.0000001	.0000022	.0000150	.0000572	.000158	.000356	000696	.001227	.002001	.003066
6		.0000001	.0000008	.0000038	.000013	.000036	000081	.000164	.000300	.000511
7				9000002	.000001	.000003	.000008	.000019	.000039	.000073
8							.000001	.000002	.000004	.000009
9										.000001

x	$\mu t = 2.0$	$\mu t = 3.0$	$\mu t = 4.0$	$\mu t = 5.0$	$\mu t = 6.0$	$\mu t = 7.0$	$\mu t = 8.0$	$\mu t = 9.0$	$\mu t = 10.0$
0	.135335	.049787	.018316	.006738	.002479	.000912	.000335	.000123	.000045
1	.270671	.149361	.073263	.033690	.014873	.006383	.002684	.001111	.000454
2	.270671	.224042	.146525	.084224	.044618	.022341	.010735	.004998	.002270
3	.180447	.224042	.195367	.140374	.089235	.052129	.028626	.014994	.007567
4	.090224	.168031	.195367	.175467	.133853	.091226	.057252	.033737	.018917
5	.036089	.100819	.156293	.175467	.160623	.127717	.091604	.060727	.037833
6	.012030	.050409	.104196	.146223	.160623	.149003	.122138	.091090	.063055
7	.003437	.021604	.059540	.104445	.137677	.149003	.139587	.117116	.090079
8	000859	.008102	.029770	.065278	.103258	.130377	.139587	.131756	.112599
9	.000191	.002701	.013231	.036266	.068838	.101405	.124077	.131756	.125110
10	.000038	.000810	.005292	.018133	.041303	.070983	.099262	.118580	.125110
11	.000007	.000221	.001925	.008242	.022529	.045171	.072190	.097020	.113736
12	.000001	.000055	.000642	.003434	.011264	.026350	.048127	.072765	.094780
13		.000013	.000197	.001321	.005199	.014188	.029616	.050376	.072908
14		.000003	.000056	.000472	.002228	.007094	.016924	.032384	.052077
15		.000001	.000015	.000157	.000891	.003311	.009026	.019431	.034718
16			.000004	.000049	.000334	.001448	.004513	.010930	.021699
17			.000001	.000014	.000118	.000596	.002124	.005786	.012764
18				.000004	.000039	.000232	.000944	.002893	.007091
19				.000001	.000012	.000085	.000397	.001370	.003732
20					.000004	.000030	.000159	.000617	.001866
21					.000001	.000010	.000061	.000264	.000889
22						.000003	.000022	.000108	.000404
23						.000001	.000008	000042	.000176
24							.000003	.000016	.000073
25							.000001	.000006	.000029
26								.000002	.000011
27								.000001	.000004
28									.000001
29									.000001

Source: E. C. Molina, *Poisson's Exponential Binomial Limit* (Princeton, N.J.: D. Van Nostrand, 1942). Reprinted by permission of Brooks/Cole Engineering Division of Wadsworth, Inc., Monterey, California 93940.

Table B–3 Standard normal distribution areas

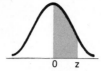

z	.00	.01	.02	.03	.04	.05	.06	.07	.08	.09
0.0	.0000	.0040	.0080	.0120	.0160	.0199	.0239	.0279	.0319	.0359
0.1	.0398	.0438	.0478	.0517	.0557	.0596	.0636	.0675	.0714	.0753
0.2	.0793	.0832	.0871	.0910	.0948	.0987	.1026	.1064	.1103	.1141
0.3	.1179	.1217	.1255	.1293	.1331	.1368	.1406	.1443	.1480	.1517
0.4	.1554	.1591	.1628	.1664	.1700	.1736	.1772	.1808	.1844	.1879
0.5	.1915	.1950	.1985	.2019	.2054	.2088	.2123	.2157	.2190	.2224
0.6	.2257	.2291	.2324	.2357	.2389	.2422	.2454	.2486	.2518	.2549
0.7	.2580	.2612	.2642	.2673	.2704	.2734	.2764	.2794	.2823	.2852
0.8	.2881	.2910	.2939	.2967	.2995	.3023	.3051	.3078	.3106	.3133
0.9	.3159	.3186	.3212	.3238	.3264	.3289	.3315	.3340	.3365	.3389
1.0	.3413	.3438	.3461	.3485	.3508	.3531	.3554	.3577	.3599	.3621
1.1	.3643	.3665	.3686	.3708	.3729	.3749	.3770	.3790	.3810	.3830
1.2	.3849	.3869	.3888	.3907	.3925	.3944	.3962	.3980	.3997	.4015
1.3	.4032	.4049	.4066	.4082	.4099	.4115	.4131	.4147	.4162	.4177
1.4	.4192	.4207	.4222	.4236	.4251	.4265	.4279	.4292	.4306	.4319
1.5	.4332	.4345	.4357	.4370	.4382	.4394	.4406	.4418	.4429	.4441
1.6	.4452	.4463	.4474	.4484	.4495	.4505	.4515	.4525	.4535	.4545
1.7	.4554	.4564	.4573	.4582	.4591	.4599	.4608	.4616	.4625	.4633
1.8	.4641	.4649	.4656	.4664	.4671	.4678	.4686	.4693	.4699	.4706
1.9	.4713	.4719	.4726	.4732	.4738	.4744	.4750	.4756	.4761	.4767
2.0	.4772	.4778	.4783	.4788	.4793	.4798	.4803	.4808	.4812	.4817
2.1	.4821	.4826	.4830	.4834	.4838	.4842	.4846	.4850	.4854	.4857
2.2	.4861	.4864	.4868	.4871	.4875	.4878	.4881	.4884	.4887	.4890
2.3	.4893	.4896	.4898	.4901	.4904	.4906	.4909	.4911	.4913	.4916
2.4	.4918	.4920	.4922	.4925	.4927	.4929	.4931	.4932	.4934	.4936
2.5	.4938	.4940	.4941	.4943	.4945	.4946	.4948	.4949	.4951	.4952
2.6	.4953	.4955	.4956	.4957	.4959	.4960	.4961	.4962	.4963	.4964
2.7	.4965	.4966	.4967	.4968	.4969	.4970	.4971	.4972	.4973	.4974
2.8	.4974	.4975	.4976	.4977	.4977	.4978	.4979	.4979	.4980	.4981
2.9	.4981	.4982	.4982	.4983	.4984	.4984	.4985	.4985	.4986	.4986
3.0	.49865	.4987	.4987	.4988	.4988	.4989	.4989	.4989	.4990	.4990
4.0	.4999683									

Source: From J. Neter, W. Wasserman, and G. A. Whitmore, *Fundamental Statistics for Business and Economics,* Fourth Edition. Copyright © 1973 by Allyn and Bacon, Inc. Reprinted with permission.

df= n−1	Level of significance for one-tailed test					
	.10	.05	.025	.01	.005	.0005
	Level of significance for two-tailed test					
	.20	.10	.05	.02	.01	.001
1	3.078	6.314	12.706	31.821	63.657	636.619
2	1.886	2.920	4.303	6.965	9.925	31.598
3	1.638	2.353	3.182	4.541	5.841	12.941
4	1.533	2.132	2.776	3.747	4.604	8.610
5	1.476	2.015	2.571	3.365	4.032	6.859
6	1.440	1.943	2.447	3.143	3.707	5.959
7	1.415	1.895	2.365	2.998	3.499	5.405
8	1.397	1.860	2.306	2.896	3.355	5.041
9	1.383	1.833	2.262	2.821	3.250	4.781
10	1.372	1.812	2.228	2.764	3.169	4.587
11	1.363	1.796	2.201	2.718	3.106	4.437
12	1.356	1.782	2.179	2.681	3.055	4.318
13	1.350	1.771	2.160	2.650	3.012	4.221
14	1.345	1.761	2.145	2.624	2.977	4.140
15	1.341	1.753	2.131	2.602	2.947	4.073
16	1.337	1.746	2.120	2.583	2.921	4.015
17	1.333	1.740	2.110	2.567	2.898	3.965
18	1.330	1.734	2.101	2.552	2.878	3.922
19	1.328	1.729	2.093	2.539	2.861	3.883
20	1.325	1.725	2.086	2.528	2.845	3.850
21	1.323	1.721	2.080	2.518	2.831	3.819
22	1.321	1.717	2.074	2.508	2.819	3.792
23	1.319	1.714	2.069	2.500	2.807	3.767
24	1.318	1.711	2.064	2.492	2.797	3.745
25	1.316	1.708	2.060	2.485	2.787	3.725
26	1.315	1.706	2.056	2.479	2.779	3.707
27	1.314	1.703	2.052	2.473	2.771	3.690
28	1.313	1.701	2.048	2.467	2.763	3.674
29	1.311	1.699	2.045	2.462	2.756	3.659
30	1.310	1.697	2.042	2.457	2.750	3.646
40	1.303	1.684	2.021	2.423	2.704	3.551
60	1.296	1.671	2.000	2.390	2.660	3.460
120	1.289	1.658	1.980	2.358	2.617	3.373
∞	1.282	1.645	1.960	2.326	2.576	3.291

Source: Table B–4 is taken from Table III of Fisher & Yates: *Statistical Tables for Biological, Agricultural and Medical Research* published by Longman Group Ltd. London (previously published by Oliver and Boyd Ltd, Edinburgh) and by permission of the authors and publishers.

Table B–5 Critical values of the F-distribution at a 5 percent level of significance ($\alpha = 0.05$)

Degrees of freedom for numerator

	1	2	3	4	5	6	7	8	9	10	12	15	20	24	30	40	60	120	∞
1	161	200	216	225	230	234	237	239	241	242	244	246	248	249	250	251	252	253	254
2	18.5	19.0	19.2	19.2	19.3	19.3	19.4	19.4	19.4	19.4	19.4	19.4	19.4	19.5	19.5	19.5	19.5	19.5	19.5
3	10.1	9.55	9.28	9.12	9.01	8.94	8.89	8.85	8.81	8.79	8.74	8.70	8.66	8.64	8.62	8.59	8.57	8.55	8.53
4	7.71	6.94	6.59	6.39	6.26	6.16	6.09	6.04	6.00	5.96	5.91	5.86	5.80	5.77	5.75	5.72	5.69	5.66	5.63
5	6.61	5.79	5.41	5.19	5.05	4.95	4.88	4.82	4.77	4.74	4.68	4.62	4.56	4.53	4.50	4.46	4.43	4.40	4.37
6	5.99	5.14	4.76	4.53	4.39	4.28	4.21	4.15	4.10	4.06	4.00	3.94	3.87	3.84	3.81	3.77	3.74	3.70	3.67
7	5.59	4.74	4.35	4.12	3.97	3.87	3.79	3.73	3.68	3.64	3.57	3.51	3.44	3.41	3.38	3.34	3.30	3.27	3.23
8	5.32	4.46	4.07	3.84	3.69	3.58	3.50	3.44	3.39	3.35	3.28	3.22	3.15	3.12	3.08	3.04	3.01	2.97	2.93
9	5.12	4.26	3.86	3.63	3.48	3.37	3.29	3.23	3.18	3.14	3.07	3.01	2.94	2.90	2.86	2.83	2.79	2.75	2.71
10	4.96	4.10	3.71	3.48	3.33	3.22	3.14	3.07	3.02	2.98	2.91	2.85	2.77	2.74	2.70	2.66	2.62	2.58	2.54
11	4.84	3.98	3.59	3.36	3.20	3.09	3.01	2.95	2.90	2.85	2.79	2.72	2.65	2.61	2.57	2.53	2.49	2.45	2.40
12	4.75	3.89	3.49	3.26	3.11	3.00	2.91	2.85	2.80	2.75	2.69	2.62	2.54	2.51	2.47	2.43	2.38	2.34	2.30
13	4.67	3.81	3.41	3.18	3.03	2.92	2.83	2.77	2.71	2.67	2.60	2.53	2.46	2.42	2.38	2.34	2.30	2.25	2.21
14	4.60	3.74	3.34	3.11	2.96	2.85	2.76	2.70	2.65	2.60	2.53	2.46	2.39	2.35	2.31	2.27	2.22	2.18	2.13
15	4.54	3.68	3.29	3.06	2.90	2.79	2.71	2.64	2.59	2.54	2.48	2.40	2.33	2.29	2.25	2.20	2.16	2.11	2.07
16	4.49	3.63	3.24	3.01	2.85	2.74	2.66	2.59	2.54	2.49	2.42	2.35	2.28	2.24	2.19	2.15	2.11	2.06	2.01
17	4.45	3.59	3.20	2.96	2.81	2.70	2.61	2.55	2.49	2.45	2.38	2.31	2.23	2.19	2.15	2.10	2.06	2.01	1.96
18	4.41	3.55	3.16	2.93	2.77	2.66	2.58	2.51	2.46	2.41	2.34	2.27	2.19	2.15	2.11	2.06	2.02	1.97	1.92
19	4.38	3.52	3.13	2.90	2.74	2.63	2.54	2.48	2.42	2.38	2.31	2.23	2.16	2.11	2.07	2.03	1.98	1.93	1.88
20	4.35	3.49	3.10	2.87	2.71	2.60	2.51	2.45	2.39	2.35	2.28	2.20	2.12	2.08	2.04	1.99	1.95	1.90	1.84
21	4.32	3.47	3.07	2.84	2.68	2.57	2.49	2.42	2.37	2.32	2.25	2.18	2.10	2.05	2.01	1.96	1.92	1.87	1.81
22	4.30	3.44	3.05	2.82	2.66	2.55	2.46	2.40	2.34	2.30	2.23	2.15	2.07	2.03	1.98	1.94	1.89	1.84	1.78
23	4.28	3.42	3.03	2.80	2.64	2.53	2.44	2.37	2.32	2.27	2.20	2.13	2.05	2.01	1.96	1.91	1.86	1.81	1.76
24	4.26	3.40	3.01	2.78	2.62	2.51	2.42	2.36	2.30	2.25	2.18	2.11	2.03	1.98	1.94	1.89	1.84	1.79	1.73
25	4.24	3.39	2.99	2.76	2.60	2.49	2.40	2.34	2.28	2.24	2.16	2.09	2.01	1.96	1.92	1.87	1.82	1.77	1.71
30	4.17	3.32	2.92	2.69	2.53	2.42	2.33	2.27	2.21	2.16	2.09	2.01	1.93	1.89	1.84	1.79	1.74	1.68	1.62
40	4.08	3.23	2.84	2.61	2.45	2.34	2.25	2.18	2.12	2.08	2.00	1.92	1.84	1.79	1.74	1.69	1.64	1.58	1.51
60	4.00	3.15	2.76	2.53	2.37	2.25	2.17	2.10	2.04	1.99	1.92	1.84	1.75	1.70	1.65	1.59	1.53	1.47	1.39
120	3.92	3.07	2.68	2.45	2.29	2.18	2.09	2.02	1.96	1.91	1.83	1.75	1.66	1.61	1.55	1.50	1.43	1.35	1.25
∞	3.84	3.00	2.60	2.37	2.21	2.10	2.01	1.94	1.88	1.83	1.75	1.67	1.57	1.52	1.46	1.39	1.32	1.22	1.00

Degrees of freedom for denominator

Source: M. Merrington and C. M. Thompson, "Tables of Percentage Points of the Inverted Beta (F) Distribution." *Biometrika* 33 (1943), by permission of the Biometrika trustees.

Table B–5 (*concluded*)   **Critical values of the *F*-distribution at a 1 percent level of significance ($\alpha = 0.01$)**

Degrees of freedom for numerator

		1	2	3	4	5	6	7	8	9	10	12	15	20	24	30	40	60	120	∞
	1	4,052	5,000	5,403	5,625	5,764	5,859	5,928	5,982	6,023	6,056	6,106	6,157	6,209	6,235	6,261	6,287	6,313	6,339	6,366
	2	98.5	99.0	99.2	99.2	99.3	99.3	99.4	99.4	99.4	99.4	99.4	99.4	99.4	99.5	99.5	99.5	99.5	99.5	99.5
	3	34.1	30.8	29.5	28.7	28.2	27.9	27.7	27.5	27.3	27.2	27.1	26.9	26.7	26.6	26.5	26.4	26.3	26.2	26.1
	4	21.2	18.0	16.7	16.0	15.5	15.2	15.0	14.8	14.7	14.5	14.4	14.2	14.0	13.9	13.8	13.7	13.7	13.6	13.5
	5	16.3	13.3	12.1	11.4	11.0	10.7	10.5	10.3	10.2	10.1	9.89	9.72	9.55	9.47	9.38	9.29	9.20	9.11	9.02
	6	13.7	10.9	9.78	9.15	8.75	8.47	8.26	8.10	7.98	7.87	7.72	7.56	7.40	7.31	7.23	7.14	7.06	6.97	6.88
	7	12.2	9.55	8.45	7.85	7.46	7.19	6.99	6.84	6.72	6.62	6.47	6.31	6.16	6.07	5.99	5.91	5.82	5.74	5.65
	8	11.3	8.65	7.59	7.01	6.63	6.37	6.18	6.03	5.91	5.81	5.67	5.52	5.36	5.28	5.20	5.12	5.03	4.95	4.86
	9	10.6	8.02	6.99	6.42	6.06	5.80	5.61	5.47	5.35	5.26	5.11	4.96	4.81	4.73	4.65	4.57	4.48	4.40	4.31
	10	10.0	7.56	6.55	5.99	5.64	5.39	5.20	5.06	4.94	4.85	4.71	4.56	4.41	4.33	4.25	4.17	4.08	4.00	3.91
	11	9.65	7.21	6.22	5.67	5.32	5.07	4.89	4.74	4.63	4.54	4.40	4.25	4.10	4.02	3.94	3.86	3.78	3.69	3.60
	12	9.33	6.93	5.95	5.41	5.06	4.82	4.64	4.50	4.39	4.30	4.16	4.01	3.86	3.78	3.70	3.62	3.54	3.45	3.36
	13	9.07	6.70	5.74	5.21	4.86	4.62	4.44	4.30	4.19	4.10	3.96	3.82	3.66	3.59	3.51	3.43	3.34	3.25	3.17
	14	8.86	6.51	5.56	5.04	4.70	4.46	4.28	4.14	4.03	3.94	3.80	3.66	3.51	3.43	3.35	3.27	3.18	3.09	3.00
	15	8.68	6.36	5.42	4.89	4.56	4.32	4.14	4.00	3.89	3.80	3.67	3.52	3.37	3.29	3.21	3.13	3.05	2.96	2.87
	16	8.53	6.23	5.29	4.77	4.44	4.20	4.03	3.89	3.78	3.69	3.55	3.41	3.26	3.18	3.10	3.02	2.93	2.84	2.75
	17	8.40	6.11	5.19	4.67	4.34	4.10	3.93	3.79	3.68	3.59	3.46	3.31	3.16	3.08	3.00	2.92	2.83	2.75	2.65
	18	8.29	6.01	5.09	4.58	4.25	4.01	3.84	3.71	3.60	3.51	3.37	3.23	3.08	3.00	2.92	2.84	2.75	2.66	2.57
	19	8.19	5.93	5.01	4.50	4.17	3.94	3.77	3.63	3.52	3.43	3.30	3.15	3.00	2.92	2.84	2.76	2.67	2.58	2.49
	20	8.10	5.85	4.94	4.43	4.10	3.87	3.70	3.56	3.46	3.37	3.23	3.09	2.94	2.86	2.78	2.69	2.61	2.52	2.42
	21	8.02	5.78	4.87	4.37	4.04	3.81	3.64	3.51	3.40	3.31	3.17	3.03	2.88	2.80	2.72	2.64	2.55	2.46	2.36
	22	7.95	5.72	4.82	4.31	3.99	3.76	3.59	3.45	3.35	3.26	3.12	2.98	2.83	2.75	2.67	2.58	2.50	2.40	2.31
	23	7.88	5.66	4.76	4.26	3.94	3.71	3.54	3.41	3.30	3.21	3.07	2.93	2.78	2.70	2.62	2.54	2.45	2.35	2.26
	24	7.82	5.61	4.72	4.22	3.90	3.67	3.50	3.36	3.26	3.17	3.03	2.89	2.74	2.66	2.58	2.49	2.40	2.31	2.21
	25	7.77	5.57	4.68	4.18	3.86	3.63	3.46	3.32	3.22	3.13	2.99	2.85	2.70	2.62	2.53	2.45	2.36	2.27	2.17
	30	7.56	5.39	4.51	4.02	3.70	3.47	3.30	3.17	3.07	2.98	2.84	2.70	2.55	2.47	2.39	2.30	2.21	2.11	2.01
	40	7.31	5.18	4.31	3.83	3.51	3.29	3.12	2.99	2.89	2.80	2.66	2.52	2.37	2.29	2.20	2.11	2.02	1.92	1.80
	60	7.08	4.98	4.13	3.65	3.34	3.12	2.95	2.82	2.72	2.63	2.50	2.35	2.20	2.12	2.03	1.94	1.84	1.73	1.60
	120	6.85	4.79	3.95	3.48	3.17	2.96	2.79	2.66	2.56	2.47	2.34	2.19	2.03	1.95	1.86	1.76	1.66	1.53	1.38
	∞	6.63	4.61	3.78	3.32	3.02	2.80	2.64	2.51	2.41	2.32	2.18	2.04	1.88	1.79	1.70	1.59	1.47	1.32	1.00

Degrees of freedom for denominator (row labels at left)

APPENDIX C STAT CITY DATA BASE

VARIABLES

01	02	03	04	05	06	07	08	09	10
KILFOYLE	406 6TH ST	43	1	1	1	502	88264	10	1199
GROGER	405 7TH ST	43	2	1	1	819	89452	9	1060
SMAY	406 7TH ST	44	3	1	1	644	76737	8	930
POTTER	405 8TH ST	44	4	1	1	716	88671	9	1051
STOUT	406 8TH ST	45	5	1	1	442	55659	6	674
DIERCKS	405 9TH ST	45	6	1	1	340	48452	5	563
READ	406 9TH ST	46	7	1	1	372	38403	4	520
NELSON	404 6TH ST	43	8	1	1	631	93538	9	1088
HILL	403 7TH ST	43	9	1	1	796	92083	9	1101
GOLDEN	404 7TH ST	44	10	1	1	690	71316	8	909
ABLES	403 8TH ST	44	11	1	1	446	65276	7	839
CURRY	404 8TH ST	45	12	1	1	1022	94549	11	1320
RIVAS	403 9TH ST	45	13	1	1	520	54886	6	694
KNIGHT	404 9TH ST	46	14	1	1	917	109640	11	1394
HALTER	403 DIVISION ST	46	15	1	1	696	97060	9	1070
MATTHEWS	402 6TH ST	43	16	1	1	549	65679	7	817
RODRIGUEZ	401 7TH ST	43	17	1	1	683	78597	8	907
MATUS	402 7TH ST	44	18	1	1	568	63206	7	821
BROWN	401 8TH ST	44	19	1	1	634	86482	9	1016
SOLLA	402 8TH ST	45	20	1	1	542	70656	7	898
KOATIK	401 9TH ST	45	21	1	1	681	83328	8	983
UNDERWOOD	402 9TH ST	46	22	1	1	446	65397	7	797
DUNN	401 10TH ST	46	23	1	1	599	89070	9	981
KELLY	402 DIVISION ST	47	24	1	1	727	84227	9	1135
MCGIBNEY	306 6TH ST	32	25	1	1	652	79381	9	1019
SMULIN	305 7TH ST	32	26	1	1	647	99762	10	1226
BRENNER	306 7TH ST	33	27	1	1	511	63601	7	846
KLIEMAN	305 8TH ST	33	28	1	1	449	73180	7	834
JONES	306 8TH ST	34	29	1	1	598	64743	7	778
BAUER	305 9TH ST	34	30	1	1	767	86624	9	1136
CLARK	306 9TH ST	35	31	1	1	766	80786	9	995
REID	305 10TH ST	35	32	1	1	657	82577	8	984
FRENCH	306 10TH ST	36	33	1	1	503	88014	9	1004
LILLEY	304 6TH ST	32	34	1	1	882	97466	11	1303
EHREN	303 7TH ST	32	35	1	1	882	97642	10	1143
KILPATRICK	304 7TH ST	33	36	1	1	611	79019	8	894
SMITH	303 8TH ST	33	37	1	1	820	94974	9	1069
AYCOCK	304 8TH ST	34	38	1	1	655	91327	10	1204
HARRIS	303 9TH ST	34	39	1	1	514	79306	8	1035
OEKLER	304 9TH ST	35	40	1	1	683	90484	10	1105
PAGE	303 10TH ST	35	41	1	1	263	46025	4	259
NEWCOMB	304 10TH ST	36	42	1	1	616	39420	8	639
COLLINS	304 DIVISION ST	36	43	1	1	545	68259	7	876
HUGHES	302 6TH ST	32	44	1	1	687	76909	9	1147
VALENTI	301 7TH ST	32	45	1	1	626	77460	9	1063
MARKLEY	302 7TH ST	33	46	1	1	795	83364	9	1115
CRIBBS	301 8TH ST	33	47	1	1	610	83133	9	1107
BOGUS	302 8TH ST	34	48	1	1	645	79711	8	1029
CLEMA	301 9TH ST	34	49	1	1	327	45077	4	513
HAYES	302 9TH ST	35	50	1	1	666	75622	8	925
BARR	301 10TH ST	35	51	1	1	332	42456	5	556
SUMMERS	302 10TH ST	36	52	1	1	775	91355	9	1084
PHILLIPS	302 DIVISION ST	36	53	1	1	433	60875	6	751
FLYTHE	206 6TH ST	20	54	1	1	566	66700	7	802
KING	205 7TH ST	20	55	1	1	678	86592	9	1065
BRYANT	206 7TH ST	21	56	1	1	321	42926	5	432
MCLEAN	205 8TH ST	21	57	1	1	445	57500	7	852
CLOSSEY	206 8TH ST	22	58	1	1	597	99761	11	1276
BARTLEY	205 9TH ST	22	59	1	1	567	78597	8	901
COX	206 9TH ST	23	60	1	1	670	109025	11	1351
NIXON	205 10TH ST	23	61	1	1	485	63638	7	875
JOHNSTONE	206 10TH ST	24	62	1	1	573	73912	8	915
SIBLEY	205 11TH ST	24	63	1	1	653	67194	7	857
LUNN	206 11TH ST	25	64	1	1	362	60679	7	821
BOYER	204 6TH ST	20	65	1	1	623	70850	7	833
EVERHAM	203 7TH ST	20	66	1	1	661	85978	9	1071
LANDER	204 7TH ST	21	67	1	1	667	90105	9	1060
CHALMERS	203 8TH ST	21	68	1	1	305	49752	5	568
BROWNING	204 8TH ST	22	69	1	1	815	92235	9	1034
STRONG	203 9TH ST	22	70	1	1	480	60991	6	709
PRICE	204 9TH ST	23	71	1	1	696	76904	8	883
DESOSA	203 10TH ST	23	72	1	1	618	69089	7	825
DILLON	204 10TH ST	24	73	1	1	685	93546	9	1152
DELVECCHIO	203 11TH ST	24	74	1	1	732	84303	8	853
STALEY	204 11TH ST	25	75	1	1	358	45419	5	548
LEONARD	204 DIVISION ST	25	76	1	1	698	107868	11	1253
HASTINGS	202 6TH ST	20	77	1	1	781	86713	9	1021
MICHAEL	201 7TH ST	20	78	1	1	456	56732	6	732
IRELAN	202 7TH ST	21	79	1	1	615	75608	8	1004
OELZE	201 8TH ST	21	80	1	1	509	89782	9	1048
COOK	202 8TH ST	22	81	1	1	840	86669	9	1086
POKORNEY	201 9TH ST	22	82	1	1	263	48297	4	338
HAMATY	202 9TH ST	23	83	1	1	802	82304	9	1056
WARREN	201 10TH ST	23	84	1	1	461	67962	7	769
SAUNDERS	202 10TH ST	24	85	1	1	566	73351	7	864
CAGLE	201 11TH ST	24	86	1	1	880	97313	10	1194
MANLEY	202 11TH ST	25	87	1	1	506	71024	8	910
SLOOP	201 12TH ST	25	88	1	1	359	50844	5	646
RADLEY	202 DIVISION ST	25	89	1	1	837	96634	10	1146
ALLISON	106 6TH ST	7	90	1	1	402	51569	5	641
TOURSH	105 7TH ST	7	91	1	1	510	58524	8	1074
IMPERI	106 7TH ST	8	92	1	1	629	80079	9	1081
SOMMER	105 8TH ST	8	93	1	1	662	82695	9	1019
GEBHARDT	106 8TH ST	9	94	1	1	530	66490	7	954
ANGUEIRA	105 9TH ST	9	95	1	1	469	52686	5	585

01	11	12	13	14	15	16	17	18	19
KILFOYLE	88	90	56419	6	3	165	55	6	1
GROGER	82	12	23280	5	3	191	64	7	1
SMAY	66	62	33350	4	2	159	80	7	2
POTTER	78	15	45826	5	3	187	62	4	2
STOUT	54	107	40603	2	2	133	67	5	0
DIERCKS	42	24	45454	2	2	133	67	3	1
READ	30	103	19469	4	1	65	65	8	1
NELSON	80	86	35118	5	3	199	66	8	2
HILL	85	13	38822	4	3	177	59	6	1
GOLDEN	80	78	21030	3	2	152	76	4	1
ABLES	62	60	54135	3	2	151	76	4	2
CURRY	95	20	57872	6	5	234	47	3	1
RIVAS	52	14	31098	2	2	127	64	4	2
KNIGHT	101	15	28180	7	2	120	60	6	1
HALTER	78	70	23328	4	3	183	61	8	1
MATTHEWS	63	79	23094	4	3	177	59	7	2
RODRIGUEZ	72	47	32756	4	2	149	75	5	2
MATUS	64	17	55162	4	3	183	61	5	1
BROWN	84	13	49234	5	3	185	62	9	0
SOLLA	61	35	41890	2	2	137	69	6	2
KOATIK	69	19	52729	4	2	146	73	6	2
UNDERWOOD	59	49	56263	3	2	123	62	9	2
DUNN	86	52	36304	5	3	186	62	5	2
KELLY	81	18	26977	4	3	174	58	10	0
MCGIBNEY	87	60	63040	5	4	201	50	6	2
SMULIN	86	23	33693	5	4	206	52	3	1
BRENNER	61	47	52135	2	1	113	113	8	2
KLIEMAN	62	13	38567	3	2	159	80	9	1
JONES	68	19	54086	3	2	133	67	6	2
BAUER	83	14	29053	4	2	144	72	7	1
CLARK	77	25	51304	5	3	197	66	8	2
REID	81	17	31608	5	4	211	53	2	2
FRENCH	84	70	49645	4	3	183	61	4	2
LILLEY	103	11	44956	7	2	120	60	9	1
EHREN	89	55	32357	6	3	165	55	4	0
KILPATRICK	74	41	65173	3	3	175	58	5	1
SMITH	83	91	36483	4	3	178	59	7	2
AYCOCK	99	46	49754	6	4	217	54	9	1
HARRIS	74	14	45814	3	3	180	60	5	2
OEKLER	93	17	30600	6	4	220	55	7	2
PAGE	27	82	19505	1	2	120	60	8	1
NEWCOMB	65	25	21756	3	2	144	72	4	2
COLLINS	64	17	38757	4	3	183	61	5	0
HUGHES	88	16	50588	4	2	147	74	11	1
VALENTI	91	17	30237	4	2	144	72	6	2
MARKLEY	85	68	35457	4	3	172	57	8	1
CRIBBS	84	20	51671	4	2	162	81	5	2
BOGUS	73	21	29375	3	2	165	83	6	1
CLEMA	35	12	19626	4	2	120	60	7	0
HAYES	66	19	45756	4	4	211	53	8	2
BARR	35	99	41062	1	1	65	65	5	2
SUMMERS	85	13	24718	6	2	120	60	5	2
PHILLIPS	51	25	43451	1	1	65	65	6	1
FLYTHE	64	55	42770	3	2	124	62	6	1
KING	83	38	33910	5	3	187	62	5	2
BRYANT	37	11	15467	4	1	65	65	8	1
MCLEAN	59	15	23580	4	3	166	55	7	2
CLOSSEY	91	63	30830	6	4	217	54	5	1
BARTLEY	76	22	33536	4	2	147	74	4	1
COX	109	15	22026	7	3	165	55	5	1
NIXON	58	81	47583	3	2	163	82	6	1
JOHNSTONE	76	69	29501	5	3	184	61	10	2
SIBLEY	65	49	35468	3	2	134	67	5	2
LUNN	61	49	30624	4	3	167	56	6	0
BOYER	57	53	29381	3	2	123	62	4	1
EVERHAM	78	18	72453	5	3	188	63	7	2
LANDER	81	24	55787	5	4	202	51	7	2
CHALMERS	39	21	41836	1	1	65	65	4	1
BROWNING	82	59	33369	5	4	210	53	8	1
STRONG	53	44	54578	3	2	149	75	8	2
PRICE	67	27	22348	4	3	181	60	4	2
DESOSA	70	39	29910	3	2	131	66	3	2
DILLON	82	69	39119	6	4	214	54	6	2
DELVECCHIO	77	70	43249	3	2	129	65	5	1
STALEY	44	22	28812	1	1	85	85	2	1
LEONARD	100	11	46185	6	4	219	55	6	2
HASTINGS	90	53	59055	5	4	204	51	5	0
MICHAEL	54	16	60633	2	1	117	117	4	2
IRELAN	76	59	41305	4	2	160	80	9	2
OELZE	81	21	39197	4	2	150	75	6	1
COOK	87	14	33260	5	3	193	64	8	1
POKORNEY	28	13	17046	5	1	65	65	9	0
HAMATY	86	20	23483	5	3	197	66	8	1
WARREN	64	22	53950	3	2	163	82	5	2
SAUNDERS	65	19	39171	3	2	156	78	3	0
CAGLE	102	17	42818	6	4	219	55	7	0
MANLEY	73	16	37523	5	4	211	53	7	2
SLOOP	42	19	40283	2	1	99	99	2	2
RADLEY	92	19	34680	6	2	120	60	7	2
ALLISON	40	15	57205	1	1	79	79	3	2
TOURSH	70	18	44914	4	3	177	59	4	1
IMPERI	84	19	52548	5	3	190	63	8	1
SOMMER	82	22	56518	5	5	231	46	4	1
GEBHARDT	59	23	36165	3	2	158	79	8	1
ANGUEIRA	42	23	27039	1	1	75	75	6	2

01	20	21	22	23	24	25	26	27
KILFOYLE	2	6	128	21	3	2	0	5
GROGER	1	6	81	16	2	2	0	0
SMAY	1	3	95	24	3	1	0	0
POTTER	1	3	104	21	2	0	6	0
STOUT	1	2	52	26	3	2	0	1
DIERCKS	1	1	50	25	3	0	0	1
READ	1	4	107	27	3	0	4	2
NELSON	2	5	106	21	3	0	0	0
HILL	1	3	95	24	3	2	0	1
GOLDEN	2	1	60	20	3	0	4	3
ABLES	1	3	67	22	3	0	3	6
CURRY	2	3	122	20	3	2	0	5
RIVAS	2	1	54	27	3	0	0	0
KNIGHT	2	7	155	22	3	0	0	1
HALTER	2	4	90	23	3	0	1	0
MATTHEWS	2	8	92	23	3	0	1	10
RODRIGUEZ	2	4	77	19	3	1	0	0
MATUS	2	6	101	25	3	0	5	3
BROWN	2	1	101	20	3	0	0	0
SOLLA	2	1	49	25	3	0	4	0
KOATIK	2	6	72	18	3	0	0	0
UNDERWOOD	1	3	74	25	3	0	0	3
DUNN	1	3	115	23	3	0	0	2
KELLY	1	5	91	23	3	0	0	0
MCGIBNEY	2	4	90	18	3	0	5	1
SMULIN	1	5	121	24	3	1	0	9
BRENNER	2	3	45	23	3	0	0	0
KLIEMAN	2	1	64	21	3	2	0	0
JONES	1	2	75	25	3	0	0	0
BAUER	1	5	79	20	2	1	0	2
CLARK	1	4	99	20	3	0	0	0
REID	2	6	95	19	3	0	0	0
FRENCH	1	1	83	21	3	0	0	0
LILLEY	2	7	176	25	3	0	3	1
EHREN	1	3	115	19	3	0	0	0
KILPATRICK	2	4	74	25	3	2	0	3
SMITH	1	2	83	21	2	0	1	0
AYCOCK	2	9	108	18	3	0	2	0
HARRIS	1	1	68	23	2	0	1	3
OEKLER	1	8	116	19	3	0	4	1
PAGE	2	1	23	23	2	0	0	0
NEWCOMB	1	0	72	24	2	2	0	0
COLLINS	2	5	111	28	3	1	0	5
HUGHES	2	2	76	19	3	1	0	0
VALENTI	1	2	88	22	3	0	1	8
MARKLEY	2	6	75	19	3	2	0	4
CRIBBS	2	3	80	20	3	0	0	6
BOGUS	2	4	58	19	3	0	0	0
CLEMA	1	3	82	21	3	0	1	11
HAYES	2	2	96	24	3	1	0	0
BARR	2	1	22	22	2	0	6	0
SUMMERS	2	8	110	18	2	0	0	0
PHILLIPS	2	1	27	27	3	2	0	0
FLYTHE	2	3	69	23	3	0	4	0
KING	2	0	91	18	3	1	0	8
BRYANT	1	3	74	19	3	1	0	3
MCLEAN	2	5	78	20	3	2	0	0
CLOSSEY	2	8	125	21	3	1	0	3
BARTLEY	1	3	88	22	3	0	5	0
COX	1	2	157	22	3	2	0	0
NIXON	1	2	70	23	3	0	0	4
JOHNSTONE	2	5	108	22	3	0	0	0
SIBLEY	2	2	67	22	3	0	0	0
LUNN	1	4	86	22	3	1	0	0
BOYER	2	2	69	23	1	1	0	0
EVERHAM	2	5	93	19	3	1	0	0
LANDER	1	3	111	22	3	2	0	8
CHALMERS	2	1	27	27	3	1	0	2
BROWNING	2	8	108	22	3	0	0	0
STRONG	2	2	58	19	3	0	1	0
PRICE	1	2	97	24	3	0	0	0
DESOSA	1	3	79	26	3	0	4	0
DILLON	2	1	113	19	3	0	0	0
DELVECCHIO	2	1	69	23	3	2	0	4
STALEY	1	1	36	36	3	0	0	4
LEONARD	2	6	108	18	3	2	0	2
HASTINGS	2	3	91	18	3	0	0	0
MICHAEL	2	1	41	21	3	0	1	0
IRELAN	2	2	72	18	1	2	0	7
OELZE	2	6	74	19	3	2	0	1
COOK	2	4	97	19	2	0	1	2
POKORNEY	2	1	103	21	2	2	0	4
HAMATY	2	0	96	19	3	0	1	4
WARREN	1	1	73	24	3	0	0	0
SAUNDERS	1	1	51	17	3	0	0	0
CAGLE	1	1	145	24	3	0	0	0
MANLEY	1	4	107	21	2	0	3	2
SLOOP	2	2	41	21	3	1	0	3
RADLEY	1	10	146	24	3	0	1	4
ALLISON	2	1	24	24	3	1	0	0
TOURSH	2	3	99	25	3	0	0	0
IMPERI	2	3	97	19	3	0	1	0
SOMMER	2	3	106	21	3	2	0	2
GEBHARDT	2	3	57	19	3	0	0	3
ANGUEIRA	2	1	24	24	3	2	0	5

Appendix C

230

VARIABLES

01	02	03	04	05	06	07	08	09	10	11	12	13	14	15	16	17	18	19	20	21	22	23	24	25	26	27
RABIN	106 9TH ST	10	96	1	1	537	65852	6	778	55	28	52221	2	2	140	70	9	1	1	1	58	29	3	0	1	2
JOHNSON	105 10TH ST	10	97	1	1	411	59277	6	805	53	71	57553	1	1	68	68	4	2	2	0	28	28	3	0	5	0
LONGUEIRA	106 10TH ST	11	98	1	1	547	66167	6	752	53	18	57092	2	2	142	71	7	2	2	1	63	32	3	2	0	7
DEGEORGE	105 11TH ST	11	99	1	1	551	78553	8	912	74	16	22771	4	2	164	82	10	2	2	2	84	21	3	0	4	8
BARNES	106 11TH ST	12	100	1	1	819	91319	9	1061	85	84	28863	4	3	170	57	6	2	2	1	99	25	3	0	0	0
LUND	105 12TH ST	12	101	1	1	580	75224	8	957	71	21	30853	3	2	164	82	6	0	2	2	66	22	3	0	0	0
QUINTANA	106 12TH ST	13	102	1	1	504	67803	7	767	66	73	35399	3	2	165	83	9	0	2	4	64	21	3	0	4	5
SWICK	104 6TH ST	7	103	1	1	696	69735	8	948	73	15	57061	4	3	166	55	7	2	1	2	65	16	3	0	0	2
GRIMANY	103 7TH ST	7	104	1	1	487	63368	6	706	53	80	41494	2	2	136	68	9	2	2	1	51	26	3	0	0	0
GRIFFITH	104 7TH ST	8	105	1	1	563	73015	7	802	64	21	49068	3	2	147	74	6	1	2	2	50	17	3	0	0	3
CAPPER	103 8TH ST	8	106	1	1	560	78595	9	1138	85	20	32150	5	4	200	50	8	0	2	3	114	23	3	0	5	1
BECTON	104 8TH ST	9	107	1	1	559	74627	8	1072	73	19	41403	4	2	150	75	6	2	2	5	77	19	3	0	0	0
KAUFMAN	103 9TH ST	9	108	1	1	483	85407	9	1017	84	34	40120	4	2	149	75	6	2	1	5	82	21	3	0	4	6
LINTON	104 9TH ST	10	109	1	1	611	82737	8	1017	75	13	56580	4	2	161	81	5	2	1	6	91	23	3	0	4	0
NEUMAN	103 10TH ST	10	110	1	1	445	57964	6	809	52	23	15895	6	1	65	65	4	1	1	6	126	21	1	0	4	1
ALONSO	104 10TH ST	11	111	1	1	490	76738	8	973	75	14	51931	5	3	193	64	5	2	2	0	99	20	3	0	0	11
KEENBURG	103 11TH ST	11	112	1	1	462	74980	8	940	70	17	32375	3	2	158	79	3	2	2	1	56	19	3	0	4	0
GETTIS	104 11TH ST	12	113	1	1	484	83877	8	971	67	17	26802	5	5	230	46	5	1	2	4	111	22	3	2	0	0
ROTHMAN	103 12TH ST	12	114	1	1	658	79453	8	925	67	20	33910	4	3	177	59	3	1	2	5	109	27	3	2	0	2
UPTON	104 12TH ST	13	115	1	1	897	94751	9	1157	83	33	23375	5	4	214	54	5	2	1	3	106	21	3	1	0	3
STUART	104 DIVISION ST	13	116	1	1	527	76193	9	1077	83	13	31244	6	4	221	55	6	2	2	4	105	18	3	0	3	0
WEEKES	102 6TH ST	7	117	1	1	598	74658	7	809	62	37	32502	3	2	145	73	5	1	2	3	56	19	3	0	0	0
PHILLIPS	101 7TH ST	7	118	1	1	549	77118	8	1027	71	26	51411	4	3	183	61	3	2	2	4	87	22	3	0	4	3
ROTH	102 7TH ST	8	119	1	1	389	57601	6	635	57	81	15300	6	1	65	65	4	0	2	4	126	21	3	0	0	0
DEROSE	101 8TH ST	8	120	1	1	509	68227	7	908	54	87	51526	2	2	139	70	8	2	1	3	45	23	1	2	0	0
RITZIE	102 8TH ST	9	121	1	1	757	101541	10	1130	96	16	24839	5	3	194	65	4	1	1	5	96	19	1	1	0	0
MILLER	101 9TH ST	9	122	1	1	537	83132	9	1036	86	49	62205	5	3	192	64	6	2	2	6	105	21	3	0	0	3
FLETCHER	102 9TH ST	10	123	1	1	298	37598	4	483	32	15	56479	1	1	65	65	5	2	2	0	25	25	3	0	0	0
GITLOW	101 10TH ST	10	124	1	1	448	52737	6	739	53	101	60229	1	1	86	86	1	0	1	1	28	28	2	0	0	10
SUGRUE	102 10TH ST	11	125	1	1	590	73690	8	972	74	15	48674	4	2	163	82	2	2	1	4	82	21	3	2	0	0
NELSON	101 11TH ST	11	126	1	1	456	47385	5	628	40	21	26316	2	2	139	70	6	0	1	2	52	26	3	0	5	3
FLEMING	102 11TH ST	12	127	1	1	411	46942	5	633	41	21	38611	2	1	115	115	1	2	2	2	47	24	3	2	0	4
BOYD	101 12TH ST	12	128	1	1	803	87319	9	1097	84	15	24917	4	2	150	75	4	2	1	3	78	20	3	0	1	0
SCHERER	102 12TH ST	13	129	1	1	606	70174	7	874	62	21	63849	3	2	122	61	5	2	2	1	75	25	3	0	1	0
ZANCA	102 DIVISION ST	13	130	1	1	590	86491	9	1067	86	11	31286	5	3	193	64	3	2	1	4	113	23	2	2	0	4
WOODSON	505 1ST ST	48	131	2	1	607	75135	6	781	57	20	51589	2	2	134	67	6	2	1	1	53	27	3	1	0	0
RIVAS	506 1ST ST	49	132	2	1	486	62746	7	541	66	52	20234	4	3	167	56	7	1	1	4	69	17	3	0	1	0
HERN	505 2ND ST	49	133	2	1	744	101344	8	1042	67	101	50001	4	3	170	57	6	2	2	0	79	20	3	2	0	0
KATZ	506 2ND ST	50	134	2	1	498	78077	7	1005	62	25	63330	3	2	125	63	5	1	1	0	65	22	3	2	0	0
MORALES	505 3RD ST	50	135	2	1	554	78850	7	987	58	84	70831	4	3	180	60	9	0	1	1	86	22	3	1	0	0
VEYHL	506 3RD ST	51	136	2	1	884	107094	9	1228	84	13	32648	5	3	190	63	5	0	2	1	89	18	3	2	0	9
SANDT	505 4TH ST	51	137	2	1	880	94070	8	1154	74	18	83345	3	2	135	68	6	2	2	2	70	23	2	0	6	0
WARREN	506 4TH ST	52	138	2	1	754	98666	8	1184	65	61	94249	3	2	139	70	7	1	2	0	78	26	3	0	0	2
GOLDBERG	503 1ST ST	48	139	2	1	742	108917	9	1253	84	22	90435	4	2	150	75	8	2	1	3	67	17	3	2	0	0
BOLDING	504 1ST ST	49	140	2	1	401	72312	6	475	50	51	23576	6	2	120	60	4	2	1	4	129	22	2	0	6	4
BRYAN	503 2ND ST	49	141	2	1	708	84778	7	1002	62	102	54903	3	2	133	67	7	2	1	2	68	23	3	0	1	0
GREENE	504 2ND ST	50	142	2	1	702	92637	8	1034	75	12	61106	4	4	202	51	4	2	2	2	75	19	3	1	0	3
ALVAREZ	503 3RD ST	50	143	2	1	832	103104	9	1265	91	85	76450	4	2	162	81	3	2	1	0	82	21	1	0	0	0
MCADAMS	504 3RD ST	51	144	2	1	582	84689	7	914	68	20	58321	3	2	149	75	5	2	2	1	58	19	1	0	0	0
ALEXANDER	503 4TH ST	51	145	2	1	1031	131033	11	1436	101	104	83669	5	4	203	51	8	2	1	3	109	22	3	0	1	0
WILKES	504 4TH ST	52	146	2	1	870	115109	9	1279	84	21	62771	5	4	216	54	4	2	2	1	100	20	3	2	0	0
O'CONNELL	501 1ST ST	48	147	2	1	842	112528	9	1261	81	12	103709	4	3	188	63	4	1	1	5	78	20	3	0	1	5
LOWE	502 1ST ST	49	148	2	1	956	113924	9	1268	84	10	120030	4	2	156	78	6	2	1	3	85	21	2	0	0	1
LEE	501 2ND ST	49	149	2	1	733	85027	8	1070	72	18	89656	3	2	162	81	8	2	2	1	72	24	3	2	0	10
RIVET	502 2ND ST	50	150	2	1	593	79421	7	1010	65	86	69898	2	1	118	118	7	1	2	1	43	22	3	2	0	11
FOGELSON	501 3RD ST	50	151	2	1	763	90176	8	1122	72	91	60310	5	3	194	65	2	1	2	1	104	21	3	0	0	0
JONES	502 3RD ST	51	152	2	1	880	99279	8	1174	74	15	83484	3	3	172	57	5	2	1	1	75	25	3	1	0	1
RIVERS	501 4TH ST	51	153	2	1	771	88803	7	944	64	82	80217	2	2	140	70	7	1	2	1	53	27	1	2	0	3
FERNANDEZ	502 4TH ST	52	154	2	1	840	92755	7	1014	62	21	70927	4	2	161	81	11	2	2	6	80	20	2	0	6	0
ALDIN	405 1ST ST	37	155	2	1	730	94470	8	1083	75	20	56172	4	2	144	72	5	1	2	5	84	21	1	2	0	2
CAIN	406 1ST ST	38	156	2	1	1102	129056	10	1510	84	40	82770	4	3	180	60	8	2	2	4	98	25	3	1	0	0
KNIGHT	405 2ND ST	38	157	2	1	806	93774	8	1006	79	47	52841	4	2	147	74	7	2	1	2	75	19	3	0	0	0
PHELPS	406 2ND ST	39	158	2	1	1021	126065	10	1443	99	26	42314	7	2	120	60	6	1	1	9	182	26	3	0	0	0
ROMERO	405 3RD ST	39	159	2	1	970	99652	8	1089	82	15	68809	5	4	203	51	4	1	1	6	99	20	2	0	6	2
BOIX	406 3RD ST	40	160	2	1	734	89888	7	891	65	40	78086	2	2	125	63	6	2	2	1	45	23	1	0	0	0
JEFFREY	405 4TH ST	40	161	2	1	672	93422	8	1053	75	18	67110	4	3	173	58	6	1	2	3	92	23	1	1	0	11
BOBSON	406 4TH ST	41	162	2	1	785	95599	8	1189	68	23	58966	4	2	157	79	6	2	1	4	79	20	3	1	0	0
MIRANDA	405 5TH ST	41	163	2	1	692	89142	7	954	66	18	69817	4	2	151	76	7	2	1	1	77	19	3	0	1	1
INFANTE	406 5TH ST	42	164	2	1	827	103777	8	1194	70	25	68773	3	3	175	58	3	1	2	1	71	24	1	0	1	0
RASSEL	405 6TH ST	42	165	2	1	683	98453	9	1288	91	26	31189	6	4	214	54	6	2	2	3	143	24	1	0	0	0
KARD	403 1ST ST	37	166	2	1	740	84686	8	1176	72	22	108774	3	2	156	78	7	2	1	1	54	18	3	0	0	0
SHELTON	404 1ST ST	38	167	2	1	975	121338	11	1492	115	47	91910	5	4	211	53	5	2	1	3	112	22	1	0	5	0
VERHOVAN	403 2ND ST	38	168	2	1	693	83080	7	1081	61	86	52722	3	2	150	75	4	2	1	1	68	23	3	1	0	0
BOREK	404 2ND ST	39	169	2	1	781	102110	9	1237	91	44	47989	5	3	190	63	5	0	2	1	112	22	3	0	2	6
GAINES	403 3RD ST	39	170	2	1	632	86855	7	984	60	18	88572	1	1	65	65	4	1	2	0	25	25	3	0	1	0
KARSEN	404 3RD ST	40	171	2	1	1012	117967	10	1232	96	21	81374	4	2	153	77	8	1	2	4	79	20	1	0	5	1
CABILI	403 4TH ST	40	172	2	1	654	90436	8	1146	72	24	91161	4	3	175	58	6	1	1	3	95	24	3	0	0	0
KUSNICK	404 4TH ST	41	173	2	1	767	102531	8	1152	74	26	69621	4	3	179	60	6	2	1	2	112	28	3	0	4	0
ALDA	403 5TH ST	41	174	2	1	888	91706	9	1309	80	20	35840	5	4	211	53	4	1	2	8	111	22	2	0	0	2
PEEPLES	404 5TH ST	42	175	2	1	852	112325	9	1272	79	21	76434	4	2	156	78	3	1	2	3	95	24	1	0	4	2
QUEENE	403 6TH ST	42	176	2	1	664	91731	8	1097	72	41	60262	5	4	208	52	8	2	2	8	116	23	1	0	3	4
LYSINGER	401 1ST ST	37	177	2	1	639	97080	8	1121	72	48	60080	4	3	167	56	8	2	1	2	93	23	1	2	0	1
JUNCO	402 1ST ST	38	178	2	1	1101	114558	9	1247	82	17	91058	3	2	142	71	3	2	2	2	62	21	1	0	0	0
HAMBLEY	401 2ND ST	38	179	2	1	619	75661	6	845	47	43	56944	2	1	105	105	4	2	2	3	43	22	3	0	0	0
SCHWARTZ	402 2ND ST	39	180	2	1	541	70408	6	867	48	12	65255	1	1	65	65	7	2	2	1	24	24	1	0	1	2
TINDALL	401 3RD ST	39	181	2	1	945	119960	10	1476	86	32	64578	5	3	199	66	7	2	1	5	103	21	3	0	0	7
GREENBERG	402 3RD ST	40	182	2	1	818	101116	8	1123	74	15	74106	4	3	183	61	6	2	2	3	89	22	3	0	0	0
JAKUBEK	401 4TH ST	40	183	2	1	645	88804	7	935	64	18	77238	3	2	144	72	6	2	2	1	62	21	3	0	0	0
BUTLER	402 4TH ST	41	184	2	1	868	103436	9	1284	84	58	35859	5	3	190	63	5	2	1	3	93	19	3	0	0	5
DIAZ	401 5TH ST	41	185	2	1	438	56743	5	618	40	107	45174	2	1	118	118	6	2	2	1	41	21	3	0	2	7
MIRSKY	402 5TH ST	42	186	2	1	766	106234	9	1286	85	25	73136	4	3	170	57	7	2	2	5	86	22	3	0	3	3
EDMONDS	401 6TH ST	42	187	2	1	805	111800	10	1399	98	23	63817	6	3	165	55	4	2	1	4	139	23	1	0	3	4
HEACOCK	305 1ST ST	26	188	2	1	606	88689	8	1187	65	18	101290	3	2	136	68	6	2	2	1	72	24	1	0	1	0
SANDSTROM	306 1ST ST	27	189	2	1	556	76680	7	966	67	22	56692	2	2	140	70	5	0	1	1	56	28	1	0	1	5
NOVITSKY	305 2ND ST	27	190	2	1	501	69822	6	910	50	64	61570	2	2	125	63	5	2	1	2	52	26	3	0	5	0

VARIABLES

01	02	03	04	05	06	07	08	09	10	11	12	13	14	15	16	17	18	19	20	21	22	23	24	25	26	27
HEARIN	306 2ND ST	28	191	2	1	522	59099	5	627	41	64	54855	1	1	75	75	6	1	1	1	31	31	1	0	0	0
ROWE	305 3RD ST	28	192	2	1	887	93933	8	1137	68	24	89496	3	3	179	60	6	2	1	1	80	27	3	0	4	2
ROBINSON	306 3RD ST	29	193	2	1	857	113574	9	1357	85	20	87855	4	3	175	58	6	1	2	2	93	23	3	0	0	0
HOFFMAN	305 4TH ST	29	194	2	1	651	83975	8	1084	75	19	62459	4	2	144	72	7	1	2	4	73	18	3	0	0	3
BRADSTREET	306 4TH ST	30	195	2	1	728	92815	8	1207	72	16	60714	4	2	160	80	8	1	1	2	79	20	3	1	0	0
DERR	305 5TH ST	30	196	2	1	771	99445	8	1129	78	14	39201	4	2	145	73	12	1	1	4	80	20	3	0	1	3
GLADSTONE	306 5TH ST	31	197	2	1	600	82927	7	1029	64	43	38527	4	3	170	57	7	0	1	3	90	23	1	0	0	0
WAYNE	305 6TH ST	31	198	2	1	734	73395	6	764	51	66	49668	2	1	112	112	11	2	1	1	43	22	3	1	0	0
GRAY	303 1ST ST	26	199	2	1	734	107481	9	1382	74	46	95761	4	2	163	82	7	2	2	4	76	19	3	0	0	0
KAHN	304 1ST ST	27	200	2	1	777	100028	8	1153	76	24	62806	4	3	177	59	4	2	1	3	89	22	3	0	0	0
HOTCHNER	303 2ND ST	27	201	2	1	714	96955	8	1177	75	21	96448	3	2	142	71	7	2	2	1	75	25	3	1	0	0
RODALE	304 2ND ST	28	202	2	1	1089	138976	11	1560	97	19	70538	7	3	165	55	7	1	1	5	156	22	1	1	0	0
PEARL	303 3RD ST	28	203	2	1	546	62786	5	656	40	13	76402	1	1	74	74	5	2	1	1	28	28	3	2	0	2
HEMINGWAY	304 3RD ST	29	204	2	1	739	100182	8	1125	74	34	61103	3	2	132	66	8	2	2	1	65	22	3	0	4	3
CLARK	303 4TH ST	29	205	2	1	675	94743	8	1124	77	21	57598	3	2	149	75	3	2	2	4	54	18	3	1	0	1
DEAN	304 4TH ST	30	206	2	1	1018	111431	9	1225	81	14	70492	4	3	168	56	7	1	2	2	98	25	3	0	6	0
PORTER	303 5TH ST	30	207	2	1	631	69956	6	813	51	12	34286	2	1	116	116	3	2	1	1	44	22	3	0	3	0
MONTAGU	304 5TH ST	31	208	2	1	767	99957	8	1177	75	20	94297	3	2	122	61	6	1	2	1	73	24	1	0	0	0
HAMLYN	303 6TH ST	31	209	2	1	914	116695	9	1289	81	35	80920	4	2	148	74	2	2	2	2	83	21	1	2	0	4
SPINK	301 1ST ST	26	210	2	1	730	88228	8	1125	73	17	146165	3	2	121	61	7	1	2	1	63	21	2	0	0	5
MILLER	302 1ST ST	27	211	2	1	675	97688	8	1059	71	59	96618	3	2	148	74	7	2	1	2	58	19	3	0	5	5
RASMUSSON	301 2ND ST	27	212	2	1	1099	112194	9	1305	85	24	81605	3	2	142	71	7	2	1	3	72	24	3	2	0	1
ABRAMS	302 2ND ST	28	213	2	1	711	89235	7	953	66	16	117671	3	2	148	74	11	0	1	3	54	18	3	0	1	0
FIELDS	301 3RD ST	28	214	2	1	714	81858	7	1059	65	22	67797	2	1	104	104	4	2	1	1	43	22	3	0	0	0
WICKERSHAM	302 3RD ST	29	215	2	1	611	81634	7	1053	58	79	78511	4	2	163	82	8	2	2	5	95	24	1	0	2	0
STEINWAY	301 4TH ST	29	216	2	1	928	125871	10	1440	92	12	75292	5	3	195	65	5	2	2	4	94	19	3	0	6	0
FROST	302 4TH ST	30	217	2	1	513	60894	5	709	43	24	72563	2	2	129	65	7	1	1	1	50	25	1	0	0	0
BOARDMAN	301 5TH ST	30	218	2	1	785	87717	7	989	59	67	60222	2	1	112	112	6	2	2	3	34	17	3	0	0	1
HAUGEN	302 5TH ST	31	219	2	1	662	92710	8	1185	69	54	35701	4	3	178	59	3	1	2	3	96	24	3	0	0	0
BLOOMFIELD	301 6TH ST	31	220	2	1	731	115390	10	1431	100	19	60585	6	4	214	54	9	2	1	6	115	19	3	0	5	4
WEISS	205 1ST ST	14	221	2	1	733	90652	8	1121	69	25	132830	3	2	163	82	6	1	2	1	61	20	2	1	0	4
KRESSEL	208 1ST ST	15	222	2	1	754	85080	8	1112	71	15	126365	2	2	127	64	7	2	2	2	45	23	1	2	0	2
REISMAN	205 2ND ST	15	223	2	1	710	107242	9	1249	83	22	91255	4	2	150	75	8	2	1	2	60	15	3	0	0	0
BOAZ	206 2ND ST	16	224	2	1	699	93133	7	923	62	88	91208	3	2	130	65	6	1	2	1	68	23	2	0	0	4
MEAD	205 3RD ST	16	225	2	1	501	53910	8	747	69	13	30043	4	3	178	59	9	1	2	2	95	24	2	0	6	0
SPELLER	206 3RD ST	17	226	2	1	699	81235	7	1038	61	16	72187	3	2	154	77	6	1	1	1	68	23	1	2	0	6
HOFFMAN	205 4TH ST	17	227	2	1	674	85248	7	980	67	69	85373	2	1	110	110	5	2	2	1	45	23	3	0	0	0
BAXTER	206 4TH ST	18	228	2	1	639	86562	7	1064	63	117	80710	2	1	105	105	4	1	1	1	42	21	3	1	0	1
ROSS	205 5TH ST	18	229	2	1	624	94756	8	1126	76	76	49047	5	3	188	63	6	1	2	1	105	21	3	0	1	1
KING	206 5TH ST	19	230	2	1	818	104945	8	1031	67	25	66701	3	2	130	65	6	1	2	3	73	24	3	0	0	0
VIGIL	205 6TH ST	19	231	2	1	915	97712	8	1003	69	27	56476	4	3	173	58	8	2	1	3	80	20	3	0	0	9
DAVIS	203 1ST ST	14	232	2	1	763	105395	9	1356	89	38	80995	4	4	206	52	5	2	1	6	97	24	3	0	5	0
ROYCE	206 1ST ST	15	233	2	1	670	93724	8	1139	64	37	70074	3	2	138	69	4	1	2	1	65	22	3	2	0	0
JOHANSON	204 1ST ST	15	234	2	1	528	73921	6	646	52	93	166518	2	1	111	111	7	1	2	1	39	20	3	0	6	1
LANDRY	203 2ND ST	15	235	2	1	520	57906	5	632	44	25	78359	2	2	134	67	5	1	2	2	44	22	3	2	0	0
DEITZ	204 2ND ST	16	236	2	1	1054	115666	9	1254	79	59	74242	4	2	153	77	6	2	1	3	74	19	1	0	0	0
CALHOUN	203 3RD ST	16	237	2	1	950	105530	8	1047	78	66	61602	4	3	181	60	4	2	2	4	93	23	3	2	0	5
FLANAGHAN	204 3RD ST	17	238	2	1	909	100493	8	1201	80	34	72824	4	2	159	80	5	1	2	1	91	23	1	0	1	1
BROWN	203 4TH ST	17	239	2	1	371	65392	6	493	45	78	33411	6	2	120	60	6	2	2	4	126	21	2	0	3	3
COHN	204 4TH ST	18	240	2	1	506	74311	6	753	52	19	62170	3	2	133	67	5	2	2	4	62	21	3	0	4	1
KANDELL	203 5TH ST	18	241	2	1	768	91563	7	1052	62	19	47650	3	2	137	69	8	2	2	4	76	25	1	2	0	0
PFEIFFER	204 5TH ST	19	242	2	1	699	100303	8	1092	69	18	62162	3	3	181	60	4	1	2	1	72	24	1	0	4	0
MENDOZA	203 6TH ST	19	243	2	1	779	99421	8	1033	69	22	46337	4	2	156	78	5	2	1	4	78	20	1	0	0	5
RODRIGUEZ	201 1ST ST	14	244	2	1	867	100532	9	1270	88	16	107165	4	3	190	63	8	2	1	4	86	22	3	0	0	0
DEHAVEN	202 1ST ST	15	245	2	1	545	70147	6	907	52	42	140379	1	1	89	89	5	1	1	1	33	33	3	0	0	11
MARIUS	201 2ND ST	15	246	2	1	769	95962	9	1181	80	24	111260	3	2	121	61	6	2	2	1	63	21	1	0	2	0
VALDEZ	202 2ND ST	16	247	2	1	731	86440	7	977	63	35	141031	2	1	109	109	4	2	1	3	44	22	3	2	0	2
INNELLO	201 3RD ST	16	248	2	1	736	103402	9	1170	87	16	83195	3	2	155	78	9	1	2	1	59	20	1	2	0	1
PAULIN	202 3RD ST	17	249	2	1	635	77700	7	1018	63	34	51166	2	2	136	68	8	2	2	2	50	25	3	2	0	2
ELINOFF	201 4TH ST	17	250	2	1	797	106539	9	1311	82	16	76231	5	3	197	66	8	1	1	4	105	21	3	2	0	2
CHAVEZ	202 4TH ST	18	251	2	1	618	89219	7	1064	60	15	82865	2	2	134	67	4	2	2	1	37	19	1	0	0	0
MCINTYRE	201 5TH ST	18	252	2	1	1009	112314	9	1341	79	19	67588	4	3	171	57	8	0	1	4	94	24	3	0	4	4
SOUTHERN	202 5TH ST	19	253	2	1	875	100919	8	1179	67	37	80265	4	3	185	62	9	2	1	3	85	21	1	0	0	0
VICIOSO	201 6TH ST	19	254	2	1	846	99056	9	1185	89	38	40217	5	4	209	52	10	0	1	4	125	25	3	0	1	1
FLETCHER	105 1ST ST	1	255	2	1	577	73681	7	968	63	69	129329	2	2	124	62	5	1	2	1	54	27	3	1	0	1
CHEANEY	106 1ST ST	2	256	2	1	627	73920	6	863	53	76	121005	2	2	134	67	6	2	2	1	50	25	3	0	0	0
ANDRES	105 2ND ST	2	257	2	1	662	80570	7	964	59	27	64797	3	2	149	75	8	2	2	0	58	19	1	0	0	0
HOLMES	106 2ND ST	3	258	2	1	748	92419	8	1136	73	24	116203	3	2	127	64	3	2	2	4	63	21	1	0	0	10
QUINTANA	105 3RD ST	3	259	2	1	696	86765	8	1138	64	19	63368	4	3	179	60	2	2	2	3	113	28	3	1	0	3
VALLEJO	106 3RD ST	4	260	2	1	914	102334	9	1300	85	18	78070	5	3	195	65	6	2	1	3	105	21	3	0	1	0
PHILLIPS	105 4TH ST	4	261	2	1	518	70201	6	780	53	81	61120	3	2	125	63	7	0	1	4	68	23	3	0	5	7
HOLMAN	106 4TH ST	5	262	2	1	418	61077	5	667	40	15	52098	1	1	91	91	7	0	2	1	33	33	1	0	6	0
PHIPPS	105 5TH ST	5	263	2	1	766	99977	9	1256	80	10	70732	4	2	162	81	5	2	2	3	86	22	1	1	0	11
WALKER	106 5TH ST	6	264	2	1	1052	117710	10	1274	90	24	43708	6	4	220	55	5	1	2	6	135	23	3	2	0	7
RAMOS	105 6TH ST	6	265	2	1	724	85366	7	957	64	72	71569	3	2	122	61	3	2	2	1	50	17	3	0	0	0
PICKER	103 1ST ST	1	266	2	1	703	83619	7	1016	66	20	120939	2	2	143	72	7	2	2	1	55	28	3	0	0	1
HOWARD	104 1ST ST	2	267	2	1	733	88656	7	990	54	20	113297	2	1	105	105	8	2	1	1	34	17	3	0	0	0
MILLER	103 2ND ST	2	268	2	1	641	74903	7	1092	64	15	86760	3	2	129	65	5	1	1	1	69	23	1	0	5	6
DION	104 2ND ST	3	269	2	1	676	86714	7	979	64	26	122969	1	1	65	65	3	2	2	1	26	26	3	0	0	0
ERNST	103 3RD ST	3	270	2	1	378	61443	5	682	43	18	54379	2	1	94	94	5	2	1	1	32	16	3	0	4	0
KASTER	104 3RD ST	4	271	2	1	738	82477	7	969	64	14	116346	3	2	136	68	3	2	2	3	72	24	3	2	0	0
BOILEN	103 4TH ST	4	272	2	1	891	110652	9	1365	86	20	66364	6	3	165	55	6	2	2	4	137	23	1	0	0	10
HYKIN	104 4TH ST	5	273	2	1	689	70435	6	874	57	17	68609	1	1	65	65	5	2	2	1	22	22	3	0	0	0
SHEELEY	103 5TH ST	5	274	2	1	725	87374	7	959	66	14	57519	3	2	144	72	9	2	2	1	68	23	1	2	0	0
WILLIAMS	104 5TH ST	6	275	2	1	845	109625	9	1194	83	49	81038	3	2	141	71	2	2	2	1	71	24	3	0	5	0
NORDBERG	103 6TH ST	6	276	2	1	863	109886	10	1480	95	38	27835	6	5	237	47	5	0	2	6	124	21	3	0	4	7
HAMRICK	101 1ST ST	1	277	2	1	752	103468	9	1265	88	59	131063	4	2	156	78	6	2	2	4	70	18	1	0	6	4
POE	102 1ST ST	2	278	2	1	1180	124599	10	1407	92	66	105357	4	4	209	52	4	2	2	2	90	23	3	1	0	6
OBREGON	101 2ND ST	2	279	2	1	837	100652	9	1193	87	78	136427	3	2	150	75	5	1	2	1	60	20	1	2	0	0
RIVERO	102 2ND ST	3	280	2	1	871	99899	8	1043	71	39	134443	3	2	141	71	3	2	1	3	78	26	3	0	0	1
MILES	101 3RD ST	3	281	2	1	599	105640	9	1385	78	65	109434	3	2	128	64	6	2	2	1	66	22	3	0	0	7
GURSON	102 3RD ST	4	282	2	1	849	109223	9	1313	84	19	58220	5	4	201	50	6	2	1	6	107	21	3	0	5	1
CRABTREE	101 4TH ST	4	283	2	1	651	98405	8	1162	82	62	57095	4	2	155	78	7	1	2	4	71	18	1	0	4	0
LOWELL	102 4TH ST	5	284	2	1	713	96645	8	1089	74	14	68773	4	3	170	57	5	2	2	1	72	18	3	1	0	1
QUINN	101 5TH ST	5	285	2	1	749	82646	7	916	57	21	50449	3	3	177	59	5	2	1	1	82	27	3	0	0	0
LOPEZ	102 5TH ST	6	286	2	1	621	76856	6	809	53	71	50602	2	1	104	104	9	1	2	1	35	18	1	0	1	11
GRABON	101 6TH ST	6	287	2	1	802	107009	9	1293	81	17	43982	5	3	197	66	7	2	2	0	95	19	3	2	0	9

Appendix C

232

VARIABLES

01	02	03	04	05	06	07	08	09	10	11	12	13	14	15	16	17	18	19	20	21	22	23	24	25	26	27
POTTS	1205 1ST ST	83	288	3	1	152	19254	4	439	32	56	15340	3	2	120	60	5	1	2	2	56	19	2	0	0	0
ZACCARIA	1206 1ST ST	84	289	3	1	134	13917	3	363	21	30	17624	3	2	120	60	5	0	1	1	76	25	1	0	0	0
RIVERS	1205 2ND ST	84	290	3	1	250	34683	6	564	49	26	30236	2	2	121	61	5	0	2	1	48	24	2	1	0	4
KELLY	1205 3RD ST	85	291	3	1	505	50816	8	727	65	68	21086	4	3	176	59	8	2	1	5	80	20	2	2	0	0
RESNICK	1206 3RD ST	86	292	3	1	181	51823	8	769	73	81	48054	4	4	203	51	7	1	2	1	73	18	2	0	1	0
SMOTRICK	1205 4TH ST	86	293	3	1	176	34510	6	670	50	23	15318	6	1	65	65	8	0	1	6	121	20	1	0	0	0
BORMANN	1206 4TH ST	87	294	3	1	314	37339	7	735	64	22	31873	3	2	138	69	3	1	1	2	66	22	2	0	4	0
FOGELMAN	1205 PARK AVE	87	295	3	1	117	23773	5	538	42	65	25328	2	1	99	99	6	0	1	2	32	16	3	0	3	8
MARCEAU	1206 PARK AVE	88	296	3	1	196	45207	7	654	62	14	29773	3	2	128	64	6	2	2	4	68	23	1	0	5	0
CEBOLLERO	1205 5TH ST	88	297	3	1	275	51953	10	1034	99	23	33745	6	2	120	60	7	2	2	1	132	22	1	0	6	0
JONES	1206 5TH ST	89	298	3	1	668	82355	12	1195	127	88	34378	8	2	120	60	5	2	1	10	178	22	1	0	6	11
MILES	1205 6TH ST	89	299	3	1	335	60394	12	1206	118	56	25780	7	2	120	60	4	0	2	9	102	15	1	0	3	0
VALDES	1206 6TH ST	90	300	3	1	355	38718	8	786	73	15	33796	4	3	177	59	11	2	1	2	86	22	1	0	0	0
WEEDEN	1205 7TH ST	90	301	3	1	412	58924	10	967	96	14	41160	6	2	120	60	3	1	2	4	139	23	1	2	0	3
OLIVER	1206 7TH ST	91	302	3	1	275	39109	7	649	65	55	36110	2	1	104	104	3	2	2	1	37	19	2	2	0	7
HARDY	1205 8TH ST	91	303	3	1	365	35446	7	712	62	39	24357	3	2	139	70	6	2	2	1	75	25	2	0	0	0
LORING	1206 8TH ST	92	304	3	1	389	50679	9	875	83	59	30090	5	3	195	65	7	2	1	1	88	18	1	0	2	1
FLETCHER	1205 9TH ST	92	305	3	1	241	22518	4	436	32	90	17617	4	2	120	60	4	1	1	1	87	22	3	0	4	0
MCIVER	1203 1ST ST	83	306	3	1	297	37465	6	562	53	17	31869	2	1	114	114	6	1	1	1	47	24	1	0	4	1
ROBINS	1204 1ST ST	84	307	3	1	542	59316	10	992	88	63	27342	5	5	228	46	7	2	1	5	100	20	2	0	6	1
EDMOND	1203 2ND ST	84	308	3	1	261	39773	7	705	58	68	31791	4	3	180	60	6	2	2	3	94	24	1	2	0	0
BORGES	1203 3RD ST	85	309	3	1	383	41668	9	907	91	22	25638	6	2	120	60	8	1	2	1	144	24	1	2	0	0
GARCIA	1204 3RD ST	86	310	3	1	314	40706	7	750	56	46	33156	3	2	141	71	9	2	2	4	71	24	1	0	0	0
HALTZMAN	1203 4TH ST	86	311	3	1	204	33659	7	658	71	21	30922	4	2	155	78	9	1	2	2	84	21	1	0	0	0
RAY	1204 4TH ST	87	312	3	1	434	47312	9	970	83	82	27274	4	2	158	79	8	1	2	1	88	22	1	0	0	0
FLYNN	1203 PARK AVE	87	313	3	1	570	61440	11	1079	92	59	38188	6	2	120	60	6	2	1	6	143	24	2	0	3	0
BABBITT	1204 PARK AVE	88	314	3	1	504	55476	10	969	95	103	36939	7	3	165	55	3	2	1	7	153	22	1	0	0	0
ABRONSKI	1203 5TH ST	88	315	3	1	515	52979	9	904	78	50	58703	5	3	197	66	5	2	2	1	116	23	2	0	0	0
CLARK	1204 5TH ST	89	316	3	1	505	54997	10	1009	102	12	28784	6	2	120	60	6	1	1	6	122	20	1	0	1	0
HALL	1203 6TH ST	89	317	3	1	360	46118	7	722	65	30	42701	4	3	175	58	9	1	2	3	96	24	2	0	0	0
OAKLEY	1204 6TH ST	90	318	3	1	512	66265	10	982	104	22	27684	6	2	120	60	5	0	1	4	135	23	1	2	0	0
ACCETTA	1203 7TH ST	90	319	3	1	387	43258	7	689	57	25	34235	3	3	169	56	8	1	1	1	57	19	1	2	0	3
CIVANTOS	1204 7TH ST	91	320	3	1	451	60677	9	910	81	36	31665	5	3	200	67	7	0	2	1	103	21	1	0	3	4
ERIKSON	1203 8TH ST	91	321	3	1	265	34154	6	647	54	21	142251	1	1	86	86	2	2	2	1	34	34	1	2	0	0
LITMAN	1204 8TH ST	92	322	3	1	225	38124	6	573	49	19	13452	6	2	120	60	6	0	2	9	126	21	2	0	0	0
ANGSTADT	1203 9TH ST	92	323	3	1	237	24941	4	388	31	55	15853	4	2	120	60	5	0	2	2	74	19	1	0	4	2
RAYMAN	1201 1ST ST	83	324	3	1	260	32015	7	687	53	91	46033	4	3	168	56	8	2	1	3	97	24	1	0	6	0
FLEMING	1202 1ST ST	84	325	3	1	267	39854	7	692	63	22	34853	4	3	167	56	8	1	2	5	89	22	1	0	0	0
MEYER	1201 2ND ST	84	326	3	1	210	43572	8	776	65	20	25287	5	3	192	64	7	2	1	3	98	20	1	0	0	0
KAFKA	1201 3RD ST	85	327	3	1	386	60478	11	1099	112	20	41507	6	4	220	55	4	2	2	3	110	18	2	0	1	0
ROLNICK	1202 3RD ST	86	328	3	1	275	41424	8	871	73	39	22232	4	4	205	51	9	1	2	3	90	23	1	0	1	4
STEVENSON	1201 4TH ST	86	329	3	1	397	46515	9	880	82	18	25900	6	4	221	55	2	1	2	1	121	20	1	0	0	0
PHELAN	1202 4TH ST	87	330	3	1	317	33731	7	694	60	65	21138	4	3	175	58	4	2	2	2	100	25	1	0	5	1
LARSON	1201 PARK AVE	87	331	3	1	296	48179	10	1062	98	21	33285	6	5	234	47	5	2	1	3	126	21	1	0	4	5
HARANG	1202 PARK AVE	88	332	3	1	209	29694	7	647	62	25	43656	4	2	151	76	8	1	2	0	79	20	2	0	1	0
MICHENER	1201 5TH ST	88	333	3	1	278	43591	8	770	72	13	42494	3	2	135	68	5	2	2	3	70	23	1	2	0	5
RUSSELL	1202 5TH ST	89	334	3	1	294	57954	12	1230	129	20	37439	7	3	165	55	4	1	1	7	148	21	2	0	1	0
MATEO	1201 6TH ST	89	335	3	1	537	58998	10	953	108	16	43767	5	3	199	66	5	2	1	3	102	20	2	1	0	0
HARMON	1202 6TH ST	90	336	3	1	238	36302	9	899	85	53	35425	5	4	211	53	8	2	2	4	102	20	2	0	0	0
BROWN	1201 7TH ST	90	337	3	1	806	74803	12	1133	109	118	29141	7	2	120	60	6	0	1	3	144	21	1	0	0	0
DORFMAN	1202 7TH ST	91	338	3	1	491	38901	9	852	93	39	33850	4	2	155	78	6	2	2	1	91	23	1	0	6	3
KRIMSKY	1201 8TH ST	91	339	3	1	598	33376	9	817	72	21	45259	3	2	135	68	5	1	1	2	77	26	1	0	0	0
FLEISCHER	1202 8TH ST	92	340	3	1	193	25400	4	422	33	15	18485	4	2	120	60	9	0	2	4	86	22	1	2	0	3
NORMAN	1201 9TH ST	92	341	3	1	334	52662	10	1020	92	68	21742	6	4	225	56	5	2	2	8	117	20	2	0	0	0
CRAMER	1105 1ST ST	73	342	3	1	511	64779	11	1179	115	63	43665	6	2	120	60	6	0	2	4	137	23	1	0	0	0
KONWISER	1106 1ST ST	74	343	3	1	483	70465	10	1063	96	94	28525	6	4	225	56	6	1	1	6	111	19	1	0	0	0
TRESCOTT	1105 2ND ST	74	344	3	1	286	43460	8	807	69	20	28264	4	2	144	72	5	2	1	1	78	20	1	2	0	0
DARWIN	1105 3RD ST	75	345	3	1	153	25884	6	620	49	14	33872	2	2	122	61	3	1	2	1	47	24	1	1	0	0
LANIER	1106 3RD ST	76	346	3	1	139	43320	9	862	82	39	37524	4	2	155	78	7	1	2	3	79	20	1	0	1	3
ANGONES	1105 4TH ST	76	347	3	1	194	31384	5	501	42	14	42778	1	1	65	65	7	2	2	1	30	30	2	0	1	0
CRUZ	1106 4TH ST	77	348	3	1	468	46224	9	869	85	77	54212	5	4	207	52	2	2	1	10	113	23	2	0	0	0
MITTLER	1105 PARK AVE	77	349	3	1	438	49312	10	1032	90	17	39124	6	4	224	56	8	1	1	8	153	26	3	0	4	0
SIEGEL	1106 PARK AVE	78	350	3	1	355	58520	10	951	99	89	23889	6	4	216	54	7	1	2	9	104	17	3	0	0	0
PEREZ	1105 5TH ST	78	351	3	1	296	43243	7	758	72	16	29330	3	2	148	74	5	2	1	4	55	18	1	0	1	0
COYLE	1106 5TH ST	79	352	3	1	414	63591	12	1164	120	91	45473	7	3	165	55	6	1	1	3	160	23	2	2	0	10
HOLMES	1105 6TH ST	79	353	3	1	640	55411	9	904	85	21	44163	5	4	209	52	6	1	1	0	111	22	1	0	5	0
DOLANSKY	1106 6TH ST	80	354	3	1	165	25936	5	544	46	20	16117	5	1	65	65	3	1	2	4	120	24	1	0	0	7
MANNING	1105 7TH ST	80	355	3	1	443	51830	8	784	69	27	23584	4	2	161	81	11	1	2	7	89	22	1	2	0	10
SIKES	1106 7TH ST	81	356	3	1	201	46074	9	888	78	18	25999	6	2	120	60	7	1	2	3	121	20	1	2	0	0
TRENZADO	1105 8TH ST	81	357	3	1	586	65599	10	1090	95	20	33578	6	4	214	54	12	2	1	1	134	22	1	2	0	0
GARCIA	1106 8TH ST	82	358	3	1	570	57602	9	875	80	98	21129	6	4	223	56	4	1	1	4	106	18	1	0	0	0
GRIMALDO	1105 9TH ST	82	359	3	1	249	31471	5	524	40	73	18890	4	1	65	65	2	0	2	1	78	20	1	2	0	0
ALDERSON	1103 1ST ST	73	360	3	1	470	46375	7	681	67	18	28224	3	2	140	70	4	2	2	2	82	27	2	0	0	0
GETTLEMAN	1104 1ST ST	74	361	3	1	593	52062	10	998	98	12	37448	6	2	120	60	4	2	1	3	138	23	1	2	0	0
OLIN	1103 2ND ST	74	362	3	1	328	38569	7	666	55	24	42293	3	2	162	81	7	1	1	4	59	20	2	0	0	0
WILLIAMS	1103 3RD ST	75	363	3	1	426	58335	10	1011	99	54	34926	5	3	186	62	6	1	1	4	105	21	1	0	0	0
STEWARD	1104 3RD ST	76	364	3	1	190	29513	6	647	54	59	25067	3	2	143	72	6	2	2	4	69	23	2	0	4	11
MANNION	1103 4TH ST	76	365	3	1	325	55102	10	949	96	13	36839	5	3	198	66	5	0	2	5	90	18	1	0	1	0
PEAKE	1104 4TH ST	77	366	3	1	55	19918	2	179	13	24	16945	2	2	120	60	6	0	2	0	60	30	3	0	0	0
BOLANOS	1103 PARK AVE	77	367	3	1	663	60424	9	906	82	57	32111	5	3	196	65	7	2	1	1	108	22	3	0	3	1
LOWE	1104 PARK AVE	78	368	3	1	297	51382	9	864	78	87	37566	4	2	159	80	6	1	2	2	67	17	2	0	1	0
SHELTON	1103 5TH ST	78	369	3	1	395	42780	8	816	78	23	42447	4	3	177	59	7	1	2	3	78	20	2	0	0	6
CARABALLO	1104 5TH ST	79	370	3	1	177	33211	7	676	61	23	35385	3	2	126	63	3	2	2	3	78	26	1	0	6	0
PARKE	1103 6TH ST	79	371	3	1	295	40735	7	739	56	23	33837	2	2	137	69	6	0	2	2	59	30	2	1	0	5
CLOUTIER	1104 6TH ST	80	372	3	1	241	30965	6	585	59	83	41366	2	2	139	70	5	1	2	1	53	27	2	0	1	3
HOLMAN	1103 7TH ST	80	373	3	1	126	14852	3	293	22	97	16222	3	1	65	65	7	0	1	4	66	22	1	0	4	0
PHINNEY	1104 7TH ST	81	374	3	1	641	59610	9	906	87	31	25648	5	3	198	66	4	1	2	3	109	22	2	0	2	0
SUAREZ	1103 8TH ST	81	375	3	1	419	47682	10	1020	100	78	31998	5	3	184	61	5	0	1	4	87	17	1	0	0	0
GOLDEN	1104 8TH ST	82	376	3	1	561	49679	8	779	75	34	26955	4	2	164	82	7	0	1	0	74	19	2	0	0	2
CLEMENTE	1103 9TH ST	82	377	3	1	386	48626	8	752	67	13	21821	4	3	178	59	7	2	2	1	86	22	2	0	5	0
KUSNICK	1101 1ST ST	73	378	3	1	183	43436	8	848	69	17	34245	3	2	164	82	3	1	1	1	63	21	1	0	0	7
RASHKINO	1102 1ST ST	74	379	3	1	475	46474	7	698	58	21	31622	3	2	159	80	5	2	2	1	57	19	1	0	0	0
TINDELL	1101 2ND ST	74	380	3	1	335	36486	6	600	54	33	36714	3	2	163	82	4	2	2	1	64	21	2	2	0	3
DASEN	1101 3RD ST	75	381	3	1	251	42338	7	719	60	16	46485	4	2	149	75	7	1	1	1	81	20	2	2	0	0
MCINTOSH	1102 3RD ST	76	382	3	1	366	38743	7	730	65	97	33414	4	3	174	58	7	2	2	7	89	22	2	0	0	3
JACOBS	1101 4TH ST	76	383	3	1	445	52551	9	895	85	46	28119	5	4	207	52	9	1	1	1	119	24	2	0	4	5
CARROLL	1102 4TH ST	77	384	3	1	524	80075	12	1198	111	14	53952	7	3	165	55	4	2	2	7	144	21	2	0	1	0
NARDIN	1101 PARK AVE	77	385	3	1	374	49217	9	887	86	47	33217	5	4	216	54	9	1	1	1	112	22	2	1	0	0
SENK	1102 PARK AVE	78	386	3	1	276	39558	8	792	72	23	32667	4	2	160	80	6	0	1	3	88	22	1	2	0	1
HAMBURGER	1101 5TH ST	78	387	3	1	368	43921	8	733	66	13	32136	5	3	187	62	9	1	2	4	103	21	2	0	0	0

```
                                                          VARIABLES
   0          0              0       0 0 0 0          0   0 0 0   0     1 1 1   1     1 1 2   2   2   2   2   2 2
   1          2              3       4 5 6 7          8   9 0 1   2     3 4 5   6     7 8 9   0   1   2   3   4 5
*****************************************************************************************************************

SHAFFER   1102 5TH ST  79 388 3 1 594  61733  9  827  76 20 28381  6 5 235  47  9 2 2  9 136 23 1  2 0 5
KATZMAN   1101 6TH ST  79 389 3 1 340  37227  7  636  54 21 31298  4 3 169  56  5 5 2  3  88 22 2  1 0 0
FAULKNER  1102 6TH ST  80 390 3 1 472  58108 11 1076 111 20 28284  6 4 220  55  5 5 5  2  4 125 51 1 0 0
CAPOTRIO  1101 7TH ST  80 391 3 1 323  37374  8  774  73 23 26049  4 3 174  58  5 2 4    91 53 1  0 0
DALBERT   1102 7TH ST  81 392 3 1 233  45444  9  796  86 85 32354  4 3 172  57  9 1 2  4  88 22 1  2 0 6

KARRASCH  1101 8TH ST  81 393 3 1 154  16009  4  436  32 11 15598  4 1  65  65  7 0 2  1  84 21 1  0 1 10
GREEN     1102 8TH ST  82 394 3 1 144  27898  4  489  44 45 29338  2 1 105 105  5 2 2  1  43 12 1  0 5
OLIVER    1101 9TH ST  82 395 3 1 632  70121 10 1035  97 15 27953  7 2 120  60  5 5 2  5 130 19 1  0 5
DONAHUE   1004 1ST ST  70 396 3 0 298   0     5  414  40 24 52318  1 1  65  65  6 5 2  1  25 25 3  0 0
PARKER    1004 1ST ST  70 397 3 0 681   0    10  869  83 17 49763  4 2 150  75  6 1 2    84 21 1  2 0

MARTIN    1004 1ST ST  70 398 3 0 467   0     6  623  52 64 31103  6 4 217  54  4 2 1  4 146 24 2  0 0
BOULIS    1004 1ST ST  70 399 3 0 443   0     6  570  48 16 37679  5 4 209  52  4 1 1  4  86 17 1  0 0
MATUSEK   1004 1ST ST  70 400 3 0 474   0     6  553  46 11 27444  6 4 215  54  8 0 1  3 118 20 1  0 0
ROWLEY    1004 1ST ST  70 401 3 0 637   0     5  859  78 31 37959  4 3 169  56  3 1 3  4  88 22 2  1 0
WEEKES    1004 1ST ST  70 402 3 0 434   0     6  593  52 18 32598  5 4 202  51  7 2 1  3 104 21 1  2 0 10

GOODING   1004 1ST ST  70 403 3 0 469   0     6  462  44 66 31758  2 2 130  65  7 1 2  1  56 28 1  0 1 0
AUSTIN    1004 1ST ST  70 404 3 0 582   0     6  571  41 57 33909  4 4 219  55  6 1 1  1 126 21 2  0 0
NOWACK    1004 1ST ST  70 405 3 0 591   0     8  678  63 88 38353  3 2 132  66  6 1 1  4  58 19 1  0 3
FLORA     1004 1ST ST  70 406 3 0 380   0     6  490  38 27 36635  2 2 116 116  6 1 1  1  44 22 1  0 5
VETTER    1003 2ND ST  70 407 3 0 356   0     5  497  41 14 30951  4 2 161  81  8 2 2    84 21 3  0 5 9

LANGSAM   1003 2ND ST  70 408 3 0 394   0     6  588  48 14 33300  7 3 165  55  4 2 0  9 152 22 2  2 0 0
DAVIS     1003 2ND ST  70 409 3 0 334   0     6  434  31 11 35005  3 2 149  75  4 2 2  3  86 22 1  0 7
ADAMS     1003 2ND ST  70 410 3 0 391   0     6  546  51 24 27779  4 3 182  61  7 1 1  4  92 23 1  0 0
GREENBURG 1003 2ND ST  70 411 3 0 525   0     6  555  46 59 27132  2 2 158  79  4 1 1  6  84 21 1  0 5
MASKIN    1003 2ND ST  70 412 3 0 546   0     7  624  54 74 27804  5 3 189  50  5 0 2  4 108 22 1  0 0

DOBBS     1003 2ND ST  70 413 3 0 406   0     8  431  38 26 34889  4 2 153  77  8 1 1  2  76 19 2  2 0 3
OVERTON   1003 2ND ST  70 414 3 0 668   0     9  867  75 48 30457  2 2 143  72  4 1 1  2  61 31 2  0 3
VEINGRAD  1003 2ND ST  70 415 3 0 284   0     5  413  37 22 26972  1 1  65  65  6 1 1  2  24 24 1  0 3
ROSENBUSH 1003 2ND ST  70 416 3 0 253   0     5  457  37 17 41549  4 3 174  58  6 1 1  4 123 21 3  2 0
SALICHS   1004 2ND ST  71 417 3 0 299   0     6  484  42 75 21512  6 2 120  60  6 1 2  3 163 27 1  0 0

MENENDEZ  1004 2ND ST  71 418 3 0 659   0    10  856  84 41 40912  2 2 136  68  3 1 1  2  50 25 2  0 3 2
STRICKLER 1004 2ND ST  71 419 3 0 296   0     5  455  37 18 38626  3 3 168  56  4 1 1  2  58 19 1  0 0
HALL      1004 2ND ST  71 420 3 0 408   0     5  491  37 16 31364  4 3 172  57  4 2 0  2 108 27 1  0 0
DUBOIS    1004 2ND ST  71 421 3 0 483   0     6  576  46 54 38785  5 3 193  64 10 1 2  4 101 20 2  0 0
GOLDFARB  1004 2ND ST  71 422 3 0 276   0     7  680  58 19 28757  2 2 147  74  4 1 2  1  67 22 1  0 0

DEITCH    1004 2ND ST  71 423 3 0 356   0     6  575  48 95 36459  3 2 142  71  7 2 2  1  74 25 2  0 1 1
KLEIN     1004 2ND ST  71 424 3 0 672   0     5  671  68 34 21362  3 2 199  66  7 2 1  2  79 20 1  0 1
MATEO     1004 2ND ST  71 425 3 0 659   0     9  811  80 16 28988  3 2 148  74  6 0 1  2  58 19 1  0 0
SQUILLACE 1004 2ND ST  71 426 3 0 352   0     6  494  46 87 26316  6 4 216  54  4 1 2  4 104 17 1  1 0
WEISSER   1004 2ND ST  71 427 3 0 625   0     8  689  66 16 30535  5 3 188  63  8 2 2  5 111 22 1  0 0

TOPPING   1003 3RD ST  71 428 3 0 344   0     5  465  37 100 44682  4 3 170  57  5 1 2  3  94 24 1  0 2 0
KAVANAUGH 1003 3RD ST  71 429 3 0 436   0     5  554  48 18 38844  3 3 184  61  7 1 2  3  72 19 1  2 0
MAINIERO  1003 3RD ST  71 430 3 0 345   0     5  313  34 28 31781  2 2 129  65  6 1 1  4  46 23 2  0 1
HARARI    1003 3RD ST  71 431 3 0 440   0     6  542  52 14 32195  4 3 192  64  6 1 1  2  81 20 2  0 0
KOHOUT    1003 3RD ST  71 432 3 0 571   0     8  718  69 93 32862  4 2 157  79  3 1 2    86 22 1  0 0

COWAN     1003 3RD ST  71 433 3 0 747   0    10  835  85 67 33033  3 2 124  62  5 2 2  3  66 22 1  0 5 2
ARONSON   1003 3RD ST  71 434 3 0 399   0     5  405  33 30 33632  3 3 194  65  5 2 2  3  99 20 3  0 6
GEISEL    1003 3RD ST  71 435 3 0 608   0     9  841  72 22 35841  4 3 155  78  4 0 1  3  84 21 1  0 6 3
ESPOLITA  1003 3RD ST  71 436 3 0 270   0     5  477  32 20 31597  1 1  65  65  3 1 1  2  25 25 1  0 4 0
SAMUELS   1003 3RD ST  71 437 3 0 499   0     7  566  53 46 15515  3 2 120  60  3 0 1  3  58 19 1  0 0

JOHNSEN   1004 3RD ST  72 438 3 0 451   0     6  584  48 25 21423  7 2 120  60  2 1 2  3 154 22 3  2 0 0
FRIEDMAN  1004 3RD ST  72 439 3 0 615   0     8  721  62 15 25971  3 2 145  73  2 1 2  1  72 18 3  0 2
JOHNSEN   1004 3RD ST  72 440 3 0 603   0     7  736  58 13 21729  2 2 141  71  4 1 1  2  65 36 1  0 0
HALLISSEY 1004 3RD ST  72 441 3 0 472   0     7  659  62 14 22881  2 2 138  69  4 1 0  2  57 20 1  1 0
PHILBRICK 1004 3RD ST  72 442 3 0 574   0     6  524  51 68 18941  7 1  65  65  4 0 2  5 151 22 1  0 0

MUTKA     1004 3RD ST  72 443 3 0 479   0     6  582  50 24 20404  5 4 203  51  5 1 2  4 113 23 2  0 0 6
VARGA     1004 3RD ST  72 444 3 0 506   0     6  514  50 23 30151  4 3 211  53  5 1 2  3 103 21 1  0 6
SLACK     1004 3RD ST  72 445 3 0 588   0     9  703  71 13 30474  4 2 148  74  5 1 2  1  68 17 1  0 2
TAYLOR    1004 3RD ST  72 446 3 0 452   0     6  562  47 23 30436  8 2 120  60  5 2 1  4 166 21 2  2 0
MIDONECK  1004 3RD ST  72 447 3 0 273   0     5  453  39 57 44565  3 2 148  74  8 1 1  4  57 19 1  2 0 4

GARDNER   1003 4TH ST  72 448 3 0 396   0     6  514  47 14 31574  5 4 209  52  5 2 2  1 117 23 1  1 0 0
PETERSON  1003 4TH ST  72 449 3 0 562   0     6  545  50 16 29863  6 2 120  60  5 2 2  6 130 22 1  2 0
IBANEZ    1003 4TH ST  72 450 3 0 406   0     8  756  66 30 32865  3 2 150  75  6 1 2  2  48 16 1  0 4
KASALTA   1003 4TH ST  72 451 3 0 487   0     6  528  48 17 24255  4 3 172  57  6 1 2  2  62 21 1  0 1
DODGE     1003 4TH ST  72 452 3 0 572   0     8  681  63 13 23786  4 2 156  78  4 0 2    86 22 1  0 0

MILNER    1003 4TH ST  72 453 3 0 534   0     9  785  75 23 26582  3 2 131  66  7 2 2  1  73 24 1  2 0 0
HARWELL   1003 4TH ST  72 454 3 0 389   0     6  587  48 47 37265  3 2 165  55  5 0 1 12 172 25 1  0 0
SACHER    1003 4TH ST  72 455 3 0 354   0     6  573  52 65 32136  6 2 120  60  5 1 2  1 120 20 1  0 0
LAMBERT   1003 4TH ST  72 456 3 0 333   0     6  426  41 22 34392  3 3 139  70  4 1 1  1  72 24 1  0 4
WRANGLER  1003 4TH ST  72 457 3 0 437   0     7  608  56 17 26552  5 3 189  63  4 1 1  3 109 22 2  1 5 3

FERRO     1003 4TH ST  72 458 3 0 448   0     6  534  47 43 34061  5 3 196  65  8 2 1  4 100 20 1  0 0
CREIG     1002 1ST ST  70 459 3 0 424   0     6  474  53 90 42806  4 2 206  52  6 2 2  3 124 21 2  0 5
BERG      1002 1ST ST  70 460 3 0 401   0     6  534  51 11 31137  5 4 203  51  6 2 1  3 108 22 1  0 0
KAHL      1002 1ST ST  70 461 3 0 387   0     6  482  48 23 49718  5 4 197  50  7 2 2  3 117 23 1  0 0
MANNE     1002 1ST ST  70 462 3 0 612   0     9  854  76 26 42231  2 2 143  72  9 2 1  2  56 28 1  0 0

WARREN    1002 1ST ST  70 463 3 0 479   0     7  619  58 12 56220  7 3 140  70  5 1 1 10  95 14 1  0 0 0
ZELIGMAN  1002 1ST ST  70 464 3 0 591   0     8  838  78 15 39875  5 3 165  55  8 1 1  1  85 28 1  2 0
KOX       1002 1ST ST  70 465 3 0 348   0     6  609  49 41 44327  4 2 151  76  3 2 1  1  77 19 2  0 0
ALLOWAY   1002 1ST ST  70 466 3 0 340   0     5  393  40 94 49898  3 3 179  60  7 2 1  1  64 21 1  1 0
PELTIER   1002 1ST ST  70 467 3 0 563   0     8  783  63 20 32548  3 3 167  56 10 2 1  4  69 23 2  1 2 6

SCHERER   1002 1ST ST  70 468 3 0 441   0     6  576  45 21 47583  2 2 134  67  6 1 2  1  53 27 2  0 0 3
VICIANA   1002 1ST ST  70 469 3 0 370   0     6  556  46 16 36907  3 2 197  66  6 1 1  2  99 20 1  0 6
TARREN    1001 2ND ST  70 470 3 0 807   0    10  900  87 36 45300  4 3 182  61  6 1 1  4  96 24 2  0 0
RICCI     1001 2ND ST  70 471 3 0 314   0     6  553  43 27 33427  2 1  86  86  5 2 2  4 117 30 2  0 0
OLIVER    1001 2ND ST  70 472 3 0 288   0     5  418  40 21 31856  1 1  65  65  5 2 1  1  31 31 2  0 5 0

LUNDY     1001 2ND ST  70 473 3 0 574   0     9  825  72 54 31398  3 2 131  66  5 0 2  0  74 25 2  0 0 0
BAY       1001 2ND ST  70 474 3 0 343   0     6  521  51 76 30138  3 2 165  55  5 1 1  3  44 22 1  0 0
ARGAIN    1001 2ND ST  70 475 3 0 559   0     6  524  51 22 42768  7 2 120  60 10 1 2  1 144 21 2  0 4
WHITE     1001 2ND ST  70 476 3 0 363   0     6  479  35 16 31367  2 2 119 119  6 1 3  2  47 21 1  0 4 0
GOELD     1001 2ND ST  70 477 3 0 379   0     6  566  48 20 29803  6 4 222  56 11 1 2  1 111 19 1  0 0

IGLESIAS  1001 2ND ST  70 478 3 0 537   0     6  581  47 30 27430  6 4 219  55  3 2 2  6  96 16 2  0 0 0
LANDMAN   1001 2ND ST  70 479 3 0 540   0     6  597  58 17 21848  4 3 193  64  3 2 1  2 106 21 3  0 6
ARMAS     1002 2ND ST  70 480 3 0 404   0     5  303  35 17 12976  5 1  65  65  3 2 1  2 101 20 2  0 1
COMAS     1002 2ND ST  71 481 3 0 312   0     6  698  66 39 28259  5 3 157  52  6 1 2  2  74 15 2  0 1 1
FOGELBERG 1002 2ND ST  71 482 3 0 654   0     9  822  73 25 25538  5 3 200  67  6 2 1  8  87 17 2  1 2

HOLMER    1002 2ND ST  71 483 3 0 406   0     8  671  64 16 25687  3 2 139  70  8 2 1  2  77 26 1  0 0 0
MCPHEE    1002 2ND ST  71 484 3 0 338   0     8  427  41 82 29990  3 2 132  66  4 2 1  2  61 20 1  3 1
ALLEN     1002 2ND ST  71 485 3 0 504   0     8  799  64 13 30802  3 2 179  60  4 1 2  2 103 26 1  0 0
ROTHKOPF  1002 2ND ST  71 486 3 0 683   0     6  693  63 23 36237  4 2 203  51  3 2 2  4 102 26 1  2 0
WALKER    1002 2ND ST  71 487 3 0 447   0     7  671  58 39 23527  4 2 162  81  7 2 2  4  81 20 2  0 9
```

Appendix C

234

```
                                                        VARIABLES
   0          0            0       0 0 0    0          0   0 1  1       1    1 1 1   1   1 1 2   2    2  2   2   2 2
   1          2            3       4 5 6    7          8   9 0  1       2    3 4 5   6   7 8 9   0   1 2  3   4   5 6 7

KARY       1002 2ND ST    71   488 3 0 427          0  6   489  50  14  31762  5 4 207  52  6 1 2  3 120 24 1  0  3 0
ARTEAGA    1002 2ND ST    71   489 3 0 503          0  8   693  76  23  25193  4 2 142  71  6 1 2  1  63 21 1  0  6 8
ALVAREZ    1002 2ND ST    71   490 3 0 433          0  8   757  66  16  27809  4 2 163  82  6 1 2  4  84 21 1  0  0 0
HALUSKA    1001 3RD ST    71   491 3 0 338          0  2   310  22  19  26430  2 2 122  61  5 1 1  2  51 26 1  0  1 0
OWENS      1001 3RD ST    71   492 3 0 647          0  8   811  76  38  28104  2 2 130  65  7 2 1  2  45 23 1  0  4 3

KEUSCO     1001 3RD ST    71   493 3 0 746          0 10   861  87  17  45631  2 2 141  71  7 1 2  1  58 29 1  0  0 0
MITES      1001 3RD ST    71   494 3 0 463          0  6   502  47  17  33673  4 4 224  56  5 0 2  3 140 23 1  0  0 8
HALPERN    1001 3RD ST    71   495 3 0 450          0  6   572  47  26  34054  5 4 201  50  5 7 2  2 118 24 1  0  0 0
SKIDELL    1001 3RD ST    71   496 3 0 429          0  7   679  59  18  38874  4 2 160  80  6 7 2  3  50 17 1  2  0 0
VIAROS     1001 3RD ST    71   497 3 0 448          0  6   538  42  15  34964  4 2 153  77  3 0 2  1  82 21 1  0  0 3

LAWSON     1001 3RD ST    71   498 3 0 424          0  6   546  44  19  41587  5 5 231  46  4 0 1  3 117 23 1  0  0 2
PURNELL    1001 3RD ST    71   499 3 0 407          0  6   474  46  97  31801  3 2 120  60  4 0 1  1 143 24 1  0  0 5
HECK       1001 3RD ST    71   500 3 0 428          0  6   546  51  20  40799  6 4 216  54  7 1 1  6 127 21 2  2  0 5
KOYNER     1002 3RD ST    72   501 3 0 568          0  6   546  51  88  26669  4 4 216  54  9 2 2  4 125 21 2  0  3 0
BRODY      1002 3RD ST    72   502 3 0 610          0  8   689  63  20  23004  4 3 197  66  7 1 2  2  84 21 1  0  1 0

PABLO      1002 3RD ST    72   503 3 0 350          0  6   591  46  17  24262  5 4 208  52  6 1 1  1 103 21 2  0  1 1
WRIGHT     1002 3RD ST    72   504 3 0 475          0  8   729  38  21  22306  3 2 138  69  3 1 1  1  70 23 1  0  1 0
ANDREWS    1002 3RD ST    72   505 3 0 315          0  5   385  40  28  25025  2 1 105 105  3 0 1  1  37 19 2  0  1 0
LYNN       1002 3RD ST    72   506 3 0 241          0  7   416  40  28  26628  2 1 112 112  4 2 1  1  49 25 1  0  6 0
HOBACK     1002 3RD ST    72   507 3 0 400          0  6   495  48  68  19607  5 2 120  60  7 0 2  6 106 21 2  0  0 0

SIMON      1002 3RD ST    72   508 3 0 516          0  6   501  43  21  43502  5 4 209  52  5 2 2  3 109 22 2  0  0 0
ADAMS      1002 3RD ST    72   509 3 0 346          0  5   476  38  75  27390  4 2 147  74  4 2 1  4  66 17 1  1  0 0
EPHRAIM    1002 3RD ST    72   510 3 0 459          0  6   552  42  21  32901  5 4 205  51  6 2 2  5 104 21 1  0  0 5
MANDERSON  1001 4TH ST    72   511 3 0 473          0  6   582  47  16  34899  3 2 157  79  8 1 1  3  90 23 2  0  1 5
HAMILTON   1001 4TH ST    72   512 3 0 614          0 10   892  80  22  28627  3 2 139  70  4 2 1  2  83 28 1  0  1 0

HUMPHREYS  1001 4TH ST    72   513 3 0 435          0  6   574  45  22  41266  3 2 160  80  7 2 2  3  64 21 1  0  1 0
DAUGHERTY  1001 4TH ST    72   514 3 0 236          0  6   370  26  16  44547  2 2 132  66  6 0 2  1  39 20 1  0  3 0
NELSON     1001 4TH ST    72   515 3 0 578          0  6   537  44 103  33009  2 2 132  66  5 2 2  1  61 31 1  0  0 0
TOLEDO     1001 4TH ST    72   516 3 0 404          0  7   617  59  13  25410  4 2 155  78  5 2 2  4  86 22 2  1  0 1
SCANLAN    1001 4TH ST    72   517 3 0 343          0  5   465  36  84  37596  3 2 148  74  6 1 1  3  54 18 2  0  6 0

GORDON     1001 4TH ST    72   518 3 0 172          0  4   363  27  19  32458  2 1 113 113  6 2 2  1  49 25 1  2  0 10
WHITFIELD  1001 4TH ST    72   519 3 0 327          0  4   464  41  14  44058  2 1 116 116  6 2 1  1  47 24 2  2  0 5
CAMERON    1001 4TH ST    72   520 3 0 572          0  8   748  67  19  39430  3 2 135  68  4 2 2  1  66 22 1  2  0 5
RULE       1001 4TH ST    72   521 3 0 609          0  9   822  80  21  27082  4 2 152  76  4 2 2  0  96 24 1  0  0 0
MAINSTER    907 1ST ST    65   522 3 1 353  49317  6   638  38  16  29644  3 2 145  73  7 0 2  0  90 30 3  0  1 0

HASTING     908 1ST ST    66   523 3 1 359  42834  7   729  60  25  51217  4 3 186  62  7 2 2  2  72 18 2  0  5 7
SCATTI      907 1ST ST    66   524 3 1 360  29113  9   885  82  15  28469  5 3 191  65  6 2 2  3 103 21 1  0  3 0
MIRAGLIA    908 2ND ST    67   525 3 1 226  26286  9   543  45  14  23632  1 2 113  65  4 2 2  2 128 28 1  0  0 0
ESPLIN      908 3RD ST    68   526 3 1 677  71141 11  1105 109  21  24351  7 2 132  66  7 2 2  2 153 22 1  0  0 0
KOI         907 4TH ST    68   527 3 1 222  33937  7   726  66  76  44147  3 2 132  66  7 2 2  2  63 21 2  0  4 3

HOWARD      908 4TH ST    69   528 3 1 315  48792  8   779  71  34  41557  5 3 199  66  9 1 2  5 102 20 2  2  0 0
GIROGOSIAN  907 PARK AVE  69   529 3 1 450  45206  7   726  62  99  45583  3 2 147  74  4 2 1  4  54 18 1  0  1 5
HOWELL      905 1ST ST    65   530 3 1 452  48152  7   720  62  18  49726  4 3 176  59  4 2 1  2  97 24 2  0  0 0
BROWN       906 1ST ST    66   531 3 1 175  37227  7   655  61  18  44212  5 3 198  66  8 1 2  8 127 25 1  0  5 1
WEAVER      905 2ND ST    66   532 3 1 332  35538  8   814  76  20  44212  5 3 198  66  8 1 2  8 127 25 1  0  5 1

SCOTT       906 2ND ST    67   533 3 1 405  57459 10   974  88  14  49325  5 4 204  51  8 0 2  4 109 22 2  0  0 0
MONROE      906 3RD ST    68   534 3 1 328  36898  7   675  56 106  25339  3 2 153  77  6 2 1  3  63 21 1  0  0 3
FLASHMAN    905 4TH ST    68   535 3 1 346  48219  8   833  73  12  37221  4 3 184  61  7 2 1  2  82 21 1  0  1 3
CAGIGAS     906 4TH ST    69   536 3 1 254  41578  7   676  59  55  35394  3 2 120  60  6 1 1  3  70 23 1  1  0 0
MCGILL      905 PARK AVE  69   537 3 1 222  44747  8   850  75  18  34667  4 3 175  58  8 1 1  9  95 24 1  0  0 0

WILLIAMS    903 1ST ST    65   538 3 1 589  71624 12  1160 125  15  42033  7 3 165  55  5 2 2  5 170 24 2  2  0 5
RODRIGUEZ   904 1ST ST    66   539 3 1 279  38791  7   737  55  17  49245  3 2 142  71  6 0 2  1  77 26 2  0  3 5
MORRIS      903 2ND ST    66   540 3 1 242  45198  8   825  70  12  36056  4 3 179  60  5 2 2  4  93 23 1  0  0 5
GENET       904 2ND ST    67   541 3 1 438  47259  8   701  80  16  40145  4 3 166  55  3 1 1  3  93 23 1  0  0 0
LOFF        904 3RD ST    68   542 3 1 390  49015  9   814  79  25  40026  5 5 230  46  8 2 2  8 132 26 1  0  0 0

ARONSON     903 4TH ST    68   543 3 1 428  54621 10   996  98  19  36756  6 4 224  56  4 1 2  4 118 20 2  2  0 1
VERA        904 4TH ST    69   544 3 1 242  34326  7   659  58  11  40988  5 3 137  69  4 2 2  6  64 21 2  0  6 11
WOODROW     903 PARK AVE  69   545 3 1 486  55412  8   854  67  37  30647  5 4 198  66  6 1 2  6 109 22 2  0  6 11
KIMBLER     901 1ST ST    65   546 3 1 453  52577  9   889  82  34  43090  5 4 210  33  6 1 2  3 116 23 1  0  0 0
ENGEL       902 1ST ST    66   547 3 1 421  56719  6   515  47  17  37621  6 4 218  55  6 2 2  1 134 22 1  0  4 2

LANE        901 2ND ST    66   548 3 1 458  66894 10   901  83  15  39624  5 3 193  64  4 0 2  1  92 18 1  2  0 6
BARNES      902 2ND ST    67   549 3 1 480  44243 10  1001  97  21  44364  5 3 196  65  6 2 2  2  74 19 1  0  5 3
RINKER      902 3RD ST    68   550 3 1 327  47819  8   851  75  16  45409  4 3 196  65  6 2 2  1  30 25 1  0  5 0
PETERSON    901 4TH ST    68   551 3 1 439  44707  7   709  61  23  44988  5 4 135  68  5 2 2  1  89 18 1  0  1 1
KERR        902 4TH ST    69   552 3 1 557  49224  8   793  70  21  46878  5 4 217  54 12 2 2  6  89 18 1  0  1 1

EAST        901 PARK AVE  69   553 3 1 151  20833  4   463  34  64  14275  5 2 120  60  5 0 1  1 103 21 1  0  3 0
LEVENBERG   807 1ST ST    62   554 3 1 461  45211  8   812  72  22  44108  4 2 152  76  6 2 2  3  61 15 1  0  5 0
PITMAN      808 1ST ST    62   555 3 1 347  43867  9   809  74  97  37546  3 2 146  73  6 2 2  2  83 21 1  0  0 0
MILLER      807 2ND ST    62   556 3 1 357  44160  8   768  71  37  43475  4 3 175  66  4 2 2  4 120 20 1  0  0 0
HOUGH       808 2ND ST    63   557 3 1 170  40685  9   852  81  18  43040  6 4 216  54  4 2 2  9 105 18 1  0  0 0

IMBER       808 3RD ST    64   558 3 1 396  38860  8   792  81  18  57597  3 3 176  59  5 1 1  3  61 20 1  0  1 0
GARCIA      807 4TH ST    64   559 3 1 375  39142  7   642  66  13  46247  3 2 150  75  7 1 1  4  57 19 1  0  0 5
DEJAMES     805 1ST ST    61   560 3 1 233  42074  8   796  80  23  46247  3 2 190  63  6 2 1  4 100 26 1  0  0 5
LAMPERT     806 1ST ST    62   561 3 1 514  51378  9   891  90  50  43705  4 3 169  56  5 1 1  5  36 16 1  0  0 0
SALIM       805 2ND ST    62   562 3 1 564  50267 10   986  95  16  43309  5 4 219  55  5 2 1  2  93 19 1  0  0 0

PURDY       806 2ND ST    63   563 3 1 507  59354 11  1137 103 138  21226  7 2 120  60  6 1 2  5 144 21 1  0  5 0
LAMODA      806 3RD ST    63   564 3 1 270  40776  9   952  85  11  50821  6 3 165  55 12 2 2  4 135 21 1  0  0 0
LOSNER      805 4TH ST    64   565 3 1 397  46147  9   708  60  66  43199  5 3 130  65 12 2 2  4  59 30 1  0  2 0
FRANTZ      803 1ST ST    61   566 3 1 344  39208  6   628  53  14  42614  2 2 142  71 12 2 2  1  94 20 1  0  0 4
COUDRIET    804 1ST ST    62   567 3 1 450  51210  6   548  49  85  43078  6 4 221  55  7 2 2  4 120 20 1  0  0 4

DOBBS       803 2ND ST    62   568 3 1 162  24862  5   552  42  28  51522  2 1 118 118 10 2 2  1  40 20 1  0  3 3
OCONNOR     804 2ND ST    63   569 3 1 331  51051  8   829  77  71  39785  6 3 205  51  4 2 2  3 103 21 1  0  0 0
NATT        804 3RD ST    64   570 3 1 504  46973 10   965 100  95  44546  6 2 120  60  4 1 1  3 125 21 1  0  5 0
LAZARUS     803 4TH ST    64   571 3 1 208  33247  6   633  51  22  30268  2 2 128  64  5 1 1  2  51 26 2  0  0 0
SALSBURY    801 1ST ST    61   572 3 1 384  47297  8   780  71  19  47162  4 2 164  82  4 0 2  4  99 25 1  0  0 0

REEVES      802 1ST ST    62   573 3 1 350  35536  7   729  66  15  49237  3 2 145  73 10 2 2  0  60 20 1  2  0 5
MENENDEZ    801 2ND ST    62   574 3 1 214  39234  7   685  62  22  54133  5 3 165  55  6 2 2  4 107 21 1  0  3 4
JACKSON     802 3RD ST    63   575 3 1 493  49758 10  1037  86  20  48159  5 4 217  54  6 2 2  4 107 21 1  0  3 4
POMER       802 3RD ST    64   576 3 1 696  65376 10   934 101  17  31458  6 2 120  60  5 1 1  4 141 24 2  0  0 5
LONGACRE    801 4TH ST    64   577 3 1 418  40771  7   707  62  15  48544  3 2 143  72  4 1 1  2  74 25 1  0  1 5

HAIMES      707 1ST ST    58   578 3 1 489  49073 10  1058  92  64  73548  5 3 187  62  5 2 1  5  87 17 1  2  0 11
PHILLIPS    708 1ST ST    59   579 3 1 330  36459  8   823  71  16  45016  4 3 182  61  3 2 2  1 112 28 2  0  1 0
JACOBS      707 2ND ST    59   580 3 1 562  68364 10   997 101  25  34903  6 4 222  56  6 2 2  6 112 19 1  0  1 1
KAY         708 2ND ST    60   581 3 1 270  50188  9   820  80  40  41974  5 3 189  63  6 1 2  2  92 18 1  2  0 0
CRUZ        705 1ST ST    58   582 3 1 452  64120  9   805  50  18  33408  4 3 168  55  6 1 2  2  65 16 2  0  2 11

EPSTEIN     706 1ST ST    59   583 3 1 474  48286  8   825  69  16  41420  4 3 172  57  4 1 1  3  92 23 2  1  0 7
MALLEY      705 2ND ST    59   584 3 1 387  46743  8   797  72  23  60473  4 3 183  61  4 2 2  2 105 20 1  0  0 0
PRICE       706 2ND ST    60   585 3 1 304  48560  8   825  75  13  50514  3 2 136  68  5 2 2  1  76 25 1  1  0 0
GARCIA      703 1ST ST    58   586 3 1 197  32592  8   819  82  91  23088  5 4 203  51  5 2 1  4 110 22 1  0  0 0
SHUMP       704 1ST ST    59   587 3 1 162  47148  9   814  84  69  45456  5 3 198  66  7 1 1  3 113 23 1  0  0 0
```

VARIABLES

01	02	03	04	05	06	07	08	09	10	11	12	13	14	15	16	17	18	19	20	21	22	23	24	25	26	27
KIMBLER	703 2ND ST	59	588	3	1	396	46928	6	494	46	21	44824	4	2	160	80	7	1	2	1	88	22	2	0	1	1
RUDOLPH	704 2ND ST	60	589	3	1	520	71731	12	1257	123	25	58090	7	3	165	55	4	0	1	7	135	19	1	1	0	0
LIPINSKY	701 1ST ST	58	590	3	1	193	29101	5	500	44	72	44993	1	1	65	65	4	2	2	1	25	25	2	0	0	0
WARD	702 1ST ST	59	591	3	1	221	42184	7	734	62	17	51039	3	2	123	62	5	2	2	2	72	24	3	1	0	3
CHAVEZ	701 2ND ST	59	592	3	1	497	41224	10	961	95	71	54468	5	4	214	54	7	2	2	4	104	21	2	0	0	0
MOSKOVITZ	702 2ND ST	60	593	3	1	177	26392	6	598	55	10	49830	2	1	105	105	10	0	2	3	39	20	2	0	0	0
SAMPSON	607 1ST ST	53	594	3	1	470	49256	7	761	67	63	36794	4	2	154	77	9	1	2	1	76	19	2	2	0	0
LOGAN	608 1ST ST	54	595	3	1	434	50269	8	804	75	20	107273	3	2	156	78	6	1	2	1	70	23	2	2	0	0
THOMAS	607 2ND ST	54	596	3	1	230	39216	7	769	61	38	40191	3	2	139	70	3	1	2	1	72	24	1	0	4	0
SPECTOR	608 2ND ST	55	597	3	1	346	47310	9	915	87	67	48075	4	3	183	61	2	1	2	2	84	21	1	0	4	5
HEATH	607 3RD ST	55	598	3	1	331	46275	9	952	81	15	50684	5	4	205	51	7	1	2	3	105	21	2	0	0	1
WYATT	608 3RD ST	56	599	3	1	366	41596	9	943	91	19	34509	4	3	173	58	6	2	2	0	90	23	1	2	0	0
HEYMAN	607 4TH ST	56	600	3	1	538	19500	8	668	65	22	19730	4	2	120	60	7	1	1	0	100	25	3	0	0	0
PERRY	608 4TH ST	57	601	3	1	388	50954	9	976	83	49	27868	5	4	208	52	5	1	2	5	121	24	1	0	0	9
SUSSMAN	605 1ST ST	53	602	3	1	89	18385	2	172	14	12	12805	1	2	120	60	7	1	1	0	19	19	3	0	0	0
MALLEY	606 1ST ST	54	603	3	1	276	33995	7	734	62	49	24343	3	2	137	69	3	1	1	2	76	25	2	2	0	0
UNGER	605 2ND ST	54	604	3	1	323	44674	7	730	65	92	37915	4	2	159	80	7	2	2	2	85	21	1	0	0	0
HOOK	606 2ND ST	55	605	3	1	337	40716	7	650	61	24	50032	4	3	173	58	6	1	1	7	97	24	1	1	0	0
COWAN	605 3RD ST	55	606	3	1	465	41429	8	748	77	41	44766	4	3	173	58	5	0	1	3	85	21	2	0	6	5
TRIBBLE	606 3RD ST	56	607	3	1	357	52540	10	972	93	22	54820	5	4	209	52	8	1	1	5	71	14	1	2	0	9
DEJESUS	605 4TH ST	56	608	3	1	290	39256	6	643	54	17	28593	2	2	136	68	9	2	2	1	48	24	2	1	0	11
WASHINGTON	606 4TH ST	57	609	3	1	596	48997	9	964	82	59	28050	5	3	194	65	6	1	2	1	104	21	2	0	0	0
BARQUIN	603 1ST ST	53	610	3	1	362	42466	7	754	62	14	29398	3	2	160	80	6	0	2	1	60	20	1	2	0	7
GARNET	604 1ST ST	54	611	3	1	287	35145	6	610	48	14	18591	6	2	120	60	7	1	2	4	120	20	1	1	0	0
AVELLO	603 2ND ST	54	612	3	1	231	42844	8	730	73	90	41959	4	3	190	63	5	2	2	2	85	21	1	2	0	2
DEITZ	604 2ND ST	55	613	3	1	358	47436	8	803	72	63	63844	4	2	145	73	5	0	2	1	83	21	1	0	0	9
ZEIGLER	603 3RD ST	55	614	3	1	397	50925	9	943	89	13	46125	4	3	181	60	5	2	2	3	89	22	2	0	4	11
BARNETT	604 3RD ST	56	615	3	1	457	42012	7	651	64	65	57483	3	2	149	75	5	2	2	2	62	21	2	0	0	0
HACH	603 4TH ST	56	616	3	1	484	50323	9	906	84	48	27071	6	4	217	54	5	2	2	4	138	23	1	0	2	4
YOUNG	604 4TH ST	57	617	3	1	611	51380	9	936	75	37	68981	4	3	184	61	5	2	1	6	75	19	1	0	0	3
COOPER	601 1ST ST	53	618	3	1	410	40858	9	976	81	15	28485	5	3	194	65	7	1	1	6	93	19	2	0	4	2
LESHAW	602 1ST ST	54	619	3	1	180	20029	4	422	33	66	31290	1	1	91	91	7	2	2	1	35	35	2	0	0	0
FLANAGHAN	601 2ND ST	54	620	3	1	337	39117	4	395	28	21	24156	1	1	65	65	4	2	2	1	29	29	2	0	0	0
KAPLAN	602 2ND ST	55	621	3	1	318	41702	8	788	73	22	45689	4	3	174	58	3	2	2	3	83	21	2	0	0	0
JACKSON	601 3RD ST	55	622	3	1	300	40127	7	697	65	84	28887	4	2	148	74	2	2	1	4	83	21	1	2	0	2
NOBLE	602 3RD ST	56	623	3	1	334	50065	8	850	75	12	76129	4	4	203	51	5	2	2	2	88	22	1	0	0	2
PERIGO	601 4TH ST	56	624	3	1	178	36624	8	792	69	88	42217	4	4	208	52	5	2	2	2	87	22	2	1	0	2
COPELAND	602 4TH ST	57	625	3	1	427	67105	10	988	100	21	36356	6	4	214	54	8	2	1	8	103	17	2	1	0	0
GARFIELD	1202 9TH ST	93	626	4	0	439	0	6	511	51	83	14081	7	2	120	60	6	0	2	3	136	19	2	0	1	3
AVERSA	1202 9TH ST	93	627	4	0	401	0	5	418	35	20	13735	6	2	120	60	4	0	2	3	122	20	2	0	6	1
LEOPOLD	1202 9TH ST	93	628	4	0	339	0	5	395	39	23	9483	4	1	65	65	7	0	2	1	93	23	2	1	0	7
YOUNG	1202 9TH ST	93	629	4	0	328	0	5	395	37	58	16014	5	2	120	60	5	1	1	4	77	15	2	0	6	0
HABER	1202 9TH ST	93	630	4	0	522	0	8	667	59	66	18640	5	2	120	60	8	0	2	5	91	18	2	0	3	7
BONWIT	1202 9TH ST	93	631	4	0	390	0	6	532	38	21	12037	5	2	120	60	5	1	1	3	95	19	2	0	0	3
JACOBSON	1202 9TH ST	93	632	4	0	282	0	4	335	24	67	15867	2	2	120	60	5	1	2	3	47	24	2	0	0	2
DELANEY	1202 9TH ST	93	633	4	0	447	0	6	519	50	23	19249	6	1	65	65	5	1	1	3	105	18	2	0	0	0
MCGHEE	1202 9TH ST	93	634	4	0	544	0	7	571	57	46	17987	4	2	120	60	8	1	2	4	90	23	2	0	0	0
GARDNER	1202 9TH ST	93	635	4	0	574	0	8	637	67	13	20943	5	4	215	54	9	2	1	1	107	21	2	0	1	3
XAVIER	1202 9TH ST	93	636	4	0	511	0	9	728	84	61	21902	4	3	165	55	8	2	1	3	89	22	2	2	0	8
FRIEDMAN	1202 9TH ST	93	637	4	0	453	0	6	520	40	29	16917	6	2	120	60	6	1	2	3	136	23	2	0	0	0
WARSHAW	1202 9TH ST	93	638	4	0	438	0	5	365	36	15	18736	2	2	120	60	12	1	1	1	43	22	2	2	0	0
PEREZ	1202 9TH ST	93	639	4	0	521	0	8	621	64	61	17430	2	2	120	60	5	0	2	3	54	27	2	0	6	0
KARNBLUH	1202 9TH ST	93	640	4	0	606	0	8	661	65	80	26000	3	2	148	74	4	2	2	1	67	22	2	0	0	0
ROLTER	1202 9TH ST	93	641	4	0	583	0	9	714	71	39	30195	5	3	191	64	11	1	1	5	111	22	2	0	0	0
MASK	1202 9TH ST	93	642	4	0	486	0	7	579	58	69	21554	4	3	183	61	6	2	2	1	93	23	2	0	1	0
JAFFE	1202 9TH ST	93	643	4	0	499	0	6	486	46	16	27589	6	2	120	60	10	2	1	1	137	23	2	0	4	6
STADLER	1202 9TH ST	93	644	4	0	645	0	9	723	67	22	16730	4	1	65	65	4	1	2	1	106	27	2	0	0	0
PICKERING	1202 9TH ST	93	645	4	0	388	0	6	511	43	70	18836	4	1	65	65	6	1	1	4	73	18	2	2	0	0
KIMBLE	1202 9TH ST	93	646	4	0	609	0	9	743	82	92	22956	4	3	180	60	5	2	2	4	81	20	2	1	0	11
UBEDA	1201 10TH ST	93	647	4	0	535	0	9	750	71	38	23346	2	2	129	65	8	2	2	1	49	25	2	0	0	0
JAHN	1201 10TH ST	93	648	4	0	441	0	6	529	43	95	21242	3	2	138	69	3	0	1	4	76	25	2	0	0	0
FLORX	1201 10TH ST	93	649	4	0	407	0	6	472	50	16	18425	5	1	65	65	8	1	1	4	91	18	2	0	1	0
LEVENBERG	1201 10TH ST	93	650	4	0	397	0	6	439	48	69	25052	6	4	221	55	9	2	2	3	108	18	3	0	0	0
HALL	1201 10TH ST	93	651	4	0	440	0	6	502	46	33	27781	4	2	155	78	6	2	2	8	78	20	2	1	0	0
STIMSON	1201 10TH ST	93	652	4	0	499	0	6	481	41	42	22901	5	3	197	66	9	2	1	6	116	23	2	2	0	2
MANN	1201 10TH ST	93	653	4	0	458	0	6	444	45	21	28133	6	4	217	54	8	2	1	6	132	22	2	0	0	3
OCONNELL	1201 10TH ST	93	654	4	0	620	0	8	657	65	22	29640	4	3	177	59	4	1	1	3	111	28	2	0	0	0
GORDON	1201 10TH ST	93	655	4	0	411	0	6	522	44	12	23675	8	2	120	60	4	2	1	8	174	22	2	2	0	0
TYCHE	1201 10TH ST	93	656	4	0	447	0	6	467	50	20	29070	7	3	165	55	4	2	1	2	156	22	3	0	0	9
CROWN	1201 10TH ST	93	657	4	0	450	0	6	472	40	20	28604	4	4	214	54	2	0	1	1	94	24	2	0	0	9
HAMLYN	1201 10TH ST	93	658	4	0	528	0	8	657	67	38	22708	2	2	144	72	4	0	1	1	54	27	2	0	4	5
CUSSO	1201 10TH ST	93	659	4	0	507	0	8	653	73	62	28077	3	2	134	67	4	1	2	1	70	23	2	2	0	0
ALEA	1201 10TH ST	93	660	4	0	567	0	8	610	68	29	23927	5	3	194	65	2	1	1	1	126	25	2	2	0	0
ESSNER	1201 10TH ST	93	661	4	0	538	0	6	436	48	16	28116	8	2	120	60	8	2	2	6	156	20	2	0	0	11
FLETCHER	1201 10TH ST	93	662	4	0	348	0	5	448	38	20	28110	1	1	93	93	4	2	1	1	33	33	2	0	0	7
BOOKER	1201 10TH ST	93	663	4	0	424	0	6	495	46	37	27962	5	4	204	51	5	2	2	4	93	19	2	2	0	10
LUCE	1201 10TH ST	93	664	4	0	470	0	6	495	46	17	27946	5	4	208	52	4	1	2	4	103	21	2	0	0	0
TIMMS	1201 10TH ST	93	665	4	0	435	0	6	455	51	77	21450	3	2	133	67	9	2	2	3	70	23	2	0	0	0
KAPLAN	1201 10TH ST	93	666	4	0	376	0	6	415	44	22	27030	2	2	130	65	7	1	2	1	51	26	2	2	0	0
CROVELLA	1202 10TH ST	94	667	4	0	398	0	6	470	44	17	8495	7	1	65	65	7	0	1	0	164	23	2	0	0	1
BOOTHE	1202 10TH ST	94	668	4	0	447	0	6	453	44	94	12964	6	2	120	60	6	1	2	6	99	17	2	0	0	0
LUIS	1202 10TH ST	94	669	4	0	499	0	7	598	54	59	6892	4	1	65	65	7	0	1	0	77	19	2	0	1	0
PEREZ	1202 10TH ST	94	670	4	0	432	0	6	471	46	30	16470	7	1	65	65	3	0	2	7	161	23	2	0	0	0
OSSIP	1202 10TH ST	94	671	4	0	495	0	7	559	55	84	15948	4	1	65	65	9	0	2	2	67	17	2	2	0	0
LOWE	1202 10TH ST	94	672	4	0	353	0	5	379	39	19	11480	5	1	65	65	5	0	2	1	119	24	2	0	0	0
PEREZ	1202 10TH ST	94	673	4	0	275	0	5	427	41	56	17565	6	3	165	55	2	0	2	9	142	24	2	2	0	1
GIRO	1202 10TH ST	94	674	4	0	582	0	8	614	69	83	14408	3	2	120	60	7	0	2	0	70	23	2	0	0	0
PHELPS	1202 10TH ST	94	675	4	0	437	0	6	470	49	24	12580	7	1	65	65	4	0	1	3	169	24	1	2	0	6
SHEPHARD	1202 10TH ST	94	676	4	0	492	0	7	600	56	15	12755	5	1	65	65	6	0	2	5	108	22	2	0	0	0
HALLSTROM	1202 10TH ST	94	677	4	0	616	0	10	788	86	17	17854	3	2	120	60	8	1	2	3	74	25	2	0	0	5
CANDELL	1202 10TH ST	94	678	4	0	359	0	5	376	32	27	13347	5	2	120	60	7	1	2	1	95	19	2	0	4	0
SMITH	1202 10TH ST	94	679	4	0	501	0	7	577	54	90	15597	5	1	65	65	5	1	2	4	93	19	2	2	0	0
JONES	1202 10TH ST	94	680	4	0	489	0	7	548	54	19	14055	4	1	65	65	7	0	2	6	83	21	2	1	0	5
THOMPSON	1202 10TH ST	94	681	4	0	432	0	6	486	49	13	15504	3	2	120	60	7	1	1	4	63	21	2	0	0	0
MALM	1202 10TH ST	94	682	4	0	310	0	5	429	39	15	14864	4	2	120	60	11	1	2	4	76	19	2	0	1	0
HEMINGWAY	1202 10TH ST	94	683	4	0	434	0	5	341	37	11	15937	4	1	65	65	5	0	2	4	80	20	2	0	0	1
STEINHAM	1202 10TH ST	94	684	4	0	667	0	8	668	59	20	12539	5	1	65	65	4	1	2	0	117	23	3	0	4	0
ODELL	1202 10TH ST	94	685	4	0	325	0	6	511	47	35	16566	4	2	120	60	6	1	2	3	97	24	1	0	0	0

VARIABLES

01	02	03	04	05	06	07	08	09	10	11	12	13	14	15	16	17	18	19	20	21	22	23	24	25	26	27
TANNER	1202 10TH ST	94	686	4	0	356	0	6	473	42	39	15272	7	2	120	60	8	1	1	5	139	20	2	0	5	0
COX	1202 10TH ST	94	687	4	0	547	0	8	647	60	97	14967	5	1	65	65	5	0	1	4	92	18	2	0	0	1
AXELROD	1201 11TH ST	94	688	4	0	572	0	8	618	62	16	16856	4	2	120	60	6	0	2	4	86	22	2	0	0	0
GFATTER	1201 11TH ST	94	689	4	0	433	0	5	431	36	22	14369	7	1	65	65	10	0	1	3	151	22	2	0	0	7
HAINES	1201 11TH ST	94	690	4	0	609	0	8	675	63	16	17302	4	2	120	60	6	0	1	2	87	22	2	2	0	2
JUNEAU	1201 11TH ST	94	691	4	0	242	0	4	317	32	10	18078	1	2	120	60	4	0	2	1	24	24	2	1	0	0
LEIDERMAN	1201 11TH ST	94	692	4	0	600	0	9	705	73	16	18489	4	1	65	65	7	1	1	4	79	20	2	2	0	4
DELACRUZ	1201 11TH ST	94	693	4	0	225	0	3	303	22	20	19419	1	2	120	60	6	1	2	1	29	29	2	0	0	0
SHEADE	1201 11TH ST	94	694	4	0	354	0	6	450	47	24	17601	4	1	65	65	5	0	2	2	70	18	3	0	0	6
LONGEN	1201 11TH ST	94	695	4	0	400	0	6	485	52	17	14592	5	2	120	60	4	1	2	4	99	20	2	0	3	3
FLIEGEL	1201 11TH ST	94	696	4	0	379	0	6	506	48	15	16149	7	2	120	60	5	1	2	3	140	20	2	2	0	0
HOLLADAY	1201 11TH ST	94	697	4	0	376	0	5	422	32	73	15739	4	2	120	60	6	1	2	1	90	23	2	0	0	0
CANFIELD	1201 11TH ST	94	698	4	0	454	0	5	379	35	19	12544	5	2	120	60	9	1	1	5	104	21	2	1	0	2
FLYNN	1201 11TH ST	94	699	4	0	509	0	8	672	67	20	11102	4	2	120	60	9	0	1	1	77	19	2	0	0	0
QUIROS	1201 11TH ST	94	700	4	0	323	0	4	288	33	20	14217	5	2	120	60	12	1	2	4	100	20	2	0	4	10
HAMILTON	1201 11TH ST	94	701	4	0	321	0	5	416	37	16	5184	8	2	120	60	10	1	1	8	176	22	2	0	0	0
HOLKO	1201 11TH ST	94	702	4	0	585	0	8	651	64	39	16006	4	1	65	65	9	0	1	2	105	26	2	0	3	0
NELSON	1201 11TH ST	94	703	4	0	240	0	5	401	38	24	23862	3	2	122	61	4	2	1	4	65	22	2	0	0	0
ALVAREZ	1201 11TH ST	94	704	4	0	429	0	6	471	43	85	13755	9	1	65	65	2	1	1	7	189	21	2	0	4	1
RODRIGUEZ	1201 11TH ST	94	705	4	0	425	0	7	578	54	80	16726	4	1	65	65	6	1	1	4	74	19	2	0	0	0
TARRUELLA	1201 11TH ST	94	706	4	0	404	0	7	590	57	71	12004	4	1	65	65	5	0	2	4	72	18	2	2	0	5
HOLLAND	1201 11TH ST	94	707	4	0	488	0	6	428	46	84	21191	6	2	120	60	11	1	2	3	134	22	2	0	0	9
SWAN	1202 11TH ST	95	708	4	0	292	0	5	380	32	86	15796	5	2	120	60	6	1	1	6	106	21	2	1	0	4
SWAN	1202 11TH ST	95	709	4	0	402	0	5	392	35	15	9416	8	1	65	65	8	1	1	6	160	20	2	1	0	1
OTERO	1202 11TH ST	95	710	4	0	409	0	6	416	51	11	9213	3	2	120	60	4	1	1	1	75	25	2	0	0	1
HINES	1202 11TH ST	95	711	4	0	449	0	6	493	55	16	10961	6	1	65	65	6	0	2	3	143	24	2	2	0	0
WAGNER	1202 11TH ST	95	712	4	0	462	0	6	475	42	18	13527	7	2	120	60	7	0	2	7	157	22	2	1	0	0
BRYANT	1202 11TH ST	95	713	4	0	493	0	7	620	53	55	9759	3	2	120	60	4	1	2	2	66	22	2	2	0	1
GORDON	1202 11TH ST	95	714	4	0	353	0	6	521	48	21	12078	6	2	120	60	5	1	1	4	110	18	1	0	1	0
APONTE	1202 11TH ST	95	715	4	0	281	0	3	215	20	24	18864	2	2	120	60	7	0	2	1	45	23	2	0	2	0
ABLE	1202 11TH ST	95	716	4	0	389	0	5	445	33	70	12364	6	3	165	55	10	1	1	8	128	21	3	0	0	0
EDDY	1202 11TH ST	95	717	4	0	505	0	8	637	73	13	17484	5	1	65	65	1	0	2	1	111	22	2	0	1	5
KIMMEL	1202 11TH ST	95	718	4	0	477	0	6	483	50	71	19359	5	2	120	60	6	1	2	1	130	26	2	0	3	3
SLACK	1202 11TH ST	95	719	4	0	146	0	3	183	20	13	17045	3	2	120	60	3	0	2	1	63	21	2	0	0	4
HOFFMANN	1202 11TH ST	95	720	4	0	455	0	5	380	39	79	16777	6	1	65	65	4	0	2	3	115	19	2	0	0	0
TOLEDO	1202 11TH ST	95	721	4	0	202	0	3	262	22	34	13329	4	2	120	60	7	1	2	6	93	23	2	0	0	0
WATKINS	1202 11TH ST	95	722	4	0	412	0	5	413	40	23	12872	4	1	65	65	7	1	2	5	77	19	2	2	0	1
MATTHIAS	1202 11TH ST	95	723	4	0	564	0	8	638	69	23	13377	3	2	120	60	7	0	2	1	67	22	1	0	3	3
STOUFFER	1202 11TH ST	95	724	4	0	518	0	8	596	62	20	17943	2	2	120	60	6	1	2	1	53	27	2	0	0	0
TOMAS	1202 11TH ST	95	725	4	0	404	0	6	477	45	110	15310	5	2	120	60	10	1	1	5	119	24	2	0	2	3
POSEDA	1202 11TH ST	95	726	4	0	467	0	7	565	59	20	16743	4	2	120	60	8	1	2	5	96	24	1	2	0	1
KING	1202 11TH ST	95	727	4	0	341	0	5	439	44	53	15458	3	2	120	60	4	0	2	4	48	16	2	0	0	0
WILLIS	1202 11TH ST	95	728	4	0	238	0	5	433	34	17	17288	4	2	120	60	10	1	2	4	107	27	2	0	2	0
GLOVER	1201 12TH ST	95	729	4	0	498	0	6	486	49	24	13937	7	2	120	60	8	0	1	5	152	22	2	2	0	5
MATTHEW	1201 12TH ST	95	730	4	0	188	0	3	247	20	19	16900	1	2	120	60	5	0	2	0	32	32	1	2	0	0
MCBRIDE	1201 12TH ST	95	731	4	0	375	0	6	432	47	97	17309	8	1	65	65	6	0	1	8	162	20	2	0	1	2
BLUM	1201 12TH ST	95	732	4	0	363	0	5	429	40	19	10795	9	2	120	60	11	1	2	2	188	21	1	2	0	6
YAEGER	1201 12TH ST	95	733	4	0	381	0	5	414	37	25	5897	8	1	65	65	6	0	1	4	127	16	2	0	0	1
MORERA	1201 12TH ST	95	734	4	0	435	0	6	482	47	54	20553	6	2	120	60	6	1	2	1	135	23	2	0	0	0
HABER	1201 12TH ST	95	735	4	0	313	0	6	498	44	27	12741	8	1	65	65	6	0	2	10	162	20	2	0	2	4
ABBEY	1201 12TH ST	95	736	4	0	307	0	5	392	38	14	18404	4	2	120	60	12	1	2	2	73	18	2	0	4	3
FLEMING	1201 12TH ST	95	737	4	0	407	0	5	407	34	58	13397	6	1	65	65	3	1	1	3	135	23	2	2	0	0
COLLINS	1201 12TH ST	95	738	4	0	445	0	6	489	45	21	12821	4	2	120	60	8	1	2	3	87	22	2	0	0	6
HECHT	1201 12TH ST	95	739	4	0	514	0	7	526	53	34	11381	2	2	120	60	9	1	1	1	50	25	2	0	0	0
PHILLIPPE	1201 12TH ST	95	740	4	0	536	0	7	523	62	12	11186	4	2	120	60	4	0	2	4	75	19	2	0	0	11
HORAN	1201 12TH ST	95	741	4	0	564	0	7	570	62	86	16674	3	1	65	65	8	0	1	2	66	22	2	0	0	0
TONKIN	1201 12TH ST	95	742	4	0	484	0	6	451	44	18	18317	6	1	65	65	8	1	2	3	133	22	2	0	0	11
WILLOUGHBY	1201 12TH ST	95	743	4	0	647	0	9	741	73	16	16473	4	2	120	60	8	1	1	5	85	21	2	2	0	0
KAHN	1201 12TH ST	95	744	4	0	330	0	5	414	38	11	4405	4	1	65	65	3	0	2	5	99	25	2	0	0	2
SMYTHE	1201 12TH ST	95	745	4	0	380	0	5	356	39	77	25044	2	1	100	100	4	2	2	2	39	20	2	2	0	6
FLOWERS	1201 12TH ST	95	746	4	0	561	0	8	652	69	29	12769	4	2	120	60	6	1	2	0	87	22	2	1	0	0
LAMM	1201 12TH ST	95	747	4	0	315	0	5	420	34	46	7691	6	2	120	60	3	0	2	4	134	22	2	0	0	0
SODERHOLM	1201 12TH ST	95	748	4	0	575	0	7	575	61	19	10380	2	2	120	60	5	0	1	2	44	22	2	1	0	0
MCBEE	1202 12TH ST	96	749	4	0	513	0	6	517	41	21	8460	7	2	120	60	5	1	1	3	127	18	2	0	0	4
BROWN	1202 12TH ST	96	750	4	0	314	0	5	380	37	33	10754	5	2	120	60	2	1	2	3	124	25	2	0	3	0
POTTER	1202 12TH ST	96	751	4	0	478	0	6	461	48	20	10530	6	2	120	60	7	1	1	4	124	21	2	0	0	0
FIREMAN	1202 12TH ST	96	752	4	0	307	0	5	393	38	21	11187	4	1	65	65	6	1	1	2	98	25	2	0	0	0
SINCLAIR	1202 12TH ST	96	753	4	0	694	0	10	826	87	18	11643	4	1	65	65	6	1	1	2	100	25	2	0	0	3
MASI	1202 12TH ST	96	754	4	0	357	0	5	422	39	27	13761	2	2	120	60	3	1	2	1	42	21	2	2	0	1
TOMAS	1202 12TH ST	96	755	4	0	532	0	7	582	56	17	10039	4	2	120	60	5	0	2	2	83	21	2	2	0	0
PHELAN	1202 12TH ST	96	756	4	0	270	0	5	427	40	39	9490	7	2	120	60	3	1	2	3	155	22	3	2	0	0
TORRES	1202 12TH ST	96	757	4	0	496	0	6	501	42	66	12194	7	1	65	65	5	0	1	7	142	20	2	0	0	0
MONROE	1202 12TH ST	96	758	4	0	381	0	6	489	44	16	10361	8	3	165	55	8	1	2	6	182	23	2	0	0	0
FLORIO	1202 12TH ST	96	759	4	0	622	0	8	646	73	77	12717	4	2	120	60	9	0	1	3	73	18	2	0	0	7
BARREIRO	1202 12TH ST	96	760	4	0	314	0	5	384	40	19	10268	8	1	65	65	5	0	2	8	162	20	2	1	0	0
KOPP	1202 12TH ST	96	761	4	0	280	0	4	331	28	60	10045	2	2	120	60	9	0	1	1	38	19	2	2	0	3
BECK	1202 12TH ST	96	762	4	0	455	0	7	577	54	28	8857	4	2	120	60	4	1	2	1	97	24	2	0	0	2
SCHUMACKER	1202 12TH ST	96	763	4	0	126	0	3	292	21	22	8571	1	2	120	60	3	0	1	2	28	28	2	0	6	5
BAKER	1202 12TH ST	96	764	4	0	143	0	3	245	19	81	8102	2	2	120	60	3	0	2	1	55	28	2	0	0	2
MARKS	1202 12TH ST	96	765	4	0	369	0	5	383	37	25	18788	7	1	65	65	7	1	2	3	154	22	2	2	0	2
GRAVER	1202 12TH ST	96	766	4	0	336	0	5	439	36	18	8772	7	2	120	60	6	0	1	7	125	18	2	1	0	7
CRAVEN	1202 12TH ST	96	767	4	0	608	0	7	594	59	62	7809	4	2	120	60	6	0	1	3	86	22	2	0	0	0
MODER	1202 12TH ST	96	768	4	0	276	0	4	365	27	37	5630	1	2	120	60	5	0	1	0	39	39	2	0	1	10
GRAVER	1202 12TH ST	96	769	4	0	441	0	6	494	48	54	6038	6	2	120	60	10	1	2	1	122	20	3	0	1	0
STEWART	1201 13TH ST	96	770	4	0	470	0	7	539	52	26	10831	4	1	65	65	5	1	2	1	80	20	2	0	0	4
ROSEN	1201 13TH ST	96	771	4	0	467	0	6	535	44	11	7158	5	2	120	60	4	1	1	8	109	22	2	0	0	0
WURST	1201 13TH ST	96	772	4	0	231	0	3	261	22	36	4322	4	2	120	60	10	0	2	1	71	18	2	0	0	0
ALLEN	1201 13TH ST	96	773	4	0	284	0	3	258	18	26	8983	2	2	120	60	3	0	2	2	41	21	2	1	0	0
SHUMP	1201 13TH ST	96	774	4	0	294	0	4	284	29	21	4886	3	2	120	60	4	0	2	1	61	20	2	0	6	6
WEISMAN	1201 13TH ST	96	775	4	0	340	0	5	378	39	18	3168	8	1	65	65	3	1	1	8	177	22	2	0	1	1
CALHOUN	1201 13TH ST	96	776	4	0	343	0	5	385	35	23	5830	4	1	65	65	6	1	1	3	87	22	2	2	0	0
STANFORD	1201 13TH ST	96	777	4	0	533	0	8	691	63	19	17585	2	2	120	60	6	1	1	1	47	24	2	0	0	0
FOX	1201 13TH ST	96	778	4	0	298	0	5	391	38	27	6476	6	1	65	65	6	0	2	3	138	23	1	0	0	0
BITTNER	1201 13TH ST	96	779	4	0	390	0	5	445	38	20	7844	6	2	120	60	5	1	2	12	114	19	2	1	0	0
HAYES	1201 13TH ST	96	780	4	0	325	0	5	418	40	41	6269	4	2	120	60	7	0	1	1	84	21	2	0	0	0
GOLDEN	1201 13TH ST	96	781	4	0	272	0	6	443	43	18	8618	5	1	65	65	4	1	2	4	108	22	2	0	1	6
OWENS	1201 13TH ST	96	782	4	0	327	0	6	466	46	10	8142	5	2	120	60	9	0	2	9	103	21	2	2	0	6
HERNANDEZ	1201 13TH ST	96	783	4	0	332	0	5	390	39	13	4919	5	2	120	60	7	0	2	0	98	20	2	0	0	0
DAVIS	1201 13TH ST	96	784	4	0	616	0	8	670	69	22	9763	3	1	65	65	4	1	2	1	73	24	2	0	4	3
KIRKPATRIK	1201 13TH ST	96	785	4	0	403	0	7	525	56	38	9765	4	2	120	60	7	1	1	4	95	24	2	0	4	0

VARIABLES

01	02	03	04	05	06	07	08	09	10	11	12	13	14	15	16	17	18	19	20	21	22	23	24	25	26	27
DODD	1201 13TH ST	96	786	4	0	413	0	6	530	42	88	9943	7	1	65	65	2	1	1	2	134	19	2	0	4	0
POWELL	1201 13TH ST	96	787	4	0	280	0	4	340	28	35	6241	5	2	120	60	3	0	1	4	114	23	2	1	0	3
IRWIN	1201 13TH ST	96	788	4	0	286	0	5	425	36	63	7548	7	2	120	60	3	1	2	3	99	14	2	2	0	1
JOHNSON	1201 13TH ST	96	789	4	0	279	0	4	285	31	54	2402	4	2	120	60	8	1	1	1	92	23	2	1	0	0
LONG	1102 9TH ST	97	790	4	0	416	0	6	494	51	13	14757	5	2	120	60	3	0	2	3	108	22	1	2	0	1
KRUGER	1102 9TH ST	97	791	4	0	558	0	7	574	57	20	12185	5	1	65	65	6	0	2	1	85	17	2	0	1	0
WOLFE	1102 9TH ST	97	792	4	0	528	0	7	582	53	19	12063	3	2	120	60	7	0	2	1	79	26	2	0	0	1
VELLANTI	1102 9TH ST	97	793	4	0	482	0	6	428	43	41	18811	7	1	65	65	1	1	2	5	153	22	2	0	0	10
NICHOLSON	1102 9TH ST	97	794	4	0	430	0	5	451	39	25	20323	5	3	198	66	11	1	1	1	112	22	2	2	0	0
HEYMAN	1102 9TH ST	97	795	4	0	333	0	5	374	40	73	9558	5	1	65	65	4	1	2	6	107	21	2	2	0	0
SPRINGER	1102 9TH ST	97	796	4	0	391	0	5	406	41	77	16200	6	2	120	60	6	0	1	4	109	18	2	0	1	10
PATTERSON	1102 9TH ST	97	797	4	0	278	0	4	392	30	19	13705	4	1	65	65	3	1	2	1	94	24	2	2	0	0
FIELDER	1102 9TH ST	97	798	4	0	384	0	5	447	38	50	19876	6	1	65	65	6	1	1	10	129	22	2	0	4	0
CAMPBELL	1102 9TH ST	97	799	4	0	252	0	3	232	21	74	19071	2	2	120	60	9	0	1	2	37	19	1	2	0	0
EARHART	1102 9TH ST	97	800	4	0	426	0	6	463	48	63	22691	6	3	165	55	7	2	2	1	129	22	2	2	0	0
NORBECK	1102 9TH ST	97	801	4	0	541	0	7	582	54	126	19261	4	1	65	65	4	1	2	2	102	26	2	0	0	0
HILL	1102 9TH ST	97	802	4	0	349	0	5	422	39	20	16371	2	2	120	60	8	0	2	1	51	26	2	0	0	2
TILLERMAN	1102 9TH ST	97	803	4	0	766	0	10	826	89	11	26377	3	2	159	80	6	2	1	1	60	20	2	0	1	0
MAST	1102 9TH ST	97	804	4	0	299	0	6	502	45	15	18096	7	1	65	65	5	0	1	5	130	19	2	0	0	0
FERNANDEZ	1102 9TH ST	97	805	4	0	366	0	5	435	37	14	22002	5	4	222	56	3	0	2	4	111	22	2	0	1	4
BARTON	1102 9TH ST	97	806	4	0	653	0	9	720	85	109	22557	4	3	176	59	5	2	2	3	89	22	2	0	5	0
HICKS	1102 9TH ST	97	807	4	0	540	0	8	690	69	78	24995	4	2	147	74	6	2	2	5	71	18	2	2	0	10
GILBERT	1102 9TH ST	97	808	4	0	411	0	6	521	46	16	17348	5	1	65	65	1	0	2	1	99	20	2	0	0	0
DRAGO	1102 9TH ST	97	809	4	0	546	0	7	538	63	35	19029	4	2	120	60	5	1	2	2	87	22	2	2	0	0
OAKES	1102 9TH ST	97	810	4	0	377	0	6	522	49	37	17715	4	2	120	60	7	0	2	2	90	23	2	1	0	0
FEY	1101 10TH ST	97	811	4	0	581	0	8	647	63	29	18795	3	2	120	60	5	0	1	4	82	27	2	2	0	0
RUSCH	1101 10TH ST	97	812	4	0	528	0	7	545	55	70	24504	4	3	168	56	9	0	2	4	85	21	2	0	0	1
PETRUCCI	1101 10TH ST	97	813	4	0	344	0	6	505	42	17	20118	5	4	201	50	5	2	2	4	101	20	2	2	0	5
ISGAR	1101 10TH ST	97	814	4	0	267	0	3	248	22	12	15322	2	2	120	60	5	0	2	0	46	23	2	2	0	7
JEFFERY	1101 10TH ST	97	815	4	0	362	0	4	281	31	11	13161	1	2	120	60	7	0	2	1	26	26	2	2	0	1
CAPMAN	1101 10TH ST	97	816	4	0	603	0	7	557	57	16	25718	4	3	179	60	8	1	1	7	97	24	2	0	5	0
KAPLAN	1101 10TH ST	97	817	4	0	449	0	7	570	60	17	21660	3	2	155	78	6	0	1	2	62	21	2	0	6	7
LEWIS	1101 10TH ST	97	818	4	0	482	0	7	562	52	24	21081	2	2	135	68	5	1	1	3	56	28	2	0	1	5
PABLO	1101 10TH ST	97	819	4	0	652	0	9	682	73	38	22992	3	2	155	78	3	1	2	3	58	19	2	0	0	0
INGRAM	1101 10TH ST	97	820	4	0	711	0	9	726	80	82	8473	5	2	120	60	7	0	2	1	89	18	2	2	0	0
MALLERY	1101 10TH ST	97	821	4	0	436	0	6	503	50	14	19498	7	1	65	65	6	1	2	9	137	20	2	2	0	5
DAVIDSON	1101 10TH ST	97	822	4	0	434	0	6	522	45	14	17777	5	1	65	65	2	0	2	3	118	24	2	0	0	0
ALEMAN	1101 10TH ST	97	823	4	0	587	0	8	644	65	101	21115	3	2	157	79	6	0	2	1	56	19	2	0	0	8
FLOOK	1101 10TH ST	97	824	4	0	460	0	6	444	44	25	18092	2	2	120	60	8	1	2	0	50	25	2	0	0	3
INGALLS	1101 10TH ST	97	825	4	0	389	0	6	473	44	14	23502	4	3	179	60	7	2	1	2	96	24	2	0	0	1
GAFF	1101 10TH ST	97	826	4	0	417	0	6	505	49	16	23257	6	2	120	60	5	2	1	6	122	20	2	0	0	0
PEXTON	1101 10TH ST	97	827	4	0	412	0	6	492	45	20	23872	6	4	222	56	7	1	2	8	125	21	1	0	1	3
OGLE	1101 10TH ST	97	828	4	0	361	0	6	456	42	82	25649	6	4	224	56	4	2	1	6	113	19	2	2	0	0
FISHER	1101 10TH ST	97	829	4	0	224	0	4	268	30	20	24567	1	1	65	65	3	2	2	2	26	26	3	2	0	2
RAMOS	1101 10TH ST	97	830	4	0	409	0	6	541	48	21	25456	7	2	120	60	8	2	1	3	164	23	2	0	0	0
JACKSON	1102 10TH ST	98	831	4	0	377	0	5	394	38	22	14573	5	1	65	65	3	1	1	5	117	23	2	0	4	0
SKINNER	1102 10TH ST	98	832	4	0	313	0	5	407	35	72	13669	6	1	65	65	5	1	2	1	120	20	2	0	2	4
NESBITT	1102 10TH ST	98	833	4	0	269	0	5	460	36	70	6095	5	1	65	65	6	0	1	1	116	23	2	0	0	0
GAINES	1102 10TH ST	98	834	4	0	670	0	9	702	73	76	10488	5	2	120	60	5	0	2	3	109	22	2	0	1	6
HINSLEY	1102 10TH ST	98	835	4	0	483	0	7	601	59	12	13661	3	2	120	60	7	0	2	5	69	23	2	0	1	6
SANCHEZ	1102 10TH ST	98	836	4	0	530	0	8	682	68	22	14572	4	1	65	65	3	0	2	1	91	23	2	0	3	0
WILLIAMS	1102 10TH ST	98	837	4	0	564	0	8	656	68	39	13276	3	2	120	60	3	0	2	2	79	26	2	0	0	0
LEVY	1102 10TH ST	98	838	4	0	329	0	5	391	33	19	10291	6	2	120	60	5	0	2	3	119	20	2	0	1	0
ALCORN	1102 10TH ST	98	839	4	0	254	0	5	422	41	24	14253	7	2	120	60	3	1	2	5	140	20	2	0	1	10
MILLER	1102 10TH ST	98	840	4	0	501	0	6	478	45	44	14305	7	1	65	65	7	1	1	3	132	19	2	2	0	5
HITT	1102 10TH ST	98	841	4	0	238	0	4	317	31	98	10857	1	2	120	60	7	1	1	0	26	26	2	0	5	0
CAMPBELL	1102 10TH ST	98	842	4	0	381	0	6	481	49	19	15166	5	2	120	60	6	1	1	5	109	22	2	0	3	3
MALLOY	1102 10TH ST	98	843	4	0	326	0	5	419	40	53	14832	5	2	120	60	5	1	2	6	105	21	2	1	0	1
BARRON	1102 10TH ST	98	844	4	0	525	0	7	556	54	59	16144	2	2	120	60	4	0	1	1	63	32	2	0	0	9
ENGLING	1102 10TH ST	98	845	4	0	379	0	6	487	45	38	11838	4	2	120	60	6	1	1	6	84	21	2	0	0	2
SPELLMAN	1102 10TH ST	98	846	4	0	418	0	5	374	41	15	16995	3	2	120	60	7	0	2	4	65	22	2	0	1	0
FOSTER	1102 10TH ST	98	847	4	0	402	0	6	482	47	95	15130	7	1	65	65	5	0	1	2	138	20	2	0	5	10
MCCARTNEY	1102 10TH ST	98	848	4	0	354	0	7	569	50	35	13721	4	2	120	60	7	0	2	2	77	19	3	0	1	0
HARRISON	1102 10TH ST	98	849	4	0	373	0	6	524	47	37	14868	7	1	65	65	3	1	2	3	120	17	2	1	0	0
STARR	1102 10TH ST	98	850	4	0	298	0	5	456	37	23	18348	5	2	120	60	8	0	1	0	115	23	2	0	0	0
LENNON	1102 10TH ST	98	851	4	0	485	0	6	510	50	68	12642	5	1	65	65	3	0	2	5	110	22	3	0	3	0
BERAN	1101 11TH ST	98	852	4	0	385	0	6	509	46	49	18609	5	2	120	60	3	1	1	0	93	19	2	0	0	0
DEAN	1101 11TH ST	98	853	4	0	320	0	4	345	31	41	15113	4	2	120	60	6	0	2	5	79	20	2	0	0	4
PORTER	1101 11TH ST	98	854	4	0	395	0	6	504	45	23	12097	7	1	65	65	5	0	1	5	150	21	2	0	0	8
DAY	1101 11TH ST	98	855	4	0	422	0	6	536	46	38	14460	4	1	65	65	5	0	1	4	90	23	2	0	0	0
TAYLOR	1101 11TH ST	98	856	4	0	418	0	6	518	44	21	16727	3	2	120	60	6	0	2	1	73	24	2	0	0	0
LEONE	1101 11TH ST	98	857	4	0	326	0	6	502	50	22	18919	6	2	120	60	4	0	1	4	143	24	2	0	0	2
DOYLE	1101 11TH ST	98	858	4	0	246	0	3	256	21	21	17052	1	2	120	60	5	1	2	1	26	26	2	2	0	0
BRUNNER	1101 11TH ST	98	859	4	0	332	0	6	455	40	15	18060	5	2	120	60	6	0	1	4	114	23	2	0	5	6
QUAT	1101 11TH ST	98	860	4	0	545	0	8	634	67	15	11499	3	2	120	60	5	1	2	4	62	21	2	0	1	3
KRAUSE	1101 11TH ST	98	861	4	0	395	0	6	467	48	17	15055	7	1	65	65	4	1	2	2	162	23	2	0	6	0
BELLOMY	1101 11TH ST	98	862	4	0	656	0	9	727	68	22	17522	3	2	120	60	7	1	1	1	56	19	2	0	0	0
BESANT	1101 11TH ST	98	863	4	0	341	0	5	335	36	25	16916	5	1	65	65	8	1	1	3	119	24	2	0	2	0
HARRIS	1101 11TH ST	98	864	4	0	386	0	5	416	35	20	15236	4	2	120	60	8	0	1	3	77	19	2	0	0	4
CALDWELL	1101 11TH ST	98	865	4	0	420	0	5	364	40	84	15200	5	1	65	65	2	0	1	4	90	18	2	0	1	1
BAINTON	1101 11TH ST	98	866	4	0	267	0	5	436	41	15	12507	4	1	65	65	5	0	2	4	90	23	2	0	0	0
IRVING	1101 11TH ST	98	867	4	0	377	0	6	454	44	41	15501	6	2	120	60	8	1	2	1	133	22	2	1	0	1
LAPPE	1101 11TH ST	98	868	4	0	450	0	6	450	47	66	16001	5	2	120	60	5	1	2	6	96	19	3	2	0	0
JAMES	1101 11TH ST	98	869	4	0	505	0	6	480	46	44	11257	4	2	120	60	10	0	1	2	79	20	2	0	4	9
COOPER	1101 11TH ST	98	870	4	0	361	0	6	445	47	11	23114	7	3	165	55	7	1	1	3	147	21	2	0	6	0
SINGER	1101 11TH ST	98	871	4	0	303	0	5	386	37	16	17569	5	2	120	60	4	0	2	8	113	23	1	0	0	0
HARRIS	1102 11TH ST	99	872	4	0	260	0	3	265	19	19	7465	3	1	65	65	8	0	1	1	59	20	2	2	0	0
BURROUGHS	1102 11TH ST	99	873	4	0	485	0	6	464	51	25	11509	6	2	120	60	7	1	2	8	97	16	2	2	0	3
JORDAN	1102 11TH ST	99	874	4	0	338	0	5	398	40	12	11410	6	2	120	60	5	0	2	8	116	19	2	1	0	0
MCKENDRICK	1102 11TH ST	99	875	4	0	335	0	5	395	37	63	13243	5	2	120	60	6	0	2	3	87	17	3	0	1	7
BURKE	1102 11TH ST	99	876	4	0	425	0	6	505	54	19	8308	5	2	120	60	9	0	2	1	116	23	2	0	1	1
HOWATCH	1102 11TH ST	99	877	4	0	239	0	5	405	36	21	12296	5	1	65	65	6	0	2	1	109	22	2	0	3	0
SANGER	1102 11TH ST	99	878	4	0	410	0	6	533	45	82	13466	3	2	120	60	6	1	2	1	50	17	2	1	0	0
MASSEE	1102 11TH ST	99	879	4	0	280	0	6	529	47	71	13903	5	1	65	65	8	0	2	4	101	20	2	0	4	0
WHITE	1102 11TH ST	99	880	4	0	373	0	5	408	40	13	11445	5	2	120	60	4	0	2	8	102	20	3	0	0	0
ALLEN	1102 11TH ST	99	881	4	0	689	0	9	760	75	78	10441	5	1	65	65	5	1	1	5	94	19	2	0	0	0
ETHAN	1102 11TH ST	99	882	4	0	416	0	6	486	44	20	18993	5	2	120	60	8	0	2	4	111	22	2	1	0	5
LUTHOR	1102 11TH ST	99	883	4	0	356	0	5	391	38	23	10885	6	2	120	60	4	1	2	6	125	21	2	0	0	0
CLARKE	1102 11TH ST	99	884	4	0	311	0	5	363	33	139	11445	6	1	65	65	7	1	1	8	107	18	2	0	0	0
WAYNE	1102 11TH ST	99	885	4	0	293	0	5	407	39	62	10873	4	2	120	60	6	1	1	3	70	18	2	2	0	11

Appendix C
238

VARIABLES

01	02	03	04	05	06	07	08	09	10	11	12	13	14	15	16	17	18	19	20	21	22	23	24	25	26	27
BALLARD	1102 11TH ST	99	886	4	0	458	0	6	488	49	18	13897	5	1	65	65	4	1	2	5	111	22	2	0	0	0
NORRIS	1102 11TH ST	99	887	4	0	372	0	5	417	40	80	16641	5	2	120	60	4	0	1	4	94	19	2	0	0	0
DIAMOND	1102 11TH ST	99	888	4	0	438	0	6	509	46	61	12870	7	2	120	60	7	1	2	2	131	19	2	0	1	2
BULL	1102 11TH ST	99	889	4	0	370	0	5	396	36	14	14590	5	1	65	65	3	0	1	6	90	18	2	0	4	3
PEZEL	1102 11TH ST	99	890	4	0	374	0	5	391	39	14	15787	3	2	120	60	7	0	1	2	69	23	2	0	3	1
ABRAMS	1102 11TH ST	99	891	4	0	145	0	4	369	27	16	10024	3	2	120	60	9	0	1	3	77	26	2	0	2	4
THOREAU	1102 11TH ST	99	892	4	0	571	0	8	638	68	36	15678	3	1	65	65	5	0	2	4	62	21	2	0	0	0
DOBSON	1101 12TH ST	99	893	4	0	341	0	5	394	41	94	10373	6	2	120	60	4	0	1	3	120	20	2	2	0	0
CAYCE	1101 12TH ST	99	894	4	0	448	0	6	483	53	20	16307	4	2	120	60	4	1	2	4	80	20	2	0	0	0
MELVILLE	1101 12TH ST	99	895	4	0	327	0	5	385	41	25	11960	7	2	120	60	4	1	1	3	161	23	2	0	0	0
POE	1101 12TH ST	99	896	4	0	290	0	4	329	29	17	15904	1	2	120	60	4	0	2	1	36	36	2	0	3	2
EMERSON	1101 12TH ST	99	897	4	0	340	0	5	373	39	26	11969	7	1	65	65	4	1	2	2	138	20	3	0	0	0
DODSON	1101 12TH ST	99	898	4	0	419	0	6	530	49	78	14765	5	2	120	60	6	1	2	3	101	20	2	1	0	0
HAWTHORNE	1101 12TH ST	99	899	4	0	221	0	3	246	22	21	15823	2	2	120	60	12	0	2	1	46	23	1	0	4	11
KAFKA	1101 12TH ST	99	900	4	0	362	0	6	492	50	15	14657	6	1	65	65	5	0	2	6	126	21	2	0	1	11
GUERAD	1101 12TH ST	99	901	4	0	510	0	7	520	53	35	15262	4	2	120	60	7	1	1	3	75	19	2	2	0	0
SETON	1101 12TH ST	99	902	4	0	223	0	4	334	30	39	18786	5	2	120	60	7	0	2	1	106	21	2	0	0	0
FREUD	1101 12TH ST	99	903	4	0	412	0	5	448	35	40	17922	7	2	120	60	5	0	1	10	114	16	2	2	0	0
HOFFMAN	1101 12TH ST	99	904	4	0	428	0	6	564	50	21	6095	2	2	120	60	10	1	1	1	53	27	2	0	0	0
WILLIAMS	1101 12TH ST	99	905	4	0	338	0	5	387	35	15	16141	7	2	120	60	2	1	2	5	149	21	2	2	0	0
KENDA	1101 12TH ST	99	906	4	0	460	0	8	644	57	30	17915	5	2	120	60	7	0	2	3	100	20	2	0	0	6
DURANT	1101 12TH ST	99	907	4	0	465	0	6	417	52	27	18539	6	1	65	65	3	0	2	4	105	18	2	2	0	9
AMES	1101 12TH ST	99	908	4	0	419	0	6	451	48	40	10614	6	2	120	60	4	1	1	1	104	17	2	0	0	0
RULE	1101 12TH ST	99	909	4	0	316	0	5	399	34	17	18430	3	1	65	65	2	0	2	1	59	20	2	0	0	6
PAULUS	1101 12TH ST	99	910	4	0	379	0	5	426	37	23	14508	7	2	120	60	7	1	2	3	156	22	2	0	1	9
SIMON	1101 12TH ST	99	911	4	0	411	0	6	501	44	16	5375	6	2	120	60	5	1	2	8	101	17	2	0	1	6
STONE	1101 12TH ST	99	912	4	0	487	0	7	600	58	57	16601	4	1	65	65	5	0	1	2	101	25	2	0	6	3
CROSBY	1102 12TH ST	100	913	4	0	332	0	5	456	33	45	7846	2	2	120	60	8	0	2	1	46	23	2	0	0	0
STILLS	1102 12TH ST	100	914	4	0	378	0	5	430	40	56	8372	3	2	120	60	3	0	1	1	70	23	2	0	5	5
NASH	1102 12TH ST	100	915	4	0	672	0	9	757	77	68	9527	4	2	120	60	5	1	2	4	87	22	3	2	0	7
COLLINS	1102 12TH ST	100	916	4	0	358	0	5	337	41	36	3714	6	2	120	60	7	1	2	1	113	19	2	1	0	3
BAEZ	1102 12TH ST	100	917	4	0	303	0	4	353	28	11	11300	4	1	65	65	4	0	1	3	69	17	2	0	0	0
DENVER	1102 12TH ST	100	918	4	0	342	0	4	333	27	17	6305	3	2	120	60	4	0	2	2	70	23	2	2	0	9
STRAUSS	1102 12TH ST	100	919	4	0	325	0	5	404	35	29	7806	6	2	120	60	6	0	2	4	120	20	2	0	0	0
GERSHWIN	1102 12TH ST	100	920	4	0	284	0	3	212	22	18	9239	4	1	65	65	9	0	2	3	96	24	2	1	0	0
LIGHTFOOT	1102 12TH ST	100	921	4	0	457	0	6	462	43	38	3976	6	2	120	60	5	0	2	0	123	21	2	0	0	0
TENILLE	1102 12TH ST	100	922	4	0	695	0	10	760	95	20	9648	4	2	120	60	7	0	2	4	79	20	2	0	0	0
CROFTS	1102 12TH ST	100	923	4	0	575	0	7	554	46	75	10483	4	1	65	65	6	0	1	5	84	21	2	0	0	0
SEALS	1102 12TH ST	100	924	4	0	439	0	5	449	31	23	9648	5	2	120	60	6	1	1	6	112	22	2	1	0	0
BATCOCK	1102 12TH ST	100	925	4	0	377	0	7	539	63	17	9194	2	2	120	60	6	0	2	1	48	24	2	0	0	0
SCARNE	1102 12TH ST	100	926	4	0	396	0	6	471	44	84	9438	7	2	120	60	7	0	1	9	172	25	2	0	1	0
GELENDER	1102 12TH ST	100	927	4	0	356	0	5	391	34	17	9449	4	2	120	60	6	0	2	4	75	19	2	0	0	7
MYER	1102 12TH ST	100	928	4	0	357	0	5	379	40	84	7711	4	2	120	60	4	1	1	2	75	19	2	1	0	1
BACH	1102 12TH ST	100	929	4	0	680	0	8	633	62	17	11873	3	2	120	60	4	1	1	1	70	23	2	0	1	0
HYMES	1102 12TH ST	100	930	4	0	380	0	6	526	45	19	7473	7	2	120	60	5	1	2	3	132	19	2	0	0	0
SORUM	1102 12TH ST	100	931	4	0	304	0	4	324	30	59	8987	4	2	120	60	7	0	2	4	99	25	2	0	0	5
MATSON	1102 12TH ST	100	932	4	0	416	0	6	468	48	24	4896	5	1	65	65	2	1	2	0	116	23	2	2	0	7
ALTHOUSE	1102 12TH ST	100	933	4	0	382	0	6	490	46	64	16051	5	1	65	65	4	0	2	5	114	23	2	1	0	3
WISE	1101 13TH ST	100	934	4	0	277	0	4	308	30	62	3509	3	1	65	65	7	0	2	1	58	19	2	0	0	3
BAUMAN	1101 13TH ST	100	935	4	0	385	0	5	349	37	48	2574	7	1	65	65	5	0	1	5	136	19	2	0	1	11
FULLER	1101 13TH ST	100	936	4	0	302	0	3	239	18	22	4842	2	2	120	60	4	0	2	1	43	22	2	1	0	0
GEORGE	1101 13TH ST	100	937	4	0	490	0	6	451	50	12	7869	8	2	120	60	6	1	2	6	126	16	2	2	0	0
JEFFERSON	1101 13TH ST	100	938	4	0	480	0	7	569	56	12	4322	5	1	65	65	5	0	2	4	104	21	2	0	4	6
MADISON	1101 13TH ST	100	939	4	0	602	0	9	740	67	16	11033	4	2	120	60	7	1	2	4	65	16	2	2	0	3
WILSON	1101 13TH ST	100	940	4	0	351	0	5	399	38	76	8637	4	2	120	60	5	1	1	3	74	19	2	0	4	4
KENNEDY	1101 13TH ST	100	941	4	0	569	0	7	585	55	33	9726	4	2	120	60	9	0	2	5	101	25	2	1	0	5
HUMPHREY	1101 13TH ST	100	942	4	0	497	0	6	475	47	10	3938	3	2	120	60	4	1	1	1	50	17	2	0	0	1
CARTER	1101 13TH ST	100	943	4	0	309	0	5	386	34	65	8197	7	1	65	65	11	1	2	7	145	21	1	0	0	0
NIXON	1101 13TH ST	100	944	4	0	463	0	6	481	46	67	8453	3	2	120	60	8	0	2	3	82	27	2	0	0	0
STEVENSON	1101 13TH ST	100	945	4	0	292	0	3	280	18	73	10898	2	2	120	60	3	1	2	2	29	15	2	0	0	0
WALKER	1101 13TH ST	100	946	4	0	263	0	4	298	31	12	4087	3	2	120	60	5	0	1	4	53	18	3	0	1	4
BACTZER	1101 13TH ST	100	947	4	0	344	0	5	388	37	25	4148	4	2	120	60	6	1	2	3	104	26	2	2	0	4
FENSKE	1101 13TH ST	100	948	4	0	555	0	6	452	47	75	10380	5	1	65	65	4	1	2	4	109	22	2	0	0	0
KIRKLEY	1101 13TH ST	100	949	4	0	359	0	5	393	39	14	11867	5	1	65	65	6	0	2	6	130	26	1	2	0	9
WEBB	1101 13TH ST	100	950	4	0	309	0	5	392	36	53	6162	4	1	65	65	5	1	2	4	89	22	2	0	0	0
GEIL	1101 13TH ST	100	951	4	0	400	0	6	457	49	19	7673	5	1	65	65	6	1	1	4	100	20	2	0	0	0
RANDLE	1101 13TH ST	100	952	4	0	512	0	7	593	62	21	8962	4	2	120	60	4	1	1	3	84	21	2	0	0	0
BATEY	1101 13TH ST	100	953	4	0	349	0	6	475	50	21	9245	6	1	65	65	8	0	1	4	121	20	2	0	0	0
BOLTON	1004 7TH ST	101	954	4	0	477	0	7	565	54	62	14832	4	1	65	65	4	1	2	2	88	22	2	0	0	0
COLON	1004 7TH ST	101	955	4	0	396	0	5	414	38	35	8055	3	2	120	60	7	1	2	1	68	23	2	0	5	3
ERICKSON	1004 7TH ST	101	956	4	0	442	0	6	498	45	11	11543	6	1	65	65	5	1	2	1	131	22	2	0	0	0
FINKE	1004 7TH ST	101	957	4	0	451	0	7	576	45	28	22379	5	3	184	61	4	2	2	4	93	19	2	0	0	0
PERKINS	1004 7TH ST	101	958	4	0	363	0	6	553	49	42	26105	7	3	165	55	11	1	2	2	150	21	2	0	0	0
SPRAGGETT	1004 7TH ST	101	959	4	0	600	0	8	681	66	11	24312	3	2	137	69	6	2	2	6	70	23	2	0	1	11
WORRALL	1004 7TH ST	101	960	4	0	473	0	6	428	45	16	19304	4	2	120	60	4	1	1	2	23	18	2	2	0	0
BIERSTEIN	1004 7TH ST	101	961	4	0	542	0	8	704	63	13	25657	4	3	178	59	9	1	2	3	86	22	3	0	1	4
ERRICO	1004 7TH ST	101	962	4	0	365	0	6	501	48	10	7852	6	2	120	60	3	0	2	3	139	23	2	2	0	1
FISKE	1004 7TH ST	101	963	4	0	263	0	5	451	38	20	28581	3	2	149	75	6	1	1	3	52	17	2	0	4	0
GEDDES	1004 7TH ST	101	964	4	0	374	0	6	503	48	22	24560	6	5	239	48	4	1	2	9	129	22	2	1	0	2
JONES	1003 8TH ST	101	965	4	0	453	0	5	400	37	43	35201	2	1	99	99	6	1	2	3	40	20	2	2	0	10
SLATER	1003 8TH ST	101	966	4	0	368	0	5	368	38	18	9963	3	2	120	60	8	0	1	1	59	20	2	0	0	0
TRIBBE	1003 8TH ST	101	967	4	0	413	0	6	430	47	21	26820	6	2	120	60	3	1	2	4	149	25	3	0	2	0
HEARST	1003 8TH ST	101	968	4	0	375	0	6	478	48	44	21518	5	4	209	52	6	1	1	4	109	22	2	2	0	4
TUBBS	1003 8TH ST	101	969	4	0	532	0	8	616	68	17	21071	2	2	126	63	9	2	1	2	44	22	3	0	4	9
BUBECK	1003 8TH ST	101	970	4	0	337	0	6	474	47	14	21258	6	4	223	56	12	1	2	9	126	21	3	0	0	3
FREEMAN	1003 8TH ST	101	971	4	0	571	0	8	689	70	92	23203	4	2	152	76	6	1	1	2	75	19	2	0	2	1
ELTON	1003 8TH ST	101	972	4	0	232	0	4	325	29	17	21765	1	1	88	88	8	2	2	1	33	33	2	1	0	0
DURBIN	1003 8TH ST	101	973	4	0	402	0	6	478	43	80	22127	4	3	190	63	11	2	2	2	74	19	2	0	4	4
OBRIEN	1003 8TH ST	101	974	4	0	471	0	6	433	52	17	21523	2	2	133	67	6	1	2	2	50	25	2	0	0	0
HOLES	1004 8TH ST	102	975	4	0	444	0	6	526	47	64	24134	5	4	205	51	9	1	2	4	118	24	2	1	0	0
HIGGINS	1004 8TH ST	102	976	4	0	475	0	6	448	45	38	28488	7	2	120	60	5	2	2	0	147	21	2	0	0	0
RAUSCHER	1004 8TH ST	102	977	4	0	264	0	5	418	37	109	21897	1	1	87	87	5	2	2	1	28	28	3	0	0	0
SWEENEY	1004 8TH ST	102	978	4	0	446	0	6	456	44	69	24772	6	4	218	55	6	2	1	6	121	20	2	0	5	10
WRIGHT	1004 8TH ST	102	979	4	0	320	0	5	369	37	14	18807	6	1	65	65	1	0	2	3	126	21	2	1	0	11
MORGAN	1004 8TH ST	102	980	4	0	682	0	9	748	77	12	22087	4	3	175	58	3	2	1	2	80	20	2	0	0	0
SHADBOLT	1004 8TH ST	102	981	4	0	647	0	9	721	77	14	21542	2	2	142	71	2	0	2	1	53	27	2	0	0	11
BISHOP	1004 8TH ST	102	982	4	0	242	0	4	332	26	20	22970	3	2	158	79	6	0	2	2	59	20	2	0	0	0
RATZLAFF	1004 8TH ST	102	983	4	0	454	0	7	540	53	65	18909	4	2	120	60	5	0	2	4	75	19	2	0	0	3
CHARI	1004 8TH ST	102	984	4	0	441	0	6	487	40	19	24413	6	2	120	60	5	1	2	1	110	18	3	0	0	0
PALMER	1004 8TH ST	102	985	4	0	584	0	8	667	63	18	23211	3	2	128	64	6	1	1	3	63	21	2	0	0	2

Stat City data base

239

```
                                                     VARIABLES
 0        0         0      0 0 0 0      0 1 1  1   1    1 1  1  1 1 1 2  2   2  2  2 2 2
 1        2         3      4 5 6 7      8 9 0  1   2    3 4  5  6 7 8 9  0   1  2  3 4 5 6 7
*********************************************************************************************
STEINER    1003 9TH  ST  102  986 4 0 569 0 7 533 57  26 23533 4 2 156 78  9 0 2 1  89 22 2 0 0  0
JACOBS     1003 9TH  ST  102  987 4 0 437 0 6 503 48  16 20488 5 3 185 62  5 1 1 6  89 18 2 0 1  0
ROSE       1003 9TH  ST  102  988 4 0 461 0 6 443 50  23 19205 6 2 120 60  8 1 1 6 129 22 1 0 4  0
MURPHY     1003 9TH  ST  102  989 4 0 590 0 6 676 60  21 22009 5 3 186 62  5 0 2 1 100 20 2 0 5  0
HORN       1003 9TH  ST  102  990 4 0 474 0 8 669 67  21 19681 4 2 120 60  6 1 1 3  73 18 2 0 3 11

PUGH       1003 9TH  ST  102  991 4 0 580 0 8 644 68  18 20249 4 4 207 52  5 1 1 1  88 22 2 0 4  1
MATTHEWS   1003 9TH  ST  102  992 4 0 519 0 8 502 47  15 33790 5 4 211 53  7 1 2 3 104 21 2 0 6  0
WAGNER     1003 9TH  ST  102  993 4 0 557 0 9 700 74  46 25260 3 2 122 61 10 2 2 3  69 23 2 0 0  0
WARD       1003 9TH  ST  102  994 4 0 388 0 7 406 36  14 34964 3 1  65 65  8 2 2 0  23 23 2 0 6  0
EVANSTON   1003 9TH  ST  102  995 4 0 483 0 7 585 52  69 23331 5 3 190 63  6 0 2 0  95 19 2 1 0  0

KRISTOL    1004 9TH  ST  103  996 4 0 491 0 7 598 56  10 18157 4 1  65 65  4 1 2 4 102 26 2 0 0  1
SINGER     1004 9TH  ST  103  997 4 0 355 0 5 429 36  16 17201 3 2 120 60  5 1 1 1  51 17 2 0 1  0
ENGELBARDT 1004 9TH  ST  103  998 4 0 338 0 6 463 44  88 21591 6 4 225 56  7 2 1 1 111 19 2 0 0  0
WILL       1004 9TH  ST  103  999 4 0 339 0 6 397 38 120 17691 2 2 120 60  4 1 1 6 127 21 2 0 3  0
WERNICK    1004 9TH  ST  103 1000 4 0 333 0 5 362 38 107 20070 2 1  99 99  4 1 2 1  34 17 2 0 0  9

GILBERT    1004 9TH  ST  103 1001 4 0 321 0 6 506 49  19 20197 8 2 120 60  8 0 2 2 151 19 2 1 0  0
PARSONS    1004 9TH  ST  103 1002 4 0 514 0 6 647 63  78 20517 3 2 162 81  8 0 2 1  81 20 2 0 0 16
LOGAN      1004 9TH  ST  103 1003 4 0 318 0 5 393 35  58 19710 6 2 120 60  9 1 1 3 141 24 2 0 0  0
EISELEY    1004 9TH  ST  103 1004 4 0 467 0 6 429 45  21 36149 5 4 201 50  4 2 1 1 103 21 2 0 4 10
KISSINGER  1004 9TH  ST  103 1005 4 0 406 0 6 453 48  12 20075 3 2 128 64  6 2 1 1  64 21 2 0 0  1

THURBER    1004 9TH  ST  103 1006 4 0 462 0 6 445 45  42 21581 7 2 120 60  8 1 1 0 136 19 2 2 0  0
CRAWFORD   1004 9TH  ST  103 1007 4 0 388 0 6 518 46  18 25900 5 4 207 52  8 1 1 2 111 22 2 0 5  0
PARTON     1004 9TH  ST  103 1008 4 0 431 0 7 552 54  20 20216 4 3 181 60  8 0 2 5  95 24 2 5 0  0
HOLMES     1003 10TH ST  103 1009 4 0 270 0 5 254 21 134 19477 7 1  65 65  7 0 2 5 126 25 2 0 0  0
ERNST      1003 10TH ST  103 1010 4 0 360 0 6 518 50  24 17010 7 1  65 65  2 0 1 1 152 22 2 2 0  5

BATSON     1003 10TH ST  103 1011 4 0 270 0 5 396 36  78 16007 5 1  65 65  6 0 2 1  98 20 3 0 0  0
HAUSER     1003 10TH ST  103 1012 4 0 371 0 5 504 36  78 23411 5 4 205 51  9 2 0 2 109 22 2 0 0  6
GHANDI     1003 10TH ST  103 1013 4 0 352 0 5 420 36  16 20760 4 3 167 56  8 1 2 0  69 23 2 0 5  6
VASHOLZ    1003 10TH ST  103 1014 4 0 615 0 9 799 82  32 24059 5 2 157 79  8 1 2 3  86 22 2 0 0  0
BLANDINA   1003 10TH ST  103 1015 4 0 412 0 6 504 46  22 18300 5 2 120 60  6 1 2 6  99 20 3 2 0  0

GREENFELD  1003 10TH ST  103 1016 4 0 396 0 6 512 46  19 24800 5 3 185 62  5 2 1 9 105 21 2 0 0  0
FUNK       1004 10TH ST  104 1017 4 0 403 0 5 391 41  16 10581 3 2 120 60  5 1 2 4  46 15 2 0 1 11
BOLCH      1004 10TH ST  104 1018 4 0 404 0 7 616 53  46 14985 4 1  65 65  5 1 2 1  78 20 2 1 0  0
GELDER     1004 10TH ST  104 1019 4 0 522 0 8 549 54  87 15404 3 2 120 60  6 1 1 3  68 20 2 0 0  0
PORTER     1004 10TH ST  104 1020 4 0 565 0 8 653 69  31 10569 5 2 120 60  6 1 1 3  97 19 2 0 0  0

HOWARD     1004 10TH ST  104 1021 4 0 310 0 4 315 28  12 14160 1 2 120 60  5 1 2 1  36 36 3 3 0  0
WINCHESTER 1004 10TH ST  104 1022 4 0 579 0 9 735 82  18 15789 4 2 120 60  5 1 2 3 111 28 2 2 0  0
INGE       1004 10TH ST  104 1023 4 0 291 0 4 301 28  19 18889 3 2 120 60  5 1 2 3  70 25 2 0 0  0
KRESS      1004 10TH ST  104 1024 4 0 458 0 7 568 53  28 12822 6 2 120 60  3 1 1 9  59 30 2 1 0  0
ADAMS      1004 10TH ST  104 1025 4 0 455 0 6 510 42  91 17580 6 2 120 60  3 1 1 9 115 19 1 1 0  0

SHEERAN    1004 10TH ST  104 1026 4 0 393 0 6 486 43  74 13041 3 1  65 65  5 0 1 4  75 25 2 0 0  0
COLE       1004 10TH ST  104 1027 4 0 607 0 6 605 67  24 21026 3 2 152 76  5 0 1 8  83 21 1 0 0  0
WARDER     1003 11TH ST  104 1028 4 0 270 0 4 315 30  26 15240 1 2 120 60  3 0 2 1  43 23 2 0 4 11
LETULI     1003 11TH ST  104 1029 4 0 618 0 8 668 65  78 15212 4 2 120 60  4 0 2 3  53 23 2 0 0  6
ATCHLEY    1003 11TH ST  104 1030 4 0 641 0 8 678 64  20 15273 4 2 120 60  4 0 2 4  67 17 2 2 0  6

BIXLER     1003 11TH ST  104 1031 4 0 326 0 5 404 35  11 14461 5 1  65 65  6 1 2 4  96 19 2 0 5  0
GROTH      1003 11TH ST  104 1032 4 0 429 0 6 471 46  37 18651 7 1  65 65  6 1 2 6 110 16 1 0 0  0
SIDNEY     1003 11TH ST  104 1033 4 0 417 0 6 408 41  20 14215 6 2 120 60  6 1 2 6 138 23 1 0 9  2
MAROONE    1003 11TH ST  104 1034 4 0 425 0 6 487 43  16 15387 6 2 120 60  7 1 2 2  77 19 2 0 0  3
MEYENDORFF 1003 11TH ST  104 1035 4 0 392 0 6 472 42  92 17928 4 2 120 60 12 0 2 2  80 20 1 0 3  3

BOYLE      1003 11TH ST  104 1036 4 0 550 0 6 443 50  17 16847 7 1  65 65  7 0 2 3 182 26 2 0 0  5
WINFREY    1003 11TH ST  104 1037 4 0 380 0 6 497 50  92 16336 5 2 120 60  4 0 2 4 149 25 2 0 6  9
TAYLOR     1004 11TH ST  104 1038 4 0 444 0 5 437 38  79 15340 6 2 120 60  3 0 2 4 149 25 2 0 6  9
SHERHAG    1004 11TH ST  105 1039 4 0 263 0 3 200 21  78  7228 5 1  65 65  3 0 2 1  79 16 2 0 3  6
RUSSELL    1004 11TH ST  105 1040 4 0 446 0 6 519 49  23 13837 7 1  65 65  3 1 1 9 146 21 2 0 6  6

BRENNAN    1004 11TH ST  105 1041 4 0 425 0 6 503 47  21 10987 6 1  65 65 10 0 1 6 114 19 2 0 0  0
NIVEN      1004 11TH ST  105 1042 4 0 399 0 6 460 46  16  9930 5 2 120 60  6 1 2 4  97 19 2 3 0  0
COCO       1004 11TH ST  105 1043 4 0 531 0 8 626 69  36  8987 4 2 120 60  6 1 1 2  69 17 2 0 0  0
FALK       1004 11TH ST  105 1044 4 0 435 0 6 460 47  63 11892 4 2 120 60  6 1 2 1 106 21 2 0 2  0
BAILEY     1004 11TH ST  105 1045 4 0 507 0 6 490 48  14  9780 5 1  65 65  5 0 2 8 101 20 1 0 0  0

NEUMAN     1004 11TH ST  105 1046 4 0 556 0 7 532 59  98 11424 3 1  65 65  4 1 1 4  73 24 2 2 0 11
GOULEY     1004 11TH ST  105 1047 4 0 250 0 3 236 21  25 10659 2 2 120 60  5 1 1 1 103 21 2 0 4  3
WAYNE      1004 11TH ST  105 1048 4 0 566 0 8 618 63  14  9099 5 2 120 60  6 1 1 3  34 17 2 0 0  0
HEPBURN    1003 12TH ST  105 1049 4 0 368 0 5 411 41 117  9417 3 1  65 65  6 1 1 2 119 20 2 2 0  0
WOODWARD   1003 12TH ST  105 1050 4 0 527 0 8 575 60  16  8664 3 2 120 60  4 1 2 0  59 20 2 2 0  0

UGGAMS     1003 12TH ST  105 1051 4 0 571 0 8 571 63  12 12125 3 1  65 65  5 1 2 1  70 23 2 0 0  0
DENNIS     1003 12TH ST  105 1052 4 0 343 0 5 423 37  59  8618 2 2 120 60  5 1 2 4 124 21 2 0 2  2
PECK       1003 12TH ST  105 1053 4 0 355 0 6 490 49  67 12661 4 1  65 65  5 1 2 1  86 22 2 4 0  7
HAYES      1003 12TH ST  105 1054 4 0 302 0 5 424 39  14  9089 8 2 120 60  5 1 2 4 158 20 2 1 0  0
BENNY      1003 12TH ST  105 1055 4 0 404 0 5 369 37  15 13371 6 1  65 65  5 1 2 8 138 23 2 0 5 10

MARX       1003 12TH ST  105 1056 4 0 294 0 4 342 27  18  6620 4 1  65 65  7 1 2 4  87 22 2 0 0  0
PRESLEY    1003 12TH ST  105 1057 4 0 227 0 4 279 29  10  9842 2 2 120 60  7 1 2 1  36 18 2 0 1 11
ALLEN      1003 12TH ST  105 1058 4 0 251 0 3 252 19  17 19917 2 2 120 60  3 0 2 2  50 25 2 0 0  7
BENEDICT   1004 12TH ST  106 1059 4 0 300 0 6 415 35  42  5853 2 2 120 60  6 1 1 4  41 21 2 2 0  0
POMEROX    1004 12TH ST  106 1060 4 0 395 0 6 484 54  21  7288 8 1  65 65  5 1 1 4 160 21 3 2 0  7

MCMANUS    1004 12TH ST  106 1061 4 0 430 0 5 403 40  26  1399 4 2 120 60  3 0 2 7  98 25 2 0 1  0
SADAT      1004 12TH ST  106 1062 4 0 299 0 5 300 27  16  7021 3 2 120 60  4 1 1 4 101 20 2 0 0  0
KROLL      1004 12TH ST  106 1063 4 0 298 0 5 370 34 107  7717 4 2 120 60 10 1 2 9  98 20 2 0 6  0
MCCAULEY   1004 12TH ST  106 1064 4 0 433 0 8 648 55  16  7941 3 2 120 60  9 1 2 4 102 20 2 0 6  0
PETERS     1004 12TH ST  106 1065 4 0 182 0 3 202 20  18  5469 3 2 120 60  9 1 1 3  75 25 2 0 0  0

RESTON     1004 12TH ST  106 1066 4 0 466 0 7 577 55  45  8048 5 1  65 65  5 1 1 3  95 19 2 0 0  0
MCKAY      1004 12TH ST  106 1067 4 0 179 0 3 301 34  44 10811 1 2 120 60  7 0 2 1  32 32 2 0 5  1
FROELICH   1004 12TH ST  106 1068 4 0 154 0 3 239 20  19  6273 3 2 120 60  7 0 2 2  72 24 2 0 0  0
MORRIS     1004 12TH ST  106 1069 4 0 506 0 9 632 60  22  8148 3 2 120 60  4 1 2 2  67 22 2 0 0  5
DEANE      1004 12TH ST  106 1070 4 0 307 0 4 361 30  27  7936 1 2 120 60  4 1 2 1  35 35 2 0 0  5

BURGESS    1003 13TH ST  106 1071 4 0 549 0 6 446 44  40  5813 5 1  65 65 11 0 2 4 110 22 2 0 2  0
AMANS      1003 13TH ST  106 1072 4 0 478 0 6 693 44  20  7306 6 2 120 60  2 1 1 9 109 22 2 0 4  0
RYAN       1003 13TH ST  106 1073 4 0 359 0 5 381 38  71  7306 6 2 120 60  2 1 1 9 138 23 2 0 4  0
GREENWOOD  1003 13TH ST  106 1074 4 0 377 0 6 466 44  18 11770 6 1  65 65  6 1 2 4 116 19 2 0 1 10
PARKER     1003 13TH ST  106 1075 4 0 435 0 6 499 44  26  5175 6 1  65 65  6 1 2 4 123 21 2 0 1 10

HOBSON     1003 13TH ST  106 1076 4 0 467 0 6 413 48  20 27032 6 4 218 55  5 1 1 3 112 19 2 0 0  0
MCMINN     1003 13TH ST  106 1077 4 0 434 0 6 422 38  66  6920 4 2 120 60  4 1 1 6 169 24 2 0 7  0
CAMPBELL   1003 13TH ST  106 1078 4 0 283 0 5 387 43  91  6894 6 1  65 65  4 1 1 6 111 19 2 0 0  0
MILLS      1003 13TH ST  106 1079 4 0 377 0 5 367 39  62  6246 6 1  65 65  5 1 1 9 111 19 2 0 0  0
MALONEY    1002 7TH  ST  107 1080 4 0 352 0 5 451 35  46 13113 6 1  65 65  5 1 2 3 132 22 2 1 0  4

LAWRENCE   1002 7TH  ST  107 1081 4 0 503 0 7 568 53  17 19787 3 2 120 60  6 1 2 3  70 23 2 1 0  0
PENN       1002 7TH  ST  107 1082 4 0 305 0 5 393 32  28 19814 3 2 120 60  6 1 2 1  94 16 2 0 0  0
KNIGHT     1002 7TH  ST  107 1083 4 0 560 0 8 688 68  14 31908 5 3 196 65  2 1 2 2 111 20 2 0 1  0
TROY       1002 7TH  ST  107 1084 4 0 339 0 5 362 38  51 49410 5 3 117 117 3 2 1 2  40 20 2 0 1  0
VIGUERIE   1002 7TH  ST  107 1085 4 0 389 0 5 427 37  14 28679 5 3 189 63  3 2 1 4  94 19 2 0 1  9
```

Appendix C

VARIABLES

01	02	03	04	05	06	07	08	09	10	11	12	13	14	15	16	17	18	19	20	21	22	23	24	25	26	27
CHARLES	1002 7TH ST	107	1086	4	0	222	0	3	293	20	23	20539	5	3	191	64	5	1	2	4	88	18	2	0	0	0
SIMS	1002 7TH ST	107	1087	4	0	519	0	8	700	65	20	19207	4	2	120	60	8	0	1	4	88	22	2	0	0	0
OLSON	1002 7TH ST	107	1088	4	0	426	0	6	428	47	12	32365	6	2	120	60	6	2	1	6	125	21	2	2	0	0
KROHNFELDT	1002 7TH ST	107	1089	4	0	434	0	6	441	47	56	30697	6	4	218	55	4	2	2	4	127	21	2	0	5	1
SULK	1002 7TH ST	107	1090	4	0	412	0	6	478	46	21	24361	6	2	120	60	8	2	1	6	143	24	2	2	0	2
CLARK	1001 8TH ST	107	1091	4	0	436	0	6	466	46	21	35832	4	2	148	74	9	0	1	8	75	19	2	0	0	0
VOELL	1001 8TH ST	107	1092	4	0	525	0	7	592	56	13	26576	4	2	149	75	6	2	2	5	78	20	2	0	0	0
PEALE	1001 8TH ST	107	1093	4	0	328	0	6	445	39	71	25152	6	3	165	55	5	1	2	4	119	20	2	2	5	0
WILBUR	1001 8TH ST	107	1094	4	0	372	0	5	465	40	14	20876	4	2	156	78	6	1	1	6	80	20	2	2	0	6
SELLECK	1001 8TH ST	107	1095	4	0	405	0	6	465	45	23	27968	7	2	120	60	6	2	2	5	143	20	2	0	0	0
MEDOR	1001 8TH ST	107	1096	4	0	499	0	7	495	57	58	23083	5	3	185	62	3	1	2	5	80	16	2	2	0	1
WILLS	1001 8TH ST	107	1097	4	0	241	0	4	283	29	79	19691	2	2	120	60	5	1	2	2	39	20	2	0	6	2
COHEN	1001 8TH ST	107	1098	4	0	354	0	6	475	44	19	22100	7	3	165	55	10	1	1	7	132	19	2	2	0	5
MANDELL	1001 8TH ST	107	1099	4	0	666	0	9	719	76	38	29134	4	3	190	63	8	0	2	3	73	18	2	0	0	2
MARGOLIS	1001 8TH ST	107	1100	4	0	432	0	6	498	43	82	20653	5	4	206	52	6	1	2	3	105	21	2	2	0	0
CANE	1002 8TH ST	108	1101	4	0	358	0	6	471	48	70	17005	5	1	65	65	5	1	1	3	105	21	3	0	4	2
WORCHESTER	1002 8TH ST	108	1102	4	0	640	0	9	749	85	14	19518	4	2	120	60	7	1	1	6	94	24	2	2	0	0
TAFT	1002 8TH ST	108	1103	4	0	526	0	6	514	41	14	21607	6	2	120	60	6	1	1	4	138	23	2	0	0	0
LINCOLN	1002 8TH ST	108	1104	4	0	468	0	7	583	62	20	24259	4	2	162	81	9	2	1	1	80	20	2	2	0	2
EVSLIN	1002 8TH ST	108	1105	4	0	569	0	8	648	68	15	25132	5	3	199	66	6	1	2	5	106	21	2	0	1	11
COFFRIN	1002 8TH ST	108	1106	4	0	354	0	5	405	36	17	19336	5	1	65	65	6	0	1	1	119	24	2	2	0	2
PLESHAW	1002 8TH ST	108	1107	4	0	677	0	9	709	77	19	31397	4	2	158	79	4	2	2	0	70	18	2	1	0	8
DAHLIN	1002 8TH ST	108	1108	4	0	552	0	8	673	62	19	19621	2	2	120	60	6	1	1	2	49	25	2	2	0	4
DEWITT	1002 8TH ST	108	1109	4	0	550	0	8	636	68	76	20826	4	2	157	79	10	2	1	8	82	21	2	0	0	0
HOWARD	1002 8TH ST	108	1110	4	0	432	0	6	470	44	13	23902	7	3	165	55	5	0	2	7	134	19	2	1	0	0
KRAFT	1001 9TH ST	108	1111	4	0	504	0	8	613	67	94	23180	5	3	197	66	3	2	1	3	93	19	1	0	6	10
WELSH	1001 9TH ST	108	1112	4	0	469	0	6	503	45	34	22603	8	2	120	60	3	1	2	2	144	18	2	0	4	5
CHESTERTON	1001 9TH ST	108	1113	4	0	433	0	6	517	48	19	19245	5	1	65	65	4	0	1	4	112	22	3	1	0	0
QUADE	1001 9TH ST	108	1114	4	0	334	0	6	500	39	66	18105	5	1	65	65	8	1	1	6	95	19	2	0	0	0
DILLON	1001 9TH ST	108	1115	4	0	358	0	6	509	52	90	22379	7	2	120	60	7	2	1	5	142	20	2	0	4	1
COHEN	1001 9TH ST	108	1116	4	0	423	0	6	452	48	16	34432	4	3	165	55	11	0	2	1	98	25	2	0	0	1
STUART	1001 9TH ST	108	1117	4	0	478	0	6	465	41	82	30937	5	3	189	63	7	2	2	0	104	21	2	0	0	9
FUDIM	1001 9TH ST	108	1118	4	0	344	0	5	404	34	72	20851	4	3	173	58	10	2	1	6	86	22	1	0	6	0
HUGO	1001 9TH ST	108	1119	4	0	539	0	7	596	54	64	23802	4	3	168	56	8	2	2	2	85	21	2	0	1	5
FERNANDEZ	1001 9TH ST	108	1120	4	0	373	0	5	360	42	16	4519	3	2	120	60	8	0	1	1	78	26	2	0	0	0
SILER	1001 9TH ST	108	1121	4	0	373	0	6	528	45	21	22564	6	4	216	54	5	2	1	6	114	19	2	0	0	0
ETONS	1002 9TH ST	109	1122	4	0	389	0	6	466	47	84	14900	5	1	65	65	4	1	2	3	110	22	2	0	3	3
TIMOTHY	1002 9TH ST	109	1123	4	0	265	0	4	310	29	102	16711	1	2	120	60	7	0	2	1	29	29	2	0	0	0
PEREZ	1002 9TH ST	109	1124	4	0	357	0	5	432	38	28	14333	2	2	120	60	4	0	1	2	57	29	2	0	3	0
GREGORY	1002 9TH ST	109	1125	4	0	423	0	6	482	51	16	17724	5	2	120	60	9	1	1	4	128	26	2	2	0	9
SHOUMATOFF	1002 9TH ST	109	1126	4	0	325	0	6	476	47	19	22005	4	2	162	81	6	2	1	4	87	22	2	0	1	0
HEFFRON	1002 9TH ST	109	1127	4	0	517	0	8	673	52	17	18341	4	2	120	60	7	1	2	2	73	18	2	0	5	1
DEFARIA	1002 9TH ST	109	1128	4	0	609	0	8	678	68	24	26511	5	3	199	66	7	2	2	1	114	23	2	0	4	2
PETERSON	1002 9TH ST	109	1129	4	0	416	0	7	543	57	13	18771	4	1	65	65	6	1	2	6	66	17	2	0	0	0
FRITTS	1002 9TH ST	109	1130	4	0	344	0	6	521	42	21	19307	6	2	120	60	4	0	2	6	124	21	2	2	0	5
MCCABE	1002 9TH ST	109	1131	4	0	334	0	5	384	35	49	24604	2	1	105	105	5	1	2	1	39	20	2	2	0	0
ALLEN	1002 9TH ST	109	1132	4	0	233	0	4	363	28	18	19179	5	2	120	60	7	1	2	4	108	22	2	0	0	11
KAPLAN	1001 10TH ST	109	1133	4	0	495	0	6	526	50	15	20791	7	2	120	60	2	1	2	5	129	18	2	0	0	5
HOOVER	1001 10TH ST	109	1134	4	0	468	0	6	492	37	17	17395	8	1	65	65	4	1	2	6	163	20	2	2	0	0
TURNER	1001 10TH ST	109	1135	4	0	327	0	4	340	26	28	21870	1	1	65	65	6	0	2	0	23	23	2	1	0	0
OGRADY	1001 10TH ST	109	1136	4	0	424	0	6	458	43	105	23923	8	2	120	60	5	2	2	8	162	20	2	0	5	1
STORM	1001 10TH ST	109	1137	4	0	351	0	6	506	47	38	27784	5	4	206	52	6	2	2	5	102	20	2	2	0	0
GREGG	1001 10TH ST	109	1138	4	0	369	0	6	497	50	14	18481	5	1	65	65	6	0	2	4	116	23	2	0	0	10
WIRTA	1001 10TH ST	109	1139	4	0	421	0	6	469	46	28	18924	5	2	120	60	4	0	2	3	95	19	2	0	1	4
ROWLAND	1001 10TH ST	109	1140	4	0	519	0	6	420	51	19	20883	5	4	208	52	3	1	2	5	122	24	1	0	1	1
LUBITZ	1001 10TH ST	109	1141	4	0	610	0	8	644	63	39	25317	3	2	130	65	8	2	2	1	69	23	1	2	0	4
KOGER	1001 10TH ST	109	1142	4	0	441	0	6	506	47	21	20648	3	2	158	79	4	2	1	1	60	20	2	0	0	0
LONIEWSKI	1002 10TH ST	110	1143	4	0	268	0	3	229	21	16	12770	4	2	120	60	6	1	1	0	74	19	2	0	5	0
HALBROOKS	1002 10TH ST	110	1144	4	0	345	0	6	482	46	63	12597	5	1	65	65	8	1	1	5	126	25	2	1	0	9
CALDWELL	1002 10TH ST	110	1145	4	0	526	0	7	560	59	22	13043	3	2	120	60	7	1	1	1	61	20	2	1	0	1
SUSSEX	1002 10TH ST	110	1146	4	0	233	0	4	384	26	26	18919	1	2	120	60	5	1	1	1	26	26	2	2	0	0
VICKSLEY	1002 10TH ST	110	1147	4	0	270	0	5	415	34	25	14790	6	2	120	60	7	0	1	8	119	20	2	0	4	8
JOSEPHS	1002 10TH ST	110	1148	4	0	525	0	8	609	64	60	18718	4	2	120	60	9	0	1	5	70	18	2	0	0	0
GORMAN	1002 10TH ST	110	1149	4	0	438	0	8	651	69	27	19670	4	2	120	60	5	0	2	4	87	22	2	0	0	4
RUIZ	1002 10TH ST	110	1150	4	0	491	0	8	632	68	12	12653	3	2	120	60	5	1	1	4	61	20	2	0	0	0
PURCELL	1002 10TH ST	110	1151	4	0	403	0	5	425	36	81	18639	8	2	120	60	4	0	1	6	161	20	2	2	0	0
HINDS	1002 10TH ST	110	1152	4	0	412	0	6	529	42	40	11171	7	1	65	65	4	0	2	7	154	22	2	2	0	0
HOROWITZ	1002 10TH ST	110	1153	4	0	323	0	5	433	33	25	12575	5	1	65	65	2	0	1	3	120	24	1	0	2	0
NORTON	1001 11TH ST	110	1154	4	0	551	0	9	762	65	38	19199	4	1	65	65	4	0	1	4	94	24	2	0	1	0
MIDDLETON	1001 11TH ST	110	1155	4	0	445	0	7	555	61	22	14018	3	1	65	65	7	1	2	1	73	24	2	0	0	0
BASSETT	1001 11TH ST	110	1156	4	0	493	0	6	484	50	13	27039	6	4	224	56	8	2	1	4	130	22	2	0	3	9
FOSTER	1001 11TH ST	110	1157	4	0	402	0	6	393	44	29	17098	3	1	65	65	5	0	2	3	66	22	2	2	0	0
KAHN	1001 11TH ST	110	1158	4	0	261	0	4	260	29	89	7531	4	1	65	65	8	1	2	4	98	25	2	0	4	0
DUNLAN	1001 11TH ST	110	1159	4	0	417	0	6	465	49	19	18141	4	2	120	60	7	1	1	5	79	20	2	0	0	0
PRAEFKE	1001 11TH ST	110	1160	4	0	237	0	4	338	28	21	17381	3	1	65	65	6	0	2	4	81	27	2	0	5	0
OLESON	1001 11TH ST	110	1161	4	0	388	0	6	489	47	16	11075	6	2	120	60	6	0	1	0	136	23	2	2	0	4
MENDOZA	1001 11TH ST	110	1162	4	0	527	0	8	623	63	24	20610	3	2	151	76	6	2	2	1	60	20	2	2	0	8
ACEVEDO	1001 11TH ST	110	1163	4	0	583	0	8	615	64	99	19334	5	2	120	60	4	1	2	6	97	19	2	2	0	6
CLARKSON	1002 11TH ST	111	1164	4	0	206	0	3	263	22	82	7369	2	2	120	60	5	0	2	1	41	21	2	2	0	2
NEUMAN	1002 11TH ST	111	1165	4	0	282	0	3	229	24	96	11763	2	2	120	60	3	1	1	1	53	27	2	2	0	0
KATZMAN	1002 11TH ST	111	1166	4	0	448	0	6	488	46	46	9336	5	2	120	60	6	0	2	5	117	23	2	0	0	2
BAGWELL	1002 11TH ST	111	1167	4	0	372	0	5	382	34	24	17790	2	2	120	60	9	1	1	3	50	25	2	0	0	0
GREEN	1002 11TH ST	111	1168	4	0	341	0	5	452	38	22	8998	8	2	120	60	9	0	2	2	186	23	2	0	0	1
SELAWRY	1002 11TH ST	111	1169	4	0	230	0	4	279	30	51	11561	1	2	120	60	6	0	1	1	25	25	2	0	0	9
NAGEL	1002 11TH ST	111	1170	4	0	346	0	6	522	45	55	9609	4	1	65	65	4	1	2	2	84	21	2	0	5	0
HENDRICKS	1002 11TH ST	111	1171	4	0	270	0	5	412	40	81	7406	5	2	120	60	6	1	1	1	112	22	2	1	0	5
WILLIAMS	1002 11TH ST	111	1172	4	0	563	0	6	525	38	31	7983	7	2	120	60	3	1	1	9	113	16	2	0	0	7
ROOTH	1002 11TH ST	111	1173	4	0	402	0	5	429	40	13	9844	5	2	120	60	4	1	2	3	121	24	2	1	0	0
BALOUGH	1002 11TH ST	111	1174	4	0	336	0	5	412	41	68	9755	6	2	120	60	8	0	2	12	132	22	2	0	0	4
WEBB	1001 12TH ST	111	1175	4	0	339	0	5	417	34	14	10227	5	2	120	60	7	1	2	5	117	23	2	0	4	0
PRICE	1001 12TH ST	111	1176	4	0	475	0	7	523	55	85	15981	4	1	65	65	8	0	2	4	81	20	2	2	0	5
FERNANDEZ	1001 12TH ST	111	1177	4	0	513	0	7	568	51	41	19315	3	2	120	60	5	1	2	6	59	20	2	0	0	0
CARLSON	1001 12TH ST	111	1178	4	0	481	0	7	545	56	18	13035	4	1	65	65	7	0	1	3	81	20	2	2	0	0
ARMINIO	1001 12TH ST	111	1179	4	0	556	0	8	628	61	18	13622	4	2	120	60	7	0	2	2	84	21	2	0	1	9
HERNANDEZ	1001 12TH ST	111	1180	4	0	608	0	8	666	71	19	11156	3	2	120	60	7	1	2	5	76	25	3	0	4	9
DURAN	1001 12TH ST	111	1181	4	0	569	0	7	608	55	22	18194	2	2	120	60	10	1	1	2	53	27	2	0	4	4
FERGUSON	1001 12TH ST	111	1182	4	0	351	0	6	479	43	45	15918	6	2	120	60	3	1	1	8	133	22	2	0	0	0
COHEN	1001 12TH ST	111	1183	4	0	463	0	7	559	51	88	14584	5	2	120	60	5	1	2	1	81	16	2	2	0	0
HARRIS	1001 12TH ST	111	1184	4	0	424	0	5	433	35	13	6549	6	2	120	60	5	0	1	3	132	22	2	0	0	0
DELEON	1002 12TH ST	112	1185	4	0	287	0	5	350	33	13	4581	3	2	120	60	5	0	2	4	58	19	2	0	4	0

VARIABLES

```
                              0  0  0              0  0  0 0   0 0           1 1 1   1   1 1 1   1   1 1 2   2  2 2 2 2 2 2
              0 1    0 2      1  2  3              4  5  6 7   8 9 0           1 2 3   4   5 6 7   8   9 0 1   2  3 4 5 6 7
**************************************************************************************************************************
MURPHY     1002 12TH ST  112 1186 4 0 379   0  6 480 45  14  6677  6 2 120 60  6 1 1  3 130 22 2 0 0
RASKOSKY   1002 12TH ST  112 1187 4 0 307   0  4 312 31  18  6083  7 2 120 60  8 1 1  1  35 18 2 1 0 5
OSTROW     1002 12TH ST  112 1188 4 0 343   0  5 411 38 107  7504  7 2 120 60  8 1 5  1 159 33 1 2 0 5 1 0
CORONA     1002 12TH ST  112 1189 4 0 449   0  6 500 47  25  1807  5 2 120 60  8 0 5  1 125 25 1 2 0 1 0
BILLINGS   1002 12TH ST  112 1190 4 0 474   0  5 434 36  19  8126  5 2 120 60  4 1 2  1  66 22 2 0 0

PARKS      1002 12TH ST  112 1191 4 0 337   0  5 352 37  15  5322  5 2 120 60  6 1 1  4 104 21 3 0 0
FISHER     1002 12TH ST  112 1192 4 0 405   0  5 466 52  16  5786  5 1  65 65  6 1 2  4 104 21 3 0 6 6
DELGADO    1002 12TH ST  112 1193 4 0 266   0  5 365 42  22  7865  6 2 120 60  6 1 1  9 114 19 2 0 6 6
REIZEN     1002 12TH ST  112 1194 4 0 214   0  3 220 19  15  8841  6 2 120 60  6 1 1  1  28 28 2 2 0
TERCILLA   1002 12TH ST  112 1195 4 0 367   0  5 384 35  13  8256  3 2 120 60  6 0 1  4  71 24 1 2 0 4

ESCOBAR    1001 13TH ST  112 1196 4 0 281   0  4 373 26  26  5985  4 2 120 60  5 0 1  8  81 20 2 0 4 0
OWENS      1001 13TH ST  112 1197 4 0 405   0  5 423 32  91  5532  5 2 120 60  5 0 1  8  81 20 2 0 4 0
LENHARDT   1001 13TH ST  112 1198 4 0 308   0  4 344 31  39  5881  5 2 120 60  5 1 1  4 176 22 2 2 0 3
MALZONE    1001 13TH ST  112 1199 4 0 379   0  5 487 49  39  8189  5 2 120 60  6 1 5  4  96 19 2 1 0
TANNER     1001 13TH ST  112 1200 4 0 386   0  4 458 49  30  7723  5 1  65 65  7 0 2  4  93 19 2 0 0 11

STREETER   1001 13TH ST  112 1201 4 0 213   0  4 299 29  25  2974  5 1  65 65  2 0 2  3  87 17 2 0 0
BERKOWITZ  1001 13TH ST  112 1202 4 0 267   0  4 335 32 105  4739  5 1  65 65  2 0 2  4  85 28 2 0 4 0
FONTANA    1001 13TH ST  112 1203 4 0 441   0  5 504 45  11  5302  5 2 120 60  3 0 2  3 122 20 2 0 5 4
SINGLETON  1001 13TH ST  112 1204 4 0 243   0  3 210 22  38  5846  5 2 120 60  6 0 1  4  85 15 2 0 5 4
ANDERSON   1001 13TH ST  112 1205 4 0 264   0  5 428 45  89  5429  6 2 120 60  6 0 1  4 126 21 2 0 4 3

WALKER      904 9TH ST   113 1206 4 0 466   0  6 458 56  19 13222  7 1  65 65  6 1 2  2 113 16 2 3 0 0
HIROKAWA    904 9TH ST   113 1207 4 0 468   0  7 624 66  22 12451  7 1  65 65  6 1 2  4  95 24 2 0 3 0
MACHT       904 9TH ST   113 1208 4 0 504   0  8 608 65  37 14475  4 2 120 60  6 1 2  2  89 22 2 0 5 2
JIMENEZ     904 9TH ST   113 1209 4 0 379   0  8 384 34  34 10477  3 2 120 60  6 1 2  2  70 22 2 0 5 0
LINDSAY     904 9TH ST   113 1210 4 0 539   0  8 457 65  19 16988  3 2 120 60  7 0 2  4  81 27 3 0 1 0

WALLACH     904 9TH ST   113 1211 4 0 403   0  6 495 46  15 21285  5 4 202 51  5 2 2  4 110 22 2 0 1 0
SIBLEX      904 9TH ST   113 1212 4 0 487   0  7 542 46  77 22910  5 3 187 62  5 2 2  1  94 19 2 0 4 0
HILL        904 9TH ST   113 1213 4 0 382   0  6 499 42  20 21847  6 4 218 55  7 2 1  4 137 23 2 2 0 6
PACCIONE    904 9TH ST   113 1214 4 0 162   0  3 243 21  30 13461  4 2 120 60  4 2 1  2  45 23 2 2 0 0
LEHMAN      904 9TH ST   113 1215 4 0 538   0  6 487 48  20 21094  5 4 207 52  7 0 1  4 123 25 2 2 0 1

TORRES      904 9TH ST   113 1216 4 0 690   0 10 817 94  74 29141  4 3 183 61 10 0 1  4  79 20 2 0 0 3
JORGE       903 10TH ST  113 1217 4 0 267   0  5 384 39  27 20175  1 1  65 65  7 7 2  1  31 31 2 0 5 3
NATHANSON   903 10TH ST  113 1218 4 0 435   0  5 423 35  94 40836  1 1  65 65  5 2 1  1  26 26 1 2 0 6 1
EARLY       903 10TH ST  113 1219 4 0 361   0  5 521 53  18 19757  1 1  65 65  5 2 2  1  82 21 2 0 5 4
MACIEL      903 10TH ST  113 1220 4 0 561   0  8 669 64  50 27651  2 2 120 60  7 1 2  1  53 27 2 0 4 4

ZUCKERMAN   903 10TH ST  113 1221 4 0 423   0  6 490 45  72 24703  5 5 233 47  6 2 2  4 111 22 2 1 0
SANDUSKY    903 10TH ST  113 1222 4 0 365   0  6 402 36  45 20390  4 3 185 65  6 2 1  2  62 22 2 0 0
FERRE       903 10TH ST  113 1223 4 0 420   0  5 500 51  19 29086  6 3 190 56  6 3 1  4 105 18 2 0 0
MARQUEZ     903 10TH ST  113 1224 4 0 446   0  6 488 46  38 25525  6 3 169 56  6 2 0  1  79 20 2 0 0 3
GRABOWSKI   903 10TH ST  113 1225 4 0 618   0  8 666 66  53 26071  3 2 128 64  6 2 1  4  74 25 2 1 0

CORBETT     903 10TH ST  113 1226 4 0 364   0  4 418 40  79 19567  5 1  65 65  2 0 2  5  91 18 2 0 4 0
BAUER       904 10TH ST  114 1227 4 0 395   0  7 560 51  38 10312  5 1  65 65 10 0 1  4 109 27 2 0 0
DORSEY      904 10TH ST  114 1228 4 0 433   0  6 433 48  21  7967  6 1  65 65  8 1 1  3 123 21 1 0 4 0
ALVAREZ     904 10TH ST  114 1229 4 0 441   0  5 442 38  72 16968  6 1  65 65  8 1 1  3 123 21 1 0 4 11
SENN        904 10TH ST  114 1230 4 0 322   0  5 302 25  19 11648  2 2 120 60  3 0 2  1  89 18 2 0 2 3

VARGAS      904 10TH ST  114 1231 4 0 440   0  7 586 58  22 19576  3 1  65 65  3 0 2  2  68 23 2 0 4 3
BASSETT     904 10TH ST  114 1232 4 0 511   0  7 571 56  72 14331  4 1  65 65  5 2 0  4  68 23 2 0 4 0
ROSENBLUM   904 10TH ST  114 1233 4 0 656   0 10 832 79  71 13477  4 1  65 65  7 0 2  4  97 24 2 1 5 0
COOLIDGE    904 10TH ST  114 1234 4 0 475   0  7 477 49  13 17580  4 2 120 60  7 1 2  1  88 29 2 2 0 3
TERRELL     904 10TH ST  114 1235 4 0 282   0  4 295 28  19 15042  1 2 120 60  8 0 1  1  37 37 2 0 5 3

PRIETO      904 10TH ST  114 1236 4 0 469   0  8 544 52  13 19745  3 2 120 60  4 1 1  3  58 19 2 0 0 6
ADAMSON     903 11TH ST  114 1237 4 0 576   0  8 643 59  21 21549  3 2 155 78  4 1 3  1  50 17 3 1 0 0
SCHWARTZ    903 11TH ST  114 1238 4 0 453   0  6 487 42  15 13326  5 2 120 60  4 2 2  4 146 22 2 0 0 6
PROULX      903 11TH ST  114 1239 4 0 278   0  5 430 41  22 13556  2 2 120 60  4 1 1  1  74 25 2 0 2 0
WINNICK     903 11TH ST  114 1240 4 0 498   0  8 637 68  76 25241  4 3 165 55  3 2 2  1  93 23 2 0 4

RUBIN       903 11TH ST  114 1241 4 0 438   0  7 572 58  75 18150  5 2 120 60  8 0 2  3 109 22 2 0 1 0
CORDOVA     903 11TH ST  114 1242 4 0 411   0  6 471 49  19 24743  4 3 168 56  8 0 2  2  47 23 2 0 0
LO          903 11TH ST  114 1243 4 0 377   0  5 394 37  15 17542  6 1  65 65  4 1 1  1 109 18 2 0 0
KIRSCH      903 11TH ST  114 1244 4 0 352   0  5 420 39  23 16095  3 1  65 65  4 1 1  2  78 26 2 2 0 0
TALMADGE    903 11TH ST  114 1245 4 0 485   0  6 466 45  77 22413  4 3 180 60  7 2 2  4  96 24 2 1 0

MELLA       903 11TH ST  114 1246 4 0 432   0  6 494 47  58 24855  4 3 195 65  7 2 2  3  85 21 2 0 0 1
LESTER      903 11TH ST  114 1247 4 0 377   0  6 494 48  78 21735  5 4 212 53  6 2 2  3 120 24 2 1 0 0
SUTHERLAND  904 11TH ST  115 1248 4 0 437   0  6 432 48  14  5780  8 1  65 65  6 1 1  2 178 22 2 1 0 1
FOLTMAN     904 11TH ST  115 1249 4 0 113   0  2 164 13  20  5660  4 1  65 65  6 1 1 12 178 22 2 0 0
MENDOZA     904 11TH ST  115 1250 4 0 422   0  7 554 60 107 10145  4 1  65 65  5 2 1  3  84 21 2 0 11

KINSEY      904 11TH ST  115 1251 4 0 416   0  6 455 46  25 10803  5 2 120 60 10 1 2  4  96 19 2 2 0
HORNER      904 11TH ST  115 1252 4 0 393   0  6 460 48  24 10116  5 2 120 60  8 0 1  2  94 19 2 0 3
SENGUPTA    904 11TH ST  115 1253 4 0 337   0  5 419 43  93 11296  4 2 120 60  8 1 1  1 123 25 2 0 6 1
NOTARIO     904 11TH ST  115 1254 4 0 370   0  6 480 48  20 13827  1 2 120 60  5 0 2  4  51 26 2 0 7 1
JACOBSON    904 11TH ST  115 1255 4 0 353   0  5 434 40  25  9187  4 2 120 60  6 0 2 10 117 17 2 0 1 2

FOLLETT     904 11TH ST  115 1256 4 0 270   0  4 338 29  24  7492  3 2 120 60  3 0 1  7  80 21 2 2 0
HWANG       904 11TH ST  115 1257 4 0 549   0  6 667 63  21 14385  3 1  65 65  3 0 1  1  55 18 2 0 7 0
SCHNEIDER   904 11TH ST  115 1258 4 0 370   0  7 395 37  22  6473  2 2 120 60  7 1 1  1  80 22 2 0 1 7
MIDDLETON   903 12TH ST  115 1259 4 0 567   0  7 608 55  18 16406  4 2 120 60  6 1 1  1  46 23 2 0 1 5
TYLER       903 12TH ST  115 1260 4 0 512   0  7 569 58  18 10392  5 2 120 60  6 0 1  4 101 20 2 5 0 7

MCKENRY     903 12TH ST  115 1261 4 0 524   0  7 561 53  39 10925  5 2 120 60  8 1 1  3 109 21 1 0 1 3
KAPROVE     903 12TH ST  115 1262 4 0 291   0  5 256 21  14 11557  2 2 120 60  8 1 1  1  33 33 2 1 2 8
MORGAN      903 12TH ST  115 1263 4 0 352   0  5 406 35  16 18385  4 2 120 60  8 0 1  4  85 21 2 0 1 8
VASQUEZ     903 12TH ST  115 1264 4 0 514   0  7 559 58  42 12860  5 2 120 60  6 1 1  4  55 18 2 1 6 1
JACKSON     903 12TH ST  115 1265 4 0 689   0  9 700 84  38 23745  4 2 150 75 11 0 1  2  74 19 2 0 8 1

TANNER      903 12TH ST  115 1266 4 0 431   0  7 547 53  39 20685  3 2 140 70  9 1 1  4  75 25 3 0 0 0
PROVENZO    903 12TH ST  115 1267 4 0 288   0  5 408 35  47  8466  3 2 120 60  8 1 1  4 115 23 2 0 2 0
LEVERMORE   903 12TH ST  115 1268 4 0 458   0  6 586 52  24 24872  3 3 185 62  6 1 2  1  80 24 2 0 0
SEYMOUR     904 12TH ST  116 1269 4 0 398   0  6 490 48  18  4877  6 1  65 65 10 1 2  4 142 22 2 0 2 4
OYEMURA     904 12TH ST  116 1270 4 0 377   0  5 412 36  62  9179  4 1  65 65  3 0 2  3 136 23 2 0 5 0

ROSS        904 12TH ST  116 1271 4 0 405   0  5 461 33  13  6494  3 2 120 60  3 0 2  2  65 22 2 0 0 0
GREGORY     904 12TH ST  116 1272 4 0 356   0  5 422 36  20 13933  4 2 120 60  3 0 1  4 104 21 2 0 3 0
LIPMAN      904 12TH ST  116 1273 4 0 396   0  5 411 49  20  6798  4 2 120 60  6 0 2  1  68 17 2 0 3 0
VICTOR      904 12TH ST  116 1274 4 0 454   0  7 586 59  20 11148  4 1  65 65  6 1 1  2  96 24 2 0 0
FONT        904 12TH ST  116 1275 4 0 520   0  7 586 59  26 16418  4 2 120 60  7 1 1  2  83 21 1 0 3 0

DUVENHAGE   904 12TH ST  116 1276 4 0 489   0  6 515 43  37  3342  5 1  65 65  2 0 2  4 112 22 2 0 1 0
LAYTON      904 12TH ST  116 1277 4 0 445   0  5 470 37  24  6590  5 1  65 65  5 0 2  4 128 21 2 0 1 0
SIMMONS     904 12TH ST  116 1278 4 0 309   0  4 318 29  71  4664  4 2 120 60  5 0 2  4  77 19 2 0 1 0
PACUCH      904 12TH ST  116 1279 4 0 313   0  4 290 26  36  5555  4 2 120 60  5 0 2  4  45 23 2 0 1 0
FRANKLIN    903 13TH ST  116 1280 4 0 354   0  5 404 36  17  6670  5 2 120 60  8 1 1  4 110 22 2 0 1

GWYN        903 13TH ST  116 1281 4 0 372   0  5 451 37  45  6785  4 1  65 65  5 0 2  2  93 23 2 2 0
POOLEY      903 13TH ST  116 1282 4 0 301   0  4 401 41  11  5326  4 2 120 60  5 0 2  4 111 19 2 0 0
MONOSA      903 13TH ST  116 1283 4 0 271   0  4 303 27  16  7080  2 2 120 60  6 0 1  1  48 24 2 0 0
SANCHEZ     903 13TH ST  116 1284 4 0 311   0  5 529 52  20  7920  6 1  65 65  6 1 2  1 113 19 2 0 0
TOLEDO      903 13TH ST  116 1285 4 0 211   0  4 317 24  27  6671  1 2 120 60  5 1 1  1  30 30 2 0 0
```

Appendix C

```
                                                              VARIABLES
  0         0           0      0 0 0   0        0   0         1          1 1 1   1       1 1 2   2   2   2   2 2   2
  1         2           3      4 5 6   7        8   9 0 1 2   3          4 5 6   7       8 9 0   1   2   3   4   5 6   7
***********************************************************************************************************************
HERNANDEZ  903 13TH ST  116 1286 4 0   462   0 6 442 49  12  8521  6 2 120  60  6 0 1  6 140 23 2 0 5  0
LYNCH      903 13TH ST  116 1287 4 0   394   0 6 472 50  19  3102  9 1  65  65  4 0 2  7 180 20 2 0 1  1
GRAHAM     903 13TH ST  116 1288 4 0   316   0 6 487 42  50  3383  4 1  65  65  4 0 2  3  91 23 2 0 0  1
CORTINA    903 13TH ST  116 1289 4 0   453   0 6 461 48  13  8726  5 2 120  60  4 0 2  5 109 22 2 0 1  6
MEDINA     902  9TH ST  117 1290 4 0   353   0 5 382 35  15 16555  6 1  65  65  7 0 2  3 130 22 2 0 5  4

SMITH      902  9TH ST  117 1291 4 0   402   0 5 470 35  23 14120  3 2 120  60  5 1 2  2  55 18 2 2 0  0
ROHRBACK   902  9TH ST  117 1292 4 0   415   0 5 503 48  40  7480  4 1  65  65  6 0 2  2 130 22 2 2 0  0
GLASSFORD  902  9TH ST  117 1293 4 0   163   0 3 271 19  69 13723  2 2 120  60  8 0 2  2  43 22 2 2 0  0
BERGER     902  9TH ST  117 1294 4 0   436   0 6 503 40  14 14656  9 2 120  60  9 1 2  2 154 17 2 2 0  0
BOLD       902  9TH ST  117 1295 4 0   499   0 5 414 37  19 31619  4 2 152  76  4 0 2  2  70 18 2 1 0  3

HANSEN     902  9TH ST  117 1296 4 0   335   0 8 382 38  34 20703  4 3 177  59  5 1 2  4  99 25 2 0 4  2
PLANTE     902  9TH ST  117 1297 4 0   583   0 8 647 70  15 18922  5 1  65  65  6 0 1  2 109 22 2 1 0  9
KIRSCHEN   902  9TH ST  117 1298 4 0   230   0 3 221 21  75 19654  1 2 120  60  8 0 2  1  31 31 3 0 5  9
SKINNER    902  9TH ST  117 1299 4 0   305   0 3 346 29  76 29004  2 2 160  80  7 2 1  1  85 21 2 0 0  0
GOYNE      902  9TH ST  117 1300 4 0   332   0 5 418 37  18 23132  2 1 110 110  6 2 2  1  35 18 3 0 0  0

SANCHEZ    901 10TH ST  117 1301 4 0   479   0 7 566 56  17 20073  4 3 170  57  8 2 1  5  88 22 2 0 0  3
TEASLEY    901 10TH ST  117 1302 4 0   297   0 5 304 36  17 24761  2 3  95  65  7 1 1  4  34 17 2 2 0  0
GOLDFARB   901 10TH ST  117 1303 4 0   476   0 8 497 41  25 18216  2 2 120  60  7 1 2  4 122 20 2 1 0  9
LARKIN     901 10TH ST  117 1304 4 0   524   0 8 603 67  26 24422  5 3 199  66  7 1 1  4 107 21 2 0 3  0
PALMER     901 10TH ST  117 1305 4 0   439   0 6 494 53  56 22516  6 2 120  60  7 2 2  1 107 18 2 0 0 10

SOCORRO    901 10TH ST  117 1306 4 0   600   0 9 688 74  68 29034  5 3 190  63  8 0 1  9 108 22 2 1 0  0
VEIGA      901 10TH ST  117 1307 4 0   405   0 6 463 48  21 27158  6 4 218  55  5 1 2  4 127 21 2 0 5  0
FEINMEL    901 10TH ST  117 1308 4 0   386   0 6 484 43  18 28058  2 2 135  68  8 2 1  1  60 30 2 0 6  0
HAMPTON    901 10TH ST  117 1309 4 0   312   0 5 403 38  26 24029  4 2 145  73 12 1 1  1  78 20 2 0 3  0
CRANE      901 10TH ST  117 1310 4 0   325   0 5 414 38  91 28180  4 3 184  61  9 0 1  5 109 27 2 1 0  0

LAVIN      902 10TH ST  118 1311 4 0   439   0 7 589 54  60 16962  4 1  65  65  3 0 1  3  88 22 2 2 0  0
ROTH       902 10TH ST  118 1312 4 0   277   0 3 340 31 107 11936  2 2 120  60  4 1 1  3  32 16 2 0 4  0
LO         902 10TH ST  118 1313 4 0   170   0 3 226 19  21  6971  5 1  65  65  3 0 2  1  90 18 2 0 4  0
CONWAY     902 10TH ST  118 1314 4 0   438   0 6 475 47  72 14298  6 2 120  60  6 0 2  1  91 15 2 0 6  0
ABDO       902 10TH ST  118 1315 4 0   377   0 5 385 38  14 19456  7 1  65  65  7 0 2  9 155 22 2 0 6  0

HASTINGS   902 10TH ST  118 1316 4 0   261   0 3 220 19  18 12494  1 2 120  60  4 0 1  1  22 22 2 1 0  0
WULF       902 10TH ST  118 1317 4 0   452   0 6 423 40  23 26335  1 1  65  65  6 0 1  8 100 17 2 0 1  0
SHAPIRO    902 10TH ST  118 1318 4 0   350   0 5 424 38  20 14303  6 2 120  60  4 0 1  4 125 21 2 1 1  0
GUERRIER   902 10TH ST  118 1319 4 0   289   0 5 437 36  18 10349  6 1  65  65  3 0 2  4 125 21 2 1 0  0
SINGLETON  902 10TH ST  118 1320 4 0   450   0 6 470 46  10 19193  7 2 120  60  4 1 1  2 142 10 2 0 0  0

HERNANDEZ  902 10TH ST  118 1321 4 0   350   0 6 481 47  81 25493  4 4 204  51  7 2 2  3  99 25 2 0 6  0
CANETA     901 11TH ST  118 1322 4 0   599   0 8 670 66  13 16204  5 3 120  60  5 2 1  4  98 20 2 0 0  0
MILLS      901 11TH ST  118 1323 4 0   375   0 5 550 43  16 19972  5 1  65  65  2 1 1  3 145 24 2 0 0  0
KAHN       901 11TH ST  118 1324 4 0   292   0 4 456 38  33 20395  2 1 100 100  7 2 1  2  39 20 2 0 0  0
WATKINS    901 11TH ST  118 1325 4 0   407   0 6 455 46  20 13694  4 1  65  65  4 1 1  2  95 24 2 0 0  0

ROGOW      901 11TH ST  118 1326 4 0   408   0 6 501 44  56 23242  5 3 191  64  7 2 2  5  87 17 2 1 0  0
BLOCK      901 11TH ST  118 1327 4 0   400   0 6 452 31  70 18555  5 2 120  60  6 1 1  6 117 23 2 0 1  0
GUTIERREZ  901 11TH ST  118 1328 4 0   423   0 6 503 47  25 18497  5 2 120  60  6 1 1  6  88 18 2 0 3  0
PEREZ      901 11TH ST  118 1329 4 0   437   0 6 446 42  31 26012  2 2 137  69  6 1 1  6  58 29 2 0 5  0
ESPIRITO   901 11TH ST  118 1330 4 0   486   0 7 578 55  36 18369  4 2 120  60  5 1 1  5  90 23 2 0 0  0

BOLINGER   901 11TH ST  118 1331 4 0   461   0 7 564 55  73 19738  5 2 120  60  3 1 1  1 105 21 2 1 0  0
SAGE       902 11TH ST  119 1332 4 0   156   0 3 222 20  42  8999  2 2 120  60  4 1 2  2  49 25 2 0 1  5
CATTERY    902 11TH ST  119 1333 4 0   321   0 5 440 42  50  8807  2 2 120  60  4 1 2  3 145 21 2 0 1  5
FOLEY      902 11TH ST  119 1334 4 0   329   0 5 388 38  25  9257  7 2 120  60  5 1 1  4 130 22 2 0 1  3
BALDWIN    902 11TH ST  119 1335 4 0   286   0 5 398 36  39  8907  6 1  65  65  5 1 1  4 130 22 2 0 1  3

CANCELA    902 11TH ST  119 1336 4 0   234   0 4 262 27  97 10960  4 1  65  65  5 0 2  3  78 20 2 0 1  5
LEIBOWITZ  902 11TH ST  119 1337 4 0   553   0 8 629 64  19 14338  4 2 120  60  5 1 1  1 101 25 2 0 1  0
RAMIREZ    902 11TH ST  119 1338 4 0   375   0 7 542 56  23 11897  3 1  65  65  5 0 2  1  69 23 2 0 1  0
KIMBLER    902 11TH ST  119 1339 4 0   431   0 5 443 34  21 12668  6 2 120  60  5 0 2  2 141 24 2 0 0  0
DUPUIS     902 11TH ST  119 1340 4 0   408   0 7 538 60  89 10032  4 1  65  65  4 1 1  2  85 21 2 2 0  2

MORENO     902 11TH ST  119 1341 4 0   362   0 5 400 38  16 24789  4 2 149  75  7 2 1  4  83 21 2 0 6  1
QUINTERO   902 11TH ST  119 1342 4 0   407   0 6 485 51  15 16392  4 2 120  60 10 1 2  3 159 22 2 0 4  0
SMITH      901 12TH ST  119 1343 4 0   309   0 6 474 48  14 23260  3 2 144  72  7 2 2  0  51 17 2 0 4  0
KUTUN      901 12TH ST  119 1344 4 0   568   0 8 646 61  15 21308  3 2 144  72  5 2 2  3 135 22 2 0 4  4
PRIEST     901 12TH ST  119 1345 4 0   284   0 5 342 38  13 12639  6 1  65  65  2 1 2  0 112 19 2 0 6  4

HAWTHORNE  901 12TH ST  119 1346 4 0   389   0 7 412 35  26 15875  6 3 165  55  7 1 1 12 115 19 2 2 0  6
TURNER     901 12TH ST  119 1347 4 0   447   0 6 486 48  18 13580  3 1  65  65  6 1 1  1  62 21 2 0 0  0
MACMASTER  901 12TH ST  119 1348 4 0   611   0 8 629 65  19 19695  4 1  65  65  3 0 1  1  72 18 2 2 0  5
RICCI      901 12TH ST  119 1349 4 0   425   0 6 518 44  82 22455  7 2 120  60  4 1 1  9 164 23 2 0 0  6
ALANIS     901 12TH ST  119 1350 4 0   483   0 7 580 54  21 23560  4 3 167  56  4 2 2  3  74 19 2 1 0  6

DOROSKI    901 12TH ST  119 1351 4 0   257   0 4 335 29  49 20432  4 2 164  82  8 2 1  7  85 21 2 0 0  0
REEVES     901 12TH ST  119 1352 4 0   387   0 8 388 38  71  5305  2 2 120  60  6 1 1  1  47 26 2 0 0  0
LAZARUS    902 12TH ST  120 1353 4 0   539   0 8 713 69  32 15445  5 1  65  65  6 1 1  2  82 57 2 0 0  0
MARTINEZ   902 12TH ST  120 1354 4 0   464   0 6 466 49  28  8312  5 1  65  65  5 0 2  1  98 20 2 0 0  0
COTO       902 12TH ST  120 1355 4 0   347   0 5 468 38  16  6444  3 2 120  60  5 0 2  1  68 23 2 0 0  0

FEINBURG   902 12TH ST  120 1356 4 0   440   0 5 486 37  15  8544  5 2 120  60  5 1 2  6 110 22 2 2 0  0
VALDES     902 12TH ST  120 1357 4 0   381   0 5 483 47  15  3826  2 2 120  60  6 1 1  4 164 27 2 0 0  0
DEITZ      902 12TH ST  120 1358 4 0   436   0 5 410 37  29  9207  5 1  65  65  6 1 1  2 136 23 2 0 4  0
OVINNIO    902 12TH ST  120 1359 4 0   423   0 6 500 43  11  5073  6 2 120  60  9 0 2  1  54 27 2 0 0  0
WALKER     902 12TH ST  120 1360 4 0   246   0 4 291 30  49  5451  2 2 120  60  9 0 2  1  54 27 2 0 4  0

PINERO     902 12TH ST  120 1361 4 0   304   0 5 405 35  25  8998  3 1  65  65  6 1 2  2  72 24 2 2 0  0
SWANSON    902 12TH ST  120 1362 4 0   388   0 6 487 45  16  5743  8 1  65  65  6 1 2 15 182 30 2 0 3  9
COWART     902 12TH ST  120 1363 4 0   382   0 6 482 43  74  5556  8 2 120  60  7 1 2 10 160 20 2 0 3  7
PARCELL    901 13TH ST  120 1364 4 0   414   0 6 495 42  47  7145  3 2 120  60  6 1 1  1  79 25 2 0 6  7
ILNITSKY   901 13TH ST  120 1365 4 0   425   0 6 484 46  62  9715  2 2 120  60  6 1 2  2  55 28 2 0 0  0

BENBON     901 13TH ST  120 1366 4 0   374   0 6 533 44  95  8967  4 2 120  60  6 0 2  5  74 19 2 2 0  6
ROBBINS    901 13TH ST  120 1367 4 0   349   0 6 444 39  11  5503  5 1  65  65  5 0 1  4 110 22 2 0 4  0
BLEDSOE    901 13TH ST  120 1368 4 0   416   0 6 511 43  15  8603  5 2 120  60  5 0 1  4 128 26 2 0 0  4
OLSON      901 13TH ST  120 1369 4 0   153   0 3 333 22  20  6398  1 2 120  60  3 0 1  1  24 24 2 0 4  0
REDFORD    901 13TH ST  120 1370 4 0   260   0 4 355 30  40  8648  1 2 120  60  3 1 1  1  23 23 2 0 0  4

FORD       901 13TH ST  120 1371 4 0   308   0 5 413 39  16  3655  5 2 120  60  5 0 2  6 115 23 2 0 1  2
NOVACK     901 13TH ST  120 1372 4 0   269   0 5 442 37  45  7732  5 1  65  65  5 0 2  6  79 16 2 1 0  0
PRICE      901 13TH ST  120 1373 4 0   363   0 5 374 38  19  4279  6 1  65  65  5 0 2  4 123 21 2 1 0  0
```

INDEX

This book has been set VIP, in 10 and 9 point Times Roman, leaded 2 points. Part numbers are 18 point Stymie Bold. Chapter numbers are 18 and 60 point Stymie Bold, and chapter titles are 24 point Stymie Med. The size of the type page is 42 by 58 picas.

LEGEND FOR THE STAT CITY MAP

Two types of dwelling units appear on the Stat City map, houses and apartment buildings. The information recorded for each house is explained below:

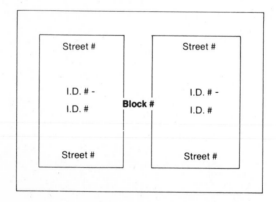

The information recorded for each apartment building is explained below:

Please note that apartment buildings in Stat City have entrances on two streets; consequently, two street numbers appear for each apartment building.